# THE STRUCTURE OF GLASS
# CATALYZED CRYSTALLIZATION OF GLASS

## STEKLOOBRAZNOE SOSTOYANIE
## KATALIZIROVANNAYA KRISTALLIZATSIYA STEKLA

## СТЕКЛООБРАЗНОЕ СОСТОЯНИЕ
## КАТАЛИЗИРОВАННАЯ КРИСТАЛЛИЗАЦИЯ СТЕКЛА

# THE STRUCTURE OF GLASS

## Volume 3

# CATALYZED CRYSTALLIZATION OF GLASS

Authorized translation from the Russian
by E. B. Uvarov, B. Sc., A.R.C.S., D.I.C., A.R.I.C.

CONSULTANTS BUREAU
NEW YORK
1964

ISBN 978-1-4684-0687-0          ISBN 978-1-4684-0685-6   (eBook)
DOI 10.1007/978-1-4684-0685-6

The original Russian text was published in 1963 by the Academy of Sciences Press in Leningrad for the I. V. Grebenshchikov Institute of Silicate Chemistry, the S. I. Vavilov (Order of Lenin) State Optical Institute, the Artistic Glass Works, and the V. I. Lenin Leningrad Institute of Electrical Engineering, as Part I of the Proceedings of the Third Conference on the Structure of Glass. [The papers comprising Part II were published in Nos. 8, 9, and 10 (1962) of *Zhurnal optiko-mekhanicheskoi promyshlennosti* (Journal of the Optico-Mechanic Industry). These papers have not been translated.]

Library of Congress Catalog Card Number 58-44503

# CONTENTS

## OTHER THREE-COMPONENT AND MULTICOMPONENT SYSTEMS

# INTRODUCTORY ADDRESS

## A. A. Lebedev

A special commission for coordination of research work on glass was set up at the 1959 conference on the glassy state. This commission, of which E. A. Porai-Koshits is the chairman, has now been transformed into the Scientific Advisory Section on Glass of the State Committee on the Coordination of Scientific Research, so that it will also have certain rights.

This commission decided to coordinate research on glass by organizing symposiums on various aspects for discussion of problems of interest in particular fields; the exchanges of views indicated the paths to be followed in subsequent work. This appears to us to be the most suitable approach.

We have met today in order to discuss problems associated with the study and practical application of the catalyzed crystallization of inorganic glasses. Many of our research organizations are now concerned with these problems; sometimes their work is not sufficiently coordinated and it is therefore necessary to introduce more clarity into these investigations.

Crystalline glass materials, known as pyroceramics or glassceramics, have appeared quite recently. The Corning Glass Works in the United States of America first produced them in 1957 on the basis of Stookey's work. Glassceramics are materials with very interesting peculiarities and properties. It is therefore not surprising that intensive work has started in many countries on investigation and production of these materials, which have good mechanical and thermal properties, high chemical resistance, and a very low coefficient of expansion.

Photosensitive glasses, containing small amounts of gold or silver and cerium, are of considerable interest. Ultraviolet irradiation of the glass results in the formation of crystal nuclei which may cause volume crystallization of the glass after suitable thermal treatment. Glasses crystallized in this manner differ very significantly from noncrystalline (nonirradiated) glass by their solubility in hydrofluoric acid. This fact can be utilized for production of orifices of any exactly reproducible shape in glass plates; this is very important, for example, in the manufacture of many electronic components.

It is highly probable that the increased mechanical and thermal strength of glassceramics is associated with their crystalline structure: the presence of intertwined crystals hinders crack growth, and this may be the cause of the increased strength. It is well known from S. N. Zhurkov's experiments that the strength of glass threads can be greatly increased if surface cracks are eliminated by etching. Considerable interest attaches in this connection to studies of the structure of glassceramics, and in particular to determinations of the form, size, and composition of the crystals.

What are the possible methods for such investigations?

In addition to the usual x-ray structural method for determination of the crystal form and size, it is possible to use the electron microscope with the aid of the usual replica and shadowing techniques, and also infrared absorption spectra.

With the aid of these methods it is possible to detect crystals in glassceramics; the crystal dimensions vary widely in accordance with the heat treatment. It is interesting to note that the crystal orientation in glassceramics is of a clearly ordered character; this appears to suggest that crystal nuclei first appear during the stage in the melting when the glass has appreciable fluidity, and they are therefore arranged in an ordered manner in the flowing viscous melt. Valuable information on the structure of glassceramics may be obtained by small-angle x-ray scattering, a method which was developed and successfully used in the Institute of Silicate Chemistry.

Determination of the chemical composition of the crystals formed, their identification, is a much more complicated matter.

Definite results are obtained by x-ray structural analysis only if the specimen contains crystals of one kind only, and if these crystals have a relatively simple structure. The situation is worse if a mixture of different crystals, of complicated structure, is present. If the specimen contains less than 10% of crystals, sometimes they cannot be identified.

Investigation of infrared absorption spectra is another possible method for identification of the crystals formed in glassceramics. However, another difficulty arises here: whereas x-ray analysis is too coarse, infrared spectral analysis is excessively sensitive. Therefore infrared spectra are often too complex to allow conclusive identification of the crystals actually present in the specimen. The results obtained in investigations of glass-ceramics by x-ray structural analysis and by infrared spectroscopy cannot as yet be fully reconciled because of the complexity of the data. However, there is no doubt that these difficulties will be overcome in the future, and it will then be possible to exercise better control over the conditions for production of glassceramics with the desired structure and properties.

Today, when this new trend in the silicate industry, associated with studies and production of a new material (intermediate between glass and ceramics), is developing so rapidly, it is particularly important to ensure active contacts between workers in industry and in the scientific research institutes. It is the purpose of the present symposium to establish such contacts. Mutual knowledge of the results achieved and exchange of experience and views will undoubtedly stimulate research ideas and assist better coordination and greater purposefulness in the work.

*General Aspects of the Crystallization of Glass*

# SEQUENCE OF SEPARATION OF CRYSTALLINE PHASES
# OF DIFFERENT COMPOSITIONS FROM SILICATE MELTS

## N. A. Toropov

The production of highly effective crystalline glass materials by catalyzed crystallization is a problem which is attracting the attention of numerous specialists in various countries. These crystalline glass materials have a fine-grained uniform structure consisting of small crystals of irregular, distorted form, often aggregated into spherulites, with residual glass interlayers cementing the crystalline glassceramic concretion.

The use of glass melts of various chemical compositions, molten metallurgical slags (according to I. I. Kitaigorodskii and K. T. Bondarev), or molten rocks has been recommended for production of crystalline glass materials.

The fundamental chemical system for production of glassceramics is usually chosen in the composition regions corresponding in the phase diagrams to the concentration regions where phase separation effects or a fairly pronounced tendency to such effects are observed, or in regions close to the phase-separation region. The catalysts used for controlled crystallization processes, giving rise to enormous numbers of nucleation centers in the original glass mass, are finely divided gold, silver, or platinum, or the oxides of titanium, chromium, cerium, vanadium, nickel, and zirconium, or certain sulfides or fluorides of heavy or transition metals. A definite similarity of lattice structure between the substance introduced as the nucleation agent and the crystals, the growth of which determines the structure of the final product, is regarded as especially important.

Glasses containing catalysts are annealed under definite temperature conditions; this is characteristic for crystalline glass materials of the vitroceramic type [1], for slag glassceramics, and for articles made from molten and annealed basalt.

While the formation of nuclei or crystallization centers is determined primarily by the structural conditions and to a certain extent by the patterns of particular plane networks in the space lattices of the crystal phase and the nucleus, during the subsequent crystallization the decisive role is played by the phase relationships of the system, which are described by the phase diagram and determine the nature of the processes taking place in the system.

An examination of literature data [2] shows that data on the synthesis of glassceramics from typical glass-forming systems can be represented roughly as follows.

a) Stookey's method. Glasses in the $Li_2O$-$Al_2O_3$-$SiO_2$ system; gold, silver, or copper as the crystallization centers; primary crystallization at 600°, secondary at 900°.

b) Nucleation by $TiO_2$. Glasses in the $Li_2O$-$Al_2O_3$-$SiO_2$ and $MgO$-$Al_2O_3$-$SiO_2$ systems are used.

c) Formation of nuclei from platinum compounds, introduced into glasses in the $LiO_2$-$SiO_2$ or $Li_2O$ (12)-$MgO$ (10-20)-$Al_2O_3$ (0-10)-$SiO_2$ (80-90) systems; percentages by weight are given in parentheses. The primary crystalline phases observed are $Li_2O \cdot 2SiO_2$, $Li_2O \cdot SiO_2$, $SiO_2$, $MgOSiO_2$; the secondary phases are $Li_2O \cdot Al_2O_3 \cdot 2SiO_2$, $Li_2O \cdot Al_2O_3 \cdot 4SiO_2$.

d) Without seeding — glasses in the $Li_2O$ (4)-$MgO$ (15-23)-$Al_2O_3$ (15-27)-$SiO_2$ (54-62) system; eucryptite forms at 750-850° as the primary phase, and at 900-1100° there is secondary separation of spodumene.

e) The fluoride method. Fluorides are used as the nucleating agents. Glasses in the $Li_2O-Al_2O_3-SiO_2-CaF_2$ system are crystallized, and mica, spodumene, and wollastonite are obtained in the crystallization products.

f) Crystallization by addition of zirconium dioxide, $ZrO_2$, to glasses in the $Li_2O-MgO-Al_2O_3-ZrO_2-P_2O_5$ system. Eucryptite, forsterite, and dimagnesium phosphate separate out in the primary crystallization phase (650-850°), and spodumene, forsterite, zircon, and $SiO_2$ in the second (950-1100°). In addition, glasses of more complex composition, such as $Li_2O \cdot ZnO \cdot Al_2O_3 \cdot SiO_2$ or $Li_2O \cdot K_2O \cdot CaO \cdot Al_2O_3 \cdot SiO_2 \cdot Cr_2O_3$, containing nickel oxide and vanadium pentoxide or antimony pentoxide, have been used. Promising results are obtained with the use of zinc oxide and transition elements such as $Cr^{3+}$, $V^{5+}$, and $Ni^{2+}$. Transition elements can alter their valence in glass, and repeated heating results in such changes, which facilitate the formation of crystal nuclei.

For production of high-quality ware from fused rocks, materials containing not more than 50% silica and definite proportions of other components are generally used. A synthetic batch containing quartz sand, dolomite, and other materials is melted for production of decorative facing tiles and ware of complex form.

According to the Czechoslovak worker A. Pelikan, the best results are obtained from raw materials of composition in the following ranges: $SiO_2 = 43.5-47.0\%$, $Al_2O_3 = 11.0-13.0\%$, $Fe_2O_3 = 4.0-7.0\%$, $FeO = 5.0-6.0\%$, $MnO = 0.2-0.3\%$, $TiO_2 = 2.0-3.9\%$, $CaO = 10.0-12.0\%$, $MgO = 8.0-11.0\%$, $Na_2O = 2.0-3.5\%$, $K_2O = 1.0-2.0\%$.

The mineral composition of vitroceramic ware has an important influence on the technological properties and service characteristics of the ware. Plagioclases, which are of higher basicity, increase the tendency to crystallization and at the same time raise the melting point of the material. The casting properties of the material are improved by the presence of olivines and a higher content of rhombic pyroxene in the original rocks. Materials rich in olivine are very brittle and more refractory. An increased pyroxene content is most desirable, as this is the mineral which increases abrasion resistance, mechanical strength, and chemical durability. The magnetite and olivine contents should not exceed 10%. Chromite ore is used as a nucleation agent in glassceramic ware production; its crystals give rise to numerous crystallization centers because of their high melting point and of the fact that they are isostructural with magnetite.

Crystallization of basalt castings is complex in character as, according to Tsvetkov [3], minute, opaque, irregular magnetite crystals begin to separate out at approximately 1250°. Separation of the latter is accompanied by clearing of the adjacent regions of the main glassy mass as the result of loss of iron oxides. If the firing time is increased, larger crystals of the same magnetite are formed, but in the form of dendrites.

A second crystalline phase, plagioclase, begins to separate out at about 1200°. The number of crystallization centers of this mineral at this stage of the process is small. Individual plagioclase crystals or groups of crystals appear only at some isolated regions of the specimens. Most of the plagioclase crystals are twinned or even triplet concretions.

At a lower temperature (1150°) the number of plagioclase crystallization centers increases sharply. This probably corresponds to the temperature maximum at the rate of formation of these centers. The result is a network of very fine transparent plagioclase crystals.

At even lower temperatures (1100°) the continuing separation of magnetite and plagioclase is accompanied by crystallization of a third phase, pyroxene, which gives rise to a number of outstanding mechanical properties of this material.

The crystallization routes of blast furnace slags of various chemical compositions can be examined with adequate approximation with the aid of the phase diagram of the ternary $CaO-Al_2O_3-SiO_2$ system, where the composition regions of basic, acidic, aluminosilicate, and high-alumina slags are usually distinguished.

The microstructure of blast-furnace slags obtained in industry depends on the cooling conditions; their properties, important in the practical sense, vary accordingly. The nature and spatial relationships of the crystalline phases formed and the amount and distribution of the residual glass are very important here.

The author, jointly with Astreeva [4], put forward a classification of blast-furnace slags which characterizes slags by the typical crystalline phases separating out during crystallization, and hence defines their physicochemical nature. A certain amount of crystals, which make it possible to determine to which group the slag belongs, separate out even in glassy slags obtained by rapid cooling in granulation equipment.

Turning to a direct examination of the crystallization process in silicate melts, we note that, similarly to magmatic rock-forming melts, they are in most cases multicomponent systems, and after crystallization are also multiphase systems, when regarded from the viewpoint of physicochemical equilibria. Therefore, for interpretation of the conditions for formation of the structure in such materials and in development of methods for modifying them, such as catalyzed crystallization, it is necessary to take into account the main characteristics of multicomponent and multiphase systems.

Even if we exclude cases of isomorphism and solid solutions, and the formation of metastable phases (although such effects are not uncommon in the crystallization of glassceramics), the structure of a crystallized melt should, in the simplest case, reflect the crystallization of a multicomponent system by a eutectic scheme. This scheme, in a three-component system, comprises three consecutive stages: 1) separation of the first phase, in equilibrium with the residual liquid; 2) simultaneous separation of two phases in reciprocal eutectic proportions; 3) separation of a ternary eutectic.

Thus, in the simplest case of crystallization of a three-component melt it undergoes a gradual transition from a homogeneous liquid phase by formation of numerous crystallization centers of the primary phase at numerous points in its volume at the liquidus temperature; these centers grow by diffusion of the corresponding ions and molecules from the melt. This process occurs in the range between the liquidus temperature and a lower temperature at which crystallization of two different phases begins. Simultaneous crystallization of the latter occurs in the temperature range, the lower boundary of which is the ternary eutectic temperature; three phases crystallize simultaneously in the melt, the phases being in equilibrium at the eutectic point. However, the crystals do not cease to grow after the temperature has fallen below the eutectic point; our observations show that growth continues even in supercooled slags down to temperatures of the order of 900-1000°.

As the result of crystallization in accordance with the eutectic scheme discussed by us in detail, the following structural formations should be present in the structure of the solidified melt: 1) porphyritic inclusions — larger and regular crystals of the first phase; 2) microcrystalline pegmatitic intergrowths of two phases in quantitative proportions corresponding to a binary eutectic; 3) even finer pegmatitic intergrowths of three phases, constituting a ternary eutectic. However, investigations of real slags show that in practice the individual phases crystallize all at once and not in two or three stages.

In contrast to eutectic crystallization, crystallization in one stage is characterized by separation of first-generation crystals around centers formed during the first stage of the crystallization process, without development of new centers of the primary phase at the second and third stages of that process. The amount of primary-phase crystals forming the prophyritic inclusions corresponds to the total content of that phase in the entire melt; crystals of the primary phase cannot be found in the composition of the main phase surrounding the inclusions.

The formation of such a structure around the crystallization centers of the first phase must be effected by collection, by means of diffusion, of the substance constituting this phase not only during the first but also during the second and third periods of the crystallization process: the appropriate substance diffuses toward the crystallization centers of the second phase during both the second and the third periods of that process. This results in a structure in which the skeletal crystals of one phase grow quite regularly into skeletal crystals of the second phase [5]. The structure then assumes an intergrown rather than a porphyritic character. For example, calcium monoaluminate forms skeletons with accumulations of pentacalcium trialuminate crystal mass concentrated between their branches. In one specimen there were found very fine branches of monoaluminate skeletons running through the $5CaO \cdot 3Al_2O_3$ mass, so that the result resembled a micropegmatitic structure. This does not interfere with the single-stage crystallization scheme, as the branches are integral with the skeletal axes (with the same optical orientation).

Investigation of the microstructures of blast-furnace slags, the chemical composition points of which lie to the left of the CA-CAS$_2$ line and the crystallization path passes through the reaction point (.) of Rankin and Wright's 13th diagram, shows that in a number of cases the reaction of the previously formed gehlenite crystals with the liquid may not go to completion. In this case, nonequilibrium states arise: gehlenite does not dissolve completely, $5CaO \cdot 3Al_2O_3$ crystals are additionally formed, and the solid product contains four crystalline phases rather than three. These microstructure regions of slag materials occur in other silicate systems (artificial and technical) and may serve for interpretation of granitic (single-stage crystallization), porphyritic (two-stage crystallization), pegmatitic, and other structures.

The observations described above show that crystallization of a multicomponent silicate melt may proceed in accordance with a variety of schemes, and that truly effective methods of structure modification may consist of various ways of creating crystallization centers artificially, subsequent thermal treatment of the crystallized glass, elimination of previously formed metastable formations, the presence of impurities forming solid solutions with the crystalline phases, and many others. Structural investigations of the products made in glassceramic ware factories, conducted at present by the electron-microscope method in the Institute of Silicate Chemistry, are also of very great importance in the development of the theory of crystalline glass materials.

## LITERATURE CITED

1.  D. Popescu-Has and St. Lungu, Industria Usoara 4:26, 1954.
2.  I. Sawai, Glass Technology 2(6):243, December, 1961.
3.  A. I. Tsvetkov, Proceedings of the Second Conference on Experimental Mineralogy and Petrography, Izd. AN SSSR, Moscow, 1937.
4.  N. A. Toropov and O. M. Astreeva, Trudy NIITsementa, No. 2, 1949.
5.  N. A. Toropov, Chemistry of Cement, Promstroiizdat, Moscow, 1956.

# INITIAL STAGES IN THE CRYSTALLIZATION OF GLASSES
# AND FORMATION OF GLASSCERAMICS

## V. N. Filipovich

### 1. Two Kinds of Relaxation Processes in Glass. Nucleation Rate of the New Phase

It is known that glass is formed by rapid cooling of a viscous melt; therefore, it has the structure of that melt if the structural changes taking place during the cooling process are disregarded [1,2]. Thus, at relatively low temperatures $T < T_g$ ($T_g$ is the softening point), the structure of a quenched glass corresponding to the melt temperatures is a nonequilibrium one in two respects: first, with regard to the thermodynamically metaequilibrium disordered structure of the glass as a supercooled liquid, corresponding to the given temperature $T > T_g$; second, with regard to the thermodynamically equilibrium structure of the crystals* which are formed during crystallization of the glass.

Accordingly, two kinds of relaxation processes (establishment of stable or metastable equilibrium) are possible in glass: relaxation processes of the liquid type leading to the establishment of a liquid glass structure, metaequilibrium in nature for the temperature T; and relaxation processes of establishment of an equilibrium crystalline structure in stable (or metastable) crystals. The former are processes of vitrification or formation of a new glassy structure, and the latter are processes of crystallization or formation of a crystalline structure. In glasses with a very low tendency to crystallization only relaxation processes of vitrification occur, in the main. On the other hand, in glasses which crystallize very well, these processes are suppressed by relaxation crystallization processes. In the intermediate case both kinds of relaxation processes are fairly pronounced, occur simultaneously, and are superposed.

A vivid example of relaxation effects of the liquid type (vitrification) is provided by metastable demixing in certain glasses (the systems $Me_2O$-$SiO_2$, $MeO$-$SiO_2$, etc.) — phase separation of the glass into two or more glasses of different chemical composition in the temperature range below the solidus temperature (i.e., where only crystals are thermodynamically stable). Metastable phase separation in glasses is analogous to phase separation in liquids in which the components have limited miscibility. It is a reversible phase transition: it occurs at lower, and disappears at higher temperatures. The metastable phase separation effect is often complicated by crystallization processes, but it may be observed independently in certain cases, e.g., in sodium silicate or sodium borosilicate glasses [13].

In this paper we are concerned mainly with relaxation effects associated with nucleation of phases with a new structure — liquid (liquation) or crystalline (crystallization). The nucleation rate I of the new phase, i.e., the number of nuclei of critical size capable of further growth appearing per unit volume in unit time at temperature T, may be expressed as:

$$I = N^* N_1^* = \left( \bar{N} e^{-\frac{\Delta \Phi^*}{kT}} \right) \left( A e^{-\frac{\Delta \Phi_A^*}{kT}} \right),$$ (1)

---

*The formation of thermodynamically nonequilibrium crystalline structures metastable at the temperature T, with thermodynamic potential $\Phi$ lower than that of glass but higher than that of crystalline modifications, truly stable at the temperature T, is also possible.

where $N^* = \tilde{N} e^{-\frac{\Delta\Phi^*}{kT}}$ is the average number of nuclei of critical size existing per unit volume (able both to

arise and to disappear again); $N_1^* = Ae^{-\frac{\Delta\Phi_A^*}{kT}}$ is the number of elementary events of activation of critical nuclei per unit time, converting them into nuclei capable of further growth; $\tilde{N}$ is a quantity proportional to the number of places where nuclei can appear; A is a quantity which depends on the mechanism of activation of the growth of the nuclei; $\Delta\Phi^*$ is the barrier or increment of thermodynamic potential of the system in the formation of a nucleus of the new phase of the critical size; $\Delta\Phi_A^*$ is the thermodynamic potential barrier which must be surmounted to convert the critical nucleus into a state of further growth.

Equation (1) for the nucleation rate I is derived from the general theory of statistical fluctuations and is of the same form for a great variety of phase changes, differing only in the values of the parameters. In particular, for the crystallization of a simple one-component viscous liquid [3] or a glass (at $T > T_g$), we have

$$\Delta\Phi^* = \Delta\Phi_v^* + \Delta\Phi_s^*, \quad \Delta\Phi_v^* = \frac{4\pi}{3} r^{*3} (\varphi_c - \varphi_g), \quad \Delta\Phi_s^* = 4\pi r^{*2}\sigma;$$
$$\Delta\Phi^* \simeq \frac{16\pi}{3} \frac{\sigma^3}{q^2 \left(1 - \frac{T}{T_s}\right)^2}, \quad r^* \simeq \frac{2\sigma}{q\left(1 - \frac{T}{T_s}\right)},$$

(2)

where r* is the critical radius of the spherical nucleus beyond which $\Delta\Phi(r)$ decreases [$\Delta\Phi^* = \Delta\Phi(r^*)$] with further increase of the nucleus radius r; q is the heat of crystallization per unit volume of the crystal; $T_s$ is the melting point of the crystal; $\varphi_c$ and $\varphi_g$ are the thermodynamic potentials per unit volume of the crystal and glass; $\sigma$ is the specific surface free energy at the crystal-glass interface.

If a phase transition takes place in a solid (e.g., in a glass at $T < T_g$), then the mechanical stresses which increase the barrier $\Delta\Phi^*$ and influence the form of the nuclei† must be taken into account in (1).

The nucleation rate as a function of temperature has a maximum, the position $T_m$ of which can be approximately found from the condition of the minimum value of the exponent in (1), $\frac{d}{dT}\left(\frac{\Delta\Phi^* + \Delta\Phi_A^*}{kT}\right) = 0$, which, with (2) taken into account, gives $\left(\text{assuming } \frac{d\Delta\Phi_A^*}{dT} = 0\right)$

$$T_m = \frac{T_s}{3}\left(1 + \frac{\Delta\Phi_A^*}{\Delta\Phi^*}\right)\Big/\left(1 + \frac{1}{3}\frac{\Delta\Phi_A^*}{\Delta\Phi^*}\right).$$

(3)

When $\Delta\Phi_A^* = 0$, $T_m = (T_s/3)$. $T_m$ increases with increase of $\Delta\Phi_A^*$; when $\Delta\Phi_A^* = \Delta\Phi^*$, $T_m = \frac{1}{2}T_s$.

We note that, in accordance with (2), $\Delta\Phi^* \sim \sigma^3/q^2$ is strongly dependent on $\sigma$. This is associated with the fact (Ostwald's step rule) that the crystalline form with the lowest $\sigma$ at the crystal-glass interface, i.e., with the structure closest to that of the glass (or liquid), is the most likely to be formed first. These are usually high-temperature and highly symmetrical forms, generally metastable at the relatively low temperatures at which crystallization of quenched glass occurs.

Equations (1) and (2) can also be applied to the formation of crystal nuclei of a particular phase in complex multicomponent systems, but $(\varphi_c - \varphi_g)$ should be replaced by $\Delta\varphi$, the volume change of the thermodynamic potential of the entire system, calculated per unit volume of the critical nucleus formed; q is the heat liberated in the formation of a unit volume of the crystalline phase from a glass of the given composition.

It should be noted that, especially in the case of complex glasses, all the quantities in (1) depend on time, as the composition and the amount of glass alter with time as the result of crystallization.

---

†Disks or plates are the most advantageous forms for the nuclei [4]. They give the least mechanical stress energy. The acicular form is less advantageous. The spherical form is the least advantageous (although it is the most advantageous in relation to the minimum surface energy).

## 2. Initial Stages in the Crystallization of Complex Glasses

**1.** Crystallization of simple (one-component) glasses is associated only with structural rearrangements in a medium, the composition of which already satisfies the stoichiometric conditions. The formation of crystal phase nuclei in complex multicomponent glasses is associated with diffusional chemical differentiation of various atoms and structural groups, which should congregate to form regions close in composition to the future crystals (either stoichiometric or the composition of possible solid solutions). Two possible cases may be imagined.

1) Glasslike regions may form with the composition of the future crystals, without simultaneous crystallization of these regions. In other words, crystallization is preceded by a certain independent effect of primary precrystallizational demixing accompanied by a gain of thermodynamic potential. The demixing (precrystallizational phase separation) occurs by way of formation of amorphous nuclei of critical size, capable of further growth, as it is accompanied by liberation of specific heat of solution $q_s$ [analogous to the heat of crystallization q; see (2)].

Because of the low value of the surface energy σ at the interface between the amorphous phases, stepwise crystallization of this kind proceeding through preliminary microseparation into metastable glassy phases appears quite probable (see end of Section 1).

Precrystallization microphase separation is analogous to the usual metastable phase separation of the liquid type (see above), except that, first, in the usual liquid metastable phase separation the composition of the glasses deposited may differ greatly from the composition of the possible crystals and, second, it may appear and disappear reversibly without crystallization, depending on the temperature. Moreover, metastable phase separation occurs only in certain glasses in a definite composition region, whereas crystallization of glasses is possible with any composition and is irreversible below the solidus temperature. If metastable liquation were a purely precrystallization effect, it could not disappear on rise of temperature once it had occurred, culminating in crystallization (below the solidus line).

We shall distinguish between the usual metastable phase separation of the liquid type in glass and precrystallizational nucleation phase separation (if the latter occurs at all).

2) The second possibility is that regions with the composition of the future crystals, arising as the result of fluctuations, cannot have a stable existence. Their formation is associated with increase of the thermodynamic potential and absorption of heat ($q_s < 0$); therefore, they become dissipated again if they do not undergo ordering with liberation of heat q. In this case, the formation of nucleation regions is in fact simultaneous with their ordering and crystallization. Of course, ordering leads to an increase of σ, but at the same time it leads to liberation of heat of crystallization q, and the latter proves decisive [see (2)].

Both the possible nucleation mechanisms are complex cooperative processes with numerous particles in coordinated motion. Equation (1) represents the probable results of these processes without reference to the mechanism of their course in time. The magnitude of the barrier for the formation of a critical nucleus $\Delta\Phi^* = \Delta\Phi_V + \Delta\Phi_S^*$, which depends most strongly on the composition and temperature ($\Delta\Phi_V^*$ and $\Delta\Phi_S^*$ are the volume and surface components of $\Delta\Phi^*$), is of considerable significance. The activation energy of nucleus growth $\Delta\Phi_A^*$ depends less strongly on the composition and temperature and is of the same order of magnitude as the activation energy of viscous flow for glass [3], as the movement of atoms and structural units on the surface of the nucleus during its growth is similar in many respects to reorientation of structural units and bond switching [5] in the flow of a viscous liquid.

Negative values of $\Delta\Phi_V$, corresponding to volume gain of thermodynamic potential in the formation of a critical nucleus, are of the greatest interest here. Let us calculate $\Delta\Phi_V$ for certain typical cases.

The structure of glass is determined by the minimum thermodynamic potential $\Phi = E - TS + pV$ or free energy

$$F = E - TS, \qquad (4)$$

if effects associated with volume changes are disregarded. As the glass forms crystals on cooling, the tendency to form groups in which the composition and mutual arrangement of the atoms are similar to those in the future crystals must already exist in the melt. Ordered arrangement of the atoms in crystal form leads to decrease of E but to increase of the entropy term −TS, as entropy decreases with increase of order. The role of the entropy disorder term diminishes with fall of temperature, and the tendency to stoichiometric aggregation eventually culminates in crystallization of the melt. The same tendency to aggregation of definite structural elements leads to metastable phase separation (when it is possible) in glasses.

**2.** Let us examine in greater detail the thermodynamic description of the process of formation of a new phase in the case of ordinary metastable phase separation. The theoretical scheme is simplest and most clear in this case. At the same time, the results obtained can be used for describing the nucleation of a new phase in crystallization, especially if the latter is preceded by precrystallizational phase separation.

In accordance with the foregoing, we consider a glass (or melt) consisting of structural units characteristic of the future crystals. For simplicity, we consider a two-component glass of the composition $xA(1-x)B$, where A and B are the compositions of the crystals formed by crystallization of the glass.‡ The structural units are taken to be, for example, the unit cells of crystals A and B, and the structure of the glass is determined by the nature of aggregation of these units. We examine two types of mutual arrangement of the structural units: a) when the units A and B are mixed uniformly without the formation of accumulations consisting predominantly of one kind of unit; b) when there is a strong tendency in the glass toward formation of groups consisting of similar structural units.

In case "a" it is supposed that z nearest neighbors of a unit A or B comprise on the average xz neighbors A (x is the concentration of component A) and, therefore, $(1-x)z$ neighbors B — a uniform random distribution. With this assumption, and with the assumption that only neighboring units interact, we can obtain the following expression for the free energy of glass**:

$$F(x) = Nf(x), \quad f(x) = \varepsilon - Ts, \tag{5}$$

$$\varepsilon = Ux(1-x) - x(U_{AA} - U_{BB}) + \frac{U_{BB}}{2}, \quad U = U_{AB} - \frac{1}{2}(U_{AA} + U_{BB}), \tag{6}$$

$$s = -k[x \ln x + (1-x) \ln (1-x)], \quad S = Ns, \tag{7}$$

where N is the total number of structural elements A and B; $U_{AA}$, $U_{BB}$, $U_{AB} \leq 0$ are the bonding energies of the units; $\varepsilon$ and s are the energy and entropy calculated per structural unit. The entropy S is calculated from the formula $S = k \ln C_N^{Nx}$, where $C_N^{Nx} = \dfrac{N!}{(Nx)!\,(N-Nx)!}$ is the number of ways (combinations) in which Nx units of type A can be distributed among N possible places (entropy of mixing of units A and B).

Figure 1 is an $f(x)$ plot for different $kT/Uz$ (it is assumed that $U_{AA} = U_{BB}$ and the constant term $U_{BB}/2$ is omitted). The $f(x)$ plot has two minima, at $x = x_1$ and $x = x_2$, which come closer together with increase of T and finally coalesce (at $x = \frac{1}{2}$). The positions of the minima can be calculated from the condition $df/dx = 0$, which gives the equation

$$(1-2x)/\ln \frac{1-x}{x} = \frac{kT}{Uz}.$$

In Fig. 2 the roots $x_1$, $x_2$ are plotted against T. This is the phase (solubility) diagram for the model under consideration. If we take the composition x (Fig. 1) and a temperature above the height of the dome in Fig. 2, the most advantageous with regard to least free energy is separation into two phases of compositions $x_1$ and $x_2$,

---

‡For example, if we consider lithium silicate glasses in the composition range from lithium disilicate to $SiO_2$, with metastable phase separation we take A to denote, for example, $SiO_2$ and B to represent lithium disilicate $Li_2O \cdot 2SiO_2$, which crystallize in this composition range.

**Details of the calculation can be found in Becker's paper [6], which gives the calculation of the free energy of a binary metallic solid solution $xA(1-x)B$ (see also [7]). Equations (5)-(7) are most correct when the units A and B have spherical (atoms) or cubic symmetry. The influence of possible relative rotations of A and B is not taken into account in (5)-(7).

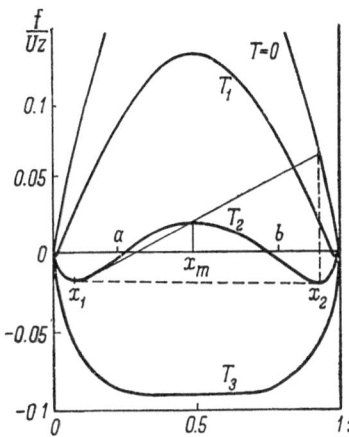

Fig. 1. Dependence of the free energy ($f$ / $Uz$) on composition and temperature for a noneutectic solution undergoing phase separation.

$$T_1 - \frac{kT}{Uz} = \frac{1}{6}; \quad T_2 - \frac{kT}{Uz} = \frac{1}{3}; \quad T_3 -$$
$$- \frac{kT}{Uz} = \frac{1}{2}.$$

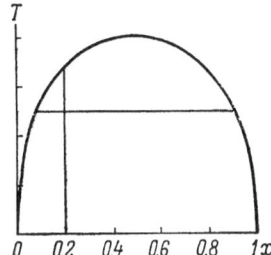

Fig. 2. Solubility (phase) diagram for a noneutectic solution.

found from the diagram in Fig. 2 by the usual method — from the abscissas of the intersections of the line for constant T with the curve separating the homophase and heterophase regions (within the curve of Fig. 2).

Let us consider the mechanism of separation of a solution (glass) of concentration x. We find the expression for the change of free energy ΔF due to occurrence of a local fluctuation of the solution concentration. It is made up of a volume component $ΔF_v$ and a surface component $ΔF_s$.

Let us find $ΔF_v$. Suppose that displacement of atoms results in the formation of a region or group in which the total number of A and B units is n and the solution concentration is x' > x (Fig. 3a). The concentration of the remaining glass also changes and becomes x" < x. From the condition that the number of units A remains constant, we have

$$Nx = nx' + (N - n) x, \text{ i.e., } x - x'' = \frac{n}{N - n}(x' - x). \quad (8)$$

The free energy change for the whole system, $ΔF_v$, is

$$ΔF_v = nf(x') + (N - n) f(x'') - Nf(x). \quad (9)$$

If N >> n, then x − x" << 1 and it may be assumed that $f(x'')$ = $f(x)$ + $(x'' - x) \partial f(x) / \partial x$. Therefore,

$$ΔF_v(x') = nΔf_v = n[f(x') - f(x) - f'(x)(x' - x)]. \quad (10)$$

If $|x' - x|$ << 1 is small, expansion of $f(x')$ by powers of x' − x gives

$$ΔF_v(x') = n \frac{1}{2} \frac{\partial^2 f}{\partial x^2}(x' - x)^2. \quad (11)$$

According to (11), the sign of $ΔF_v$ is determined by the sign of $\partial^2 f/\partial x^2$. When x < a (Fig. 1), $\partial^2 f / \partial x^2 > 0$; at the "inversion" points a and b, $\partial^2 f / \partial x^2 = 0$; when a < x < b, $\partial^2 f / \partial x^2 < 0$; and when x > b, $\partial^2 f / \partial x^2 > 0$. In other words, small fluctuations of concentration lead to decreases of free energy only in the region a < x < b, where $f(x)$ is convex with respect to the x axis. We are more interested in large fluctuations, when only Eq. (10) is valid.

The case $x_1 < x < a$. Here any fluctuations toward a decrease of x always lead, in accordance with (10), to increase of $ΔF_v$. Fluctuations toward an increase of x also initially lead to increase of $ΔF_v$. However, at x' > $x_{int}$ [$x_{int}$ is the point where the tangent to $f(x)$ at the point x cuts the $f(x)$ curve], $ΔF_v$ becomes negative and increases rapidly as x and $x_1$ approach each other. Thus, there is a free energy barrier for deposition of the phase of the composition $x_2$.

During deposition of the $x_2$ phase, the composition x of the remaining solution is displaced toward $x_1$ and the absolute value of $ΔF_v(x_2)$ diminishes until it becomes zero [the tangent at the point x = $x_1$ then touches $f(x)$ at the point $x_2$].

It is significant that in this example, with a gain of free energy, only nucleation regions of phase $x_2$, of which little should be obtained during the decomposition, are formed. This is because (Fig. 3b) the composition region x and the surrounding regions of the same composition tend to change their composition to $x_1$. If the central region attains the composition $x_1$, the regions surrounding it will become further from $x_1$ in composition and there will be an over-all loss of free energy. On the other hand, if the composition $x_2$ is becoming established in the central region, initially the process also causes an increase of free energy (transition through the barrier, see above) and subsequently the free energy of the system decreases as the composition of the central region approaches $x_2$, and that of the surrounding regions tends to $x_1$, both being advantageous.

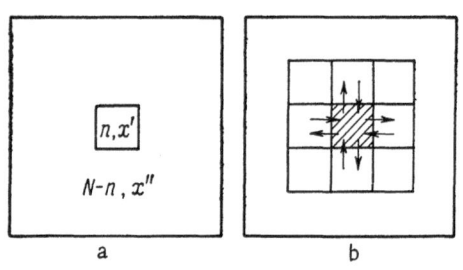

Fig. 3. Scheme (a and b) of nucleus formation during phase separation.

The case $a < x \leq x_m$ (Fig. 1). Here any fluctuations, to-ward either $x_1$ or $x_2$, proceed with decrease of free energy. There-fore, simultaneous formation of small regions of both composi-tions, $x_1$ and $x_2$, is possible.

The cases $x_m \leq x < b$ and $b < x \leq x_2$ are analogous to those examined above. In the cases $x < x_1$ and $x > x_2$, any fluc-tuations lead to increase of $\Delta F_v$ and no decomposition occurs (the homophase region is outside the curve in Fig. 2).

It was assumed above that $U_{AA} = U_{BB}$. If, for example, $|U_{AA}| > |U_{BB}|$, the term $x(U_{AA} - U_{BB})$ is negative (6) and in-creases in absolute value with increase of x. Instead of the symmetric curves of Figs. 1 and 2, we have asym-metric curves. Analysis of the possible cases of isothermal decomposition of the solution at various concentra-tions is similar to the analysis for the case $U_{AA} = U_{BB}$.

The surface part $\Delta F_s$ of the increase $\Delta F$ in the formation of a region with concentration x', according to [6], is given by the equation

$$\Delta F_s = n_s \Delta f_s = n_s U (x - x')^2, \tag{12}$$

where $n_s$ is the total number of structural units on the surface of the glassy phase separation region under con-sideration.

Equations (10) and (12) can be used for estimation of the critical size $r^*$ of the phase separation regions and of the barrier $\Delta F^*$ for formation of these regions. In the case of lithium silicate glasses in the composition range from lithium disilicate to $SiO_2$, $2r^* = (2-4)l$, where $l$ is the average linear dimension of a structural unit (cell) and $\Delta F^* = (2-4)$ kcal/mole – relatively small quantities.

We now pass to the case of solutions or glasses with a tendency to form groups predominantly of structural units A or B. If structural units A are added to a glass or liquid B, the energy of the solution first increases great-ly, as the added units A become incorporated between units B and form A−B bonds which are considerably weaker than B−B ($|U_{AA}|$, $|U_{BB}| > |U_{AB}|$). Therefore, as more A units are added, they should congregate into groups and the number of weak A−B bonds decreases. The same happens if units B are added to glass A.

At intermediate compositions the glass or solution consists of small groups, predominantly of units A and B. For these compositions, $x_B < x < x_A$, we can write (Fig. 4)

$$E = N_A z_A n_A U_{AA} + N_B z_B n_B U_{BB} + N_{AB} U_{AB} =$$
$$= N x z_A U_{AA} + N (1 - x) z_B U_{BB} + N_{AB} U_{AB}, \tag{13}$$

where, for example, $n_A$ is the number of A units in an A group; $N_A = N_x/n_A$ is the number of A groups; $z_A$ and $z_B$ are the numbers of neighboring units a given unit has in the A and B groups; $N_{AB}$ is the number of boundary pairs of structural units in the A and B groups. We assume, approximately, that $n_A$, $n_B$, $z_A$, $z_B$, and $N_{AB}$ are in-dependent of x.†† Then (13) gives a linear dependence of the energy on x. Figure 4 (T = 0) shows an E(x) plot corresponding to the above reasoning for the case $z_A U_{AA} = z_B U_{BB}$.

The expression for entropy, if we restrict ourselves only to the entropy of mixing (transposition) of A and B groups, coincides with (7), where the total number of units N should be replaced by the total number of groups $N_0 = \dfrac{N x}{n_A} + \dfrac{N (1 - x)}{n_B}$, and x by $y = \dfrac{N_A}{N_0} = \dfrac{N x}{n_A N_0} = \dfrac{x}{x + (1 - x) \dfrac{n_B}{n_A}}$. The free energy plots (5) for various temperatures are given in Fig. 4 (it is assumed that $z_A U_{AA} = z_B U_{BB}$). The phase diagram (Fig. 4) is eutectic in

††In fact, the composition, dimensions, and structure of the A and B groups depend on the concentration x. The B groups become enriched with A units as the composition shifts toward A, and vice versa. A more detailed quali-tative picture is given by Bartenev [11].

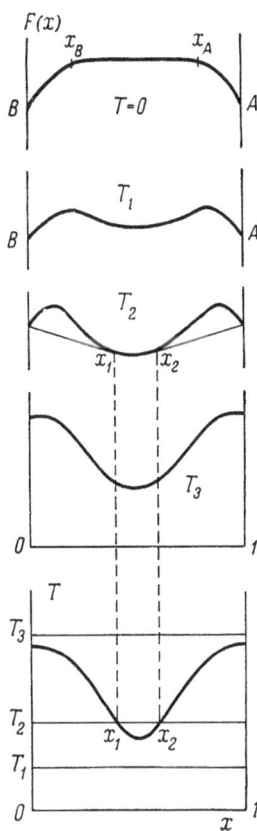

Fig. 4. Dependence of free energy on the composition and temperature for a liquating eutectic solution. $T_3 > T_2 > T_1 > 0$. The phase diagram is shown at the bottom.

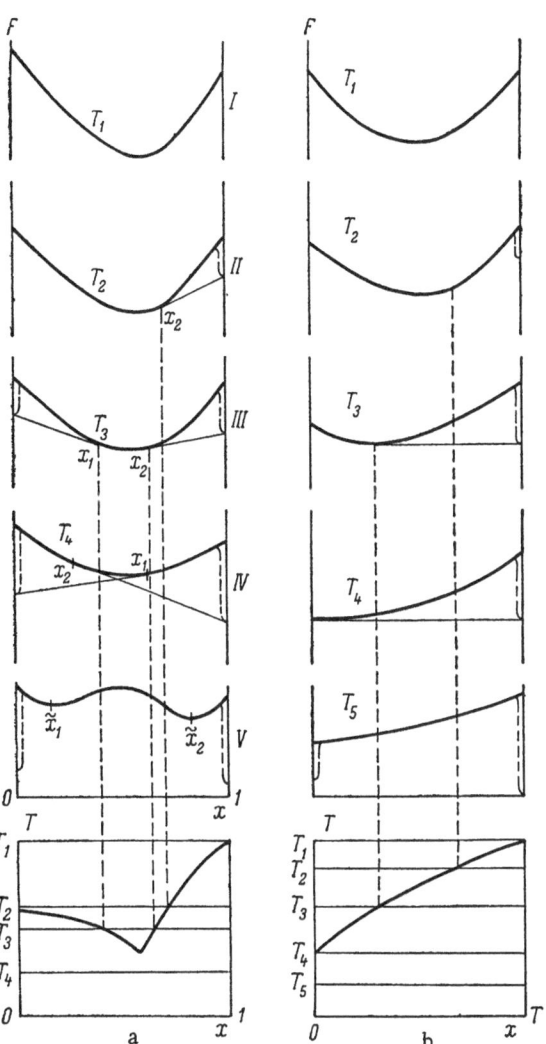

Fig. 5. Dependence of free energy on the composition and temperature in the case of crystallization without preliminary phase separation. a) With eutectic; b) without eutectic.

character and can be plotted from the positions of the points of contact between F(x) and the straight lines from the minima at x = 0 and x = 1, as in Fig. 1 for $T = T_2$. The construction scheme for determination of $\Delta F_V$ in nucleus formation is analogous to the scheme of Fig. 1, based on Eq. (10), which is of general significance.

At temperature $T_1$ a body of any composition decomposes into glasses A and B. When $T = T_2$ (Fig. 4), a glass of composition $x_1 \leq x \leq x_2$ does not undergo demixing, as this would, in this case, be associated with an increase of free energy. A glass of composition $1 > x > x_2$ separates at $T_2$ into a glass of composition $x_2$ and a glass of composition A (x = 1). The laws governing this separation are the same as for the case "a" discussed earlier, since the part of the F(x) curve for $x > x_2$ is a maximum between two minima (x = $x_2$ and x = 1), and it may be likened to the $T_2$ curve in Fig. 1. Analogous reasoning is valid for compositions $x < x_1$. At $T > T_3$ (Fig. 4), separation does not occur at all, as there is only one minimum on the F(x) curve.

It is significant that in phase separation with a solubility diagram of the eutectic type (Fig. 4), the phase of which a large amount should be obtained is the first to nucleate. It differs in this respect from liquation with a dome-shaped solubility diagram (case "a"), when the phase of which a small amount should be formed is the first to form nuclei.

15

**3.** The above scheme is most suitable for describing phase separation effects in completely liquid or glassy phases (with stable or metastable phase separation). If we assume that a new crystalline phase originates by way of an intermediate step of the metastable phase separation type (case 1, see p. 11 ), the scheme also describes this precrystallization stage of phase decomposition. With certain modifications the scheme can also be used for interpretation of effects occurring when the formation of stable regions of a new phase is impossible without simultaneous crystallization (case 2, see p. 11 ).

Figure 5 shows the phase diagram and F(x) plots for various temperatures, both in the case of a eutectic (a) and without (b). The thick lines correspond to the metastable glass phase, and the thin lines give a nominal indication of the decrease of F in crystallization of glasses with compositions close to the stoichiometric. Curve I corresponds completely to the liquid state at $T_1 > T_m$. Curve II ($T = T_2$) corresponds to the appearance at $x > x_2$ of one crystalline phase in a liquid phase of composition $x_2$, and to a completely liquid phase at $x < x_2$. Curve III corresponds to temperature $T_3$ when, dependent on the composition: 1) separation occurs into a crystalline phase A and a liquid phase of composition $x_2$ ($x_2 \leq x \leq 1$); 2) only a liquid phase is present ($x_1 < x < x_2$); 3) separation occurs into phase B and a liquid phase of composition $x_1$ ($0 \leq x \leq x_1$). Curve IV corresponds to temperature $T_4$ — below the solidus temperature (below the eutectic temperature $T_e$), when separation into the crystalline phases A and B occurs. Curve V corresponds to the more general case, when metastable phase separation of type "a" (Fig. 1) exists below the solidus line.

Let us consider in greater detail curve IV for F(x) in the region $x_2 \leq x \leq 1$ (consisting of a thick and a thin line). It may be likened to the $T_2$ curve in Fig. 1 with two minima separated by a maximum, and the earlier reasoning may be used. The formation of a crystal nucleus of phase A is represented as proceeding through the formation of an amorphous nucleus of composition $x \simeq 1$, the $\Delta F_v$ barrier associated with thermodynamic disadvantage, in the given case 2, of formation of a region of glass of stoichiometric composition being overcome. This loss, together with the surface energy $\Delta F_s$, is subsequently made good by the crystallization energy of the nucleus (transition to the thin-line curve IV).

During formation of the nuclei and crystallization of phase A, the composition of the remaining metastable glassy phase shifts to the left, and at $x < x_1$ nuclei of phase B begin to appear. This leftward shift of the composition continues until it is counterbalanced by the opposing effect of crystallization of phase B. It should be pointed out that in the composition range from $x_1$ to $x = 1$, according to Fig. 5, curve IV, nuclei of phase B ($x' = 0$) cannot be formed at all, as here $\Delta F_v(x, x' = 0) > 0$. ‡‡

The boundary of the start of formation of nuclei of phase B, $x_1$, shifts to the right with rise of temperature and with intensification of the influence of the entropy term, responsible for the depth of the minimum in the middle portion of the F(x) curve. There is an analogous boundary, $x_2$, for phase A. At low temperatures, generally speaking, formation of nuclei of both phases is possible at all compositions.

If ordinary metastable phase separation occurs (curve V, Fig. 5), because of its occurrence the compositions $\tilde{x}_1$ and $\tilde{x}_2$ of the deposited glasses approach the stoichiometric values, which facilitates the course of the initial crystallization stages. Moreover, the extent of the phase separation regions apparently determines the dimensions of the future crystals at the initial stages of crystallization. Crystallization of the metastable phase separation regions leads to changes in their composition. For example, the composition of the $\tilde{x}_2$ region shifts toward $\tilde{x}_1$. This should lead to decrease (rearrangement) of the $\tilde{x}_2$ regions.

We may note that the laws governing the formation of crystal nuclei in case 2, when there is no precrystallizational metastable phase separation, are in many respects similar to the laws governing the formation of glassy nuclei in phase separation of the eutectic type. This follows, for example, from a comparison of Figs. 4 and 5, where the curves (including the thin lines) are of the same character. From this it follows, in particular, that if precrystallizational phase separation occurs it must be of eutectic character (Fig. 4) with subsequent crystallization of stoichiometric glassy nuclei. Moreover, it may be supposed that metastable phase separation of the eutectic type, if it is detected in a given glass, is always precrystallizational irreversible phase separation

---

‡‡The sequence of phase deposition in lithium silicate glasses, studied in [8], is apparently associated with the influence of this situation.

which, having once started, does not disappear and culminates in crystallization of the glass at all temperatures below the solidus line. This question requires special experimental investigation.

## 3. The Role of the Nucleation Stage of Crystallization in the Formation of Glassceramics

Characteristic features of glassceramics are high dispersity ($\sim0.1$-$1$ $\mu$) and uniform size of crystals of different phases, joined firmly together. Microcrystallinity and size uniformity appear to be necessary conditions for a strong glassceramic, as in the case of high-strength metallic alloys. From this viewpoint, the question of the nature of the high strength of glassceramics is one form of the question of the nature of the strength of metallic alloys, which has been the subject of many investigations [9]. In any event, it is fairly obvious that a network of very fine crystals, joined directly or by means of thin glassy interlayers, hinders the development of cracks in the specimen along the grain boundaries and within the grains. At the same time, the fine crystals and glassy interlayers themselves have few defects and are very strong; their strength approaches the theoretical value, which is very high both for crystals and for glasses.

The initial (nucleation) stages of crystallization are of decisive significance in the formation of a glassceramic. With the conditions of uniformity and high dispersity taken into account, the following requirements may be formulated.

**1.** The nuclei must form throughout the volume of the glass. Glasses with surface crystallization are unsuitable for production of glassceramics. For example, sodium silicate glasses crystallize from the surface, whereas lithium silicate glasses are capable of volume crystallization, at least in the composition range between lithium metasilicate and $SiO_2$. The difference between the specific volumes or densities of the glass and the crystals formed apparently plays an important part here. A large difference between the specific volumes leads to strong elastic stresses, capable of suppressing volume nucleation and crystal growth. Nucleation is easier at the glass surface and relatively large crystals growing into the glass are formed. The crystallization is accompanied by extensive cracking of the surface layer, probably as a consequence of the stresses which hinder volume crystallization, and also because of the difference between the coefficients of thermal expansion.

For example, it is significant that in high-silica sodium silicate glasses the occurrence of metastable liquation does not cause volume crystallization [13], although the internal interface is very extensive. The crystallization of such glasses proceeds from the surface as before.

**2.** The number n of the nuclei formed at low temperatures in the region of the nucleation rate maximum (see Section 1) should be large, about one per cubic micron ($n \sim 10^{12}$ cm$^{-3}$) or more. The number of nuclei determines the number of the future crystals. A large number of nuclei can be obtained in an acceptable time (several hours or tens of hours) if at least one of the phases crystallizes rapidly enough and if sufficient of it is present.***

The nucleation rate I and, consequently, the number of crystals n depend significantly on the thermodynamic potential barrier $\Delta\Phi^*$ (or the free energy barrier $\Delta F^*$). The value of $\Delta\Phi^* = \Delta\Phi_V^* + \Delta\Phi_S^*$ may be regulated by suitable selection of the chemical composition of the crystallizing glass, as well as by the temperature (2). As the composition moves away from the stoichiometric, $\Delta\Phi^*$ (or $\Delta F^*$) increases as $\Delta\Phi_V^* \leq 0$ or $\Delta F_V^* \leq 0$ increases, and I and n therefore decrease. The lowest values of I and n (if the dependence of $\Delta\Phi_S^*$ on composition is disregarded) are obtained for intermediate compositions $x_2 \leq x \leq x_1$ (Fig. 5, curve IV). By shift of the composition toward the stoichiometric it is possible to achieve accelerated nucleation of one phase and retardation or complete suppression of nucleation of the other.

The nucleation rate also depends on the activation potential of increase of $\Delta\Phi_A^*$, which in its turn depends on the composition x. This factor may introduce significant changes into the scheme described. For example, if $\Delta\Phi_A^*$ is considerably greater for composition B (x = 0) than for composition A (x = 1), phase B and not A is deposited first even when the composition of the original glass is close to A.

***The tendency to crystallization must not be so high that the glass cannot be obtained without crystallization during the forming at all; such crystallization (with movement due to high temperatures) is always too coarse and heterogeneous owing to accidental appearance of a small number of rapidly growing crystals.

On the whole, the sequence of deposition of the crystalline phases is determined by the magnitude of the exponent $\Delta\Phi^* + \Delta\Phi_A^*$ in (1). As the different phases are deposited, $\Delta\Phi^* + \Delta\Phi_A^*$ increases and the crystallization tendency of the remaining glass diminishes.

The presence of phase separation may complicate the picture, as it leads to the appearance of glasses differing in composition from the original. For example, phase separation of type "a" (noneutectic, see Fig. 1) leads to the appearance of glass nuclei with a composition close to the composition of the crystalline phase of which a small amount should be formed, and it is this phase which may crystallize first although the composition of the original glass is nearer to that of the other crystalline phase.

The following should be noted in relation to the problem of obtaining a large number of nuclei. It is known that a microdisperse mass is characteristically formed when a melt of the eutectic composition solidifies. This is associated, in particular, with the fact that a eutectic melt before crystallization is itself a mixture of minute particles with a structure close to that of the future crystals. This view is confirmed by structural investigations of eutectic melts [10]. It is also supported by qualitative theoretical considerations [11]. For example, we showed above that the assumption of approximate additivity of energy in relation to the composition (13), which follows from the hypothesis of structural additivity in relation to A and B groupings, is sufficient for the formation of a eutectic. All this suggests that in a glass of the eutectic (or nearly eutectic) composition the conditions for simultaneous uniform formation of numerous crystal nuclei, ensuring microdisperse crystallization during sufficiently long low-temperature treatments, are satisfied most fully.

It should be pointed out that a heterogeneous structure on the atomic and molecular scale is apparently characteristic for glass melts of complex composition in general (regardless of the tendency to phase separation). These primary heterogeneities in the glass subsequently act as nucleation centers, which leads to a peculiar autocatalysis of the crystallization of complex glasses. No such primary uniform heterogeneity exists in simple one-component glasses, and in such cases uniform crystallization is hindered by the absence of nucleation centers.

**3.** Fine and uniform crystallization is by itself insufficient for the formation of a strong glassceramic. The crystals themselves must be firmly joined by concretion or by means of glassy interlayers. The principle of the minimum surface energy $\sigma$ at the interphase boundary is obeyed in the formation of nuclei of a new phase. A low value of $\sigma$ indicates, in particular, good wetting of the nucleus by the medium, i.e., effective and strong mechanical bonding with the glassy medium. Firm bonding of the crystals to the medium and absence of stresses and cracks are energetically advantageous in the formation and subsequent growth of crystals in glass, as work must be done in the formation of such stresses and cracks. The glass itself, by virtue of the principle of the minimum thermodynamic potential and statistical laws, selects the crystallization route and sequence of phase deposition which ensure the minimal mechanical stresses in the glass and the minimal amount of cracks, i.e., the maximum strength of the crystallization product.††† The only important thing is that such a route should exist for a given glass. This depends in the main on the chemical composition of the glass. In glasses of simple composition no such route may exist in practice, as the number of possible crystalline phases is small. In glasses of more complex chemical composition the deposition of a larger number of phases (stable or metastable) is possible, and there is a wider choice between the various crystallization routes ensuring minimum stresses and cracking.

A small size of the deposited crystals is important, since it is evident that the formation of cracks at the boundaries with the glass due to differences in properties between the glass and the crystals is less probable than in the case of large crystals. The high strength of the smallest crystals in conjunction with the high strength of the thin glassy interlayers ("fibers," "films") and absence or difficult development of boundary cracks lead, as already noted, to a sharp increase of the mechanical strength of the glassceramic as a whole.

---

††† This, in particular, distinguishes glassceramics from vitroceramic materials; in the production of the latter the composition and size of the crystals and the glassy matrix are predetermined, and in consequence the bond between them is weaker than in glassceramics.

During prolonged high-temperature treatments, when the crystals grow further, metastable forms are converted into more stable forms, and residual glass undergoes crystallization, the mechanical properties of glass-ceramics should deteriorate, as crack growth is inevitable. Experience in general confirms this.

All the foregoing refers to spontaneous crystallization. Catalyzed crystallization is generally used in the production of industrial glassceramics. There are three main catalytic routes, all concerned with the nucleation stage of crystallization. In the first route impurities of noble metals or other substances of low solubility and easy crystallization are added to the glass to form crystal nuclei which act as seeds for crystallization of the main phase. Here the presence of seed crystals leads either to easier deposition, on their surfaces, of the phase which crystallizes first in spontaneous crystallization or (which may prove more important) to a change in the sequence of phase deposition from the sequence in spontaneous crystallization. Apparently the latter is possible only in cases of epitaxy, with similarity of the lattices of the catalyst and the deposited phase for which the catalyst crystals act as finished nuclei.

The second route consists of irradiation of the glass by hard radiations and particles. Absorption of such radiations increases the defectiveness of the glass structure and raises its thermodynamic potential owing to absorption of energy. All this makes the glass structure less stable and more susceptible to rearrangements associated with nucleation and crystallization. More complex secondary photochemical processes leading to the formation of crystal nuclei are also possible.

The third route involves the use of glasses with phase separation. Metastable phase separation leads to appearance of primary microheterogeneity in the glass, to an enormous growth of the interfacial area between the glassy phases arising during the separation. It is believed that this surface subsequently acts as the crystallization catalyst [12]. There is as yet no direct proof of this catalysis mechanism. In this connection we may refer to sodium silicate glasses, in which metastable phase separation may be produced and prevented by increase of temperature without appreciable crystallization, which occurs at higher temperatures and at the surface. This example is a very vivid illustration of the absence of a direct connection between phase separation‡‡‡ and crystallization of glass.

It must be pointed out that inclusion of the crystallization of glasses undergoing phase separation in the class of catalyzed crystallization is itself debatable. Metastable phase separation in itself is a rightful physico-chemical property of a larger number of silicate glasses, and it depends on the chemical composition and temperature of the glass. It is preferable to speak merely of the strong influence of phase separation on the character of the subsequent crystallization rather than of some sort of catalysis of crystallization. Moreover, it is also possible that the role of metastable phase separation lies not in growth of the interfacial area, but in the formation of very small glassy regions, crystallization of which (independently of the part played by the surface) may result in the formation of very small crystals, determined by the size of the phase separation regions. In general, the significance of metastable phase separation in the formation of glassceramics is not yet clear, although many systems capable of forming good glassceramics have the property of phase separation.

In the lithium silicate system, suitable heat treatment leads to the formation of fairly strong microcrystalline products in the composition range from lithium disilicate to $SiO_2$, where metastable phase separation occurs [8, 13]. However, the best result is obtained by crystallization in the intermediate composition range from the meta- to the disilicate, where no phase separation occurs. Examples of spontaneous crystallization in the absence of phase separation or catalysts can probably be found among glasses of other systems. These are likely to be easily crystallizing glasses with a high nucleation rate I and at the same time with a low rate of linear growth in the crystals during low-temperature treatments (the temperatures of the maximum nucleation rate and maximum crystal growth rate differ sharply). In general, the presence of a large number of nuclei is the only necessary condition for formation of a microcrystalline material, independently of the manner in which this is achieved — by catalyzing additives, by irradiation of the glass in conjunction with subsequent heat treatment, by liquation, or merely in consequence of a relatively high rate I of spontaneous nucleation.

---

‡‡‡It seems that noneutectic relaxational phase separation of the liquid type occurs in this case. In contrast to the eutectic type, it is not directly associated with crystallization (see Section 2), but is an independent intermediate phase change.

Investigation of spontaneous crystallization and comparison of spontaneous and catalyzed crystallization are the main routes for elucidation of the action of crystallization catalysts and of the mechanism of formation of glassceramics. Such investigations are also of great importance for understanding the nature of the glassy state.

Summary

1. There are two types of relaxation processes in glass: vitrification and crystallization.

2. Two possible crystallization mechanisms are suggested for complex glasses: with and without precrystallizational phase separation.

3. An outline is given of a statistical thermodynamic theory of the formation of nuclei of a new phase, in relation to metastable phase separation of the noneutectic and eutectic types.

4. Certain consequences of the theory are examined in relation to regulation of the nucleation process and the rate and sequence of phase deposition in the formation of glassceramics.

5. The high mechanical strength of glassceramics is explained on the basis of the general principles of statistical physics and thermodynamics and of concepts of the mechanical strength of multicrystalline materials.

LITERATURE CITED

1. P. P. Kobeko, Amorphous Substances, Izd. AN SSSR, Moscow, 1952.
2. M. V. Vol'kenshtein and O. B. Ptitsyn, Zhur. Tekh. Fiz. 26(10):2205, 1956.
3. Ya. I. Frenkel', Statistical Physics, Izd. AN SSSR, Moscow-Leningrad, 1948; Ya. S. Umanskii, B. N. Finkel'shtein, and M. E. Blanter, Physical Principles of Metal Science, Metallurgizdat, 1949.
4. J. H. Hollomon and D. Turnbull, Advances in Metal Physics, Vol. 1 [Russian translation], Metallurgizdat, 1956.
5. R. L. Myuller, Zhur. Prik. Khim. 28(363):1077, 1955; Trudy Tomsk. Univ. 145:33, 1957.
6. R. Becker, Ann. Physik 32(1):128, 1938.
7. B. Ya. Pines, Essays on Metal Physics, Izd. Khar'kovskogo Univ., 1961.
8. A. M. Kalinina, V. N. Filipovich, V. A. Kolesova, and I. A. Bondar', this collection, p. 53.
9. R. I. Garber and I. A. Gindin, Uspekhi Fiz. Nauk 70:57, 1960.
10. V. I. Danilov, Structure and Crystallization of Liquids, Izd. AN UkrSSR, Kiev, 1956.
11. G. M. Bartenev, Thermodynamics and Structure of Solutions, Izd. AN SSSR, Moscow, 1959.
12. C. D. Stookey, Glastechn. Ber., Special volume, V. Intern. Glass Congress 32K(5):1, 1959.
13. N. S. Andreev, D. A. Goganov, E. A. Porai-Koshits, and Yu. G. Sokolov, this collection, p. 47.

# INFLUENCE OF CRYSTAL-CHEMICAL SIMILARITY

# ON HETEROGENEOUS CRYSTALLIZATION OF GLASSES

## É. M. Rabinovich

Substances which cause crystallization of a volume-homogeneous microcrystalline structure when added to glasses may be divided into two groups, differing in the mechanism of their action.

Group A includes substances capable of forming minute crystals in the glass, of low solubility in silicate glasses; these crystals act as crystallization centers for the silicates. The noble metals and copper are such substances. They have a high tendency to reduction from the ionic to the neutral state in silicate glasses; the neutral atoms cannot react chemically with the silicate melt and the solubility of the atoms in the glasses is extremely low. Therefore, introduction of even small amounts of these additives (1% in the case of copper and 0.01% in the case of platinum) results in precipitation, either after thermal treatment or without, of the free metals in the form of colloidal crystals (20-100 A), the lattice parameters of which do not differ from the parameters of the same metals in the massive state [1].

Group B comprises substances such as $TiO_2$, $ZrO_2$, etc., which are readily soluble in silicate glasses, so that large amounts (5-20% by weight) must be added. These substances rarely separate out as the independent oxides at the start of crystallization, while their salts (such as $MgTiO_3$), because of the complex formation of their own crystal structure (in contrast to Au, Ag, or Cu) and of the complex position of $Ti^{4+}$, $Zr^{4+}$, $Mo^{6+}$, and $W^{6+}$ ions in the glass structure, have no decisive advantages over silicates in relation to spontaneous formation of crystal nuclei which could then act as crystallization centers for the silicates. In conjunction with $SiO_2$, the oxides of group B form very extensive immiscibility regions (from 7 to 80% in the $TiO_2$-$SiO_2$ system) [2]. The region is usually narrowed by addition of other components, but continues to occupy a considerable position in the phase diagram. Therefore, almost any glass composition can be taken into the metastable liquation region by introduction of group B oxides. The subsequent crystallization is due to the very extensive interface between the metastable "phases" [3].

Fluorine as a crystallization catalyst belongs to group A, as it usually separates out not in the form of fluosilicates, but as fluorides [4], which are relatively insoluble in glasses although they can subsequently react chemically in the solid state with the silicates growing on them [5].

This paper is concerned with a study of the influence of crystal-chemical similarity between heterogeneous crystallization centers formed by group A substances and the primary silicate phase on crystallization of glasses [6].

Crystal-chemical similarity means similarity, within about 10%, between the linear lattice parameters of one substance and the lattice parameters, or multiples of them, of another substance [7-9]. If one ionic crystal grows on another, only direct overgrowth is possible (see figure, a); if an ionic crystal grows on a metal crystal and vice versa, direct overgrowth (figure, b) or diagonal overgrowth (figure, c) may occur; in the latter, the parameters of the one substance are equal to the parameters of the other multiplied by $\sqrt{2}$ or $\sqrt{2}/2$. If there is similarity in three pairs of linear parameters (all for direct overgrowth or two pairs for diagonal and one for direct), overgrowth of a single crystal of one substance by a single crystal of the other is possible. If two pairs of parameters are similar, crystals of the second substance may grow from the corresponding faces of the supporting crystal. Under favorable conditions similarity of one pair of parameters may be sufficient for one substance to induce crystallization of another. In the case of monoclinic and triclinic crystals, similarity is difficult to

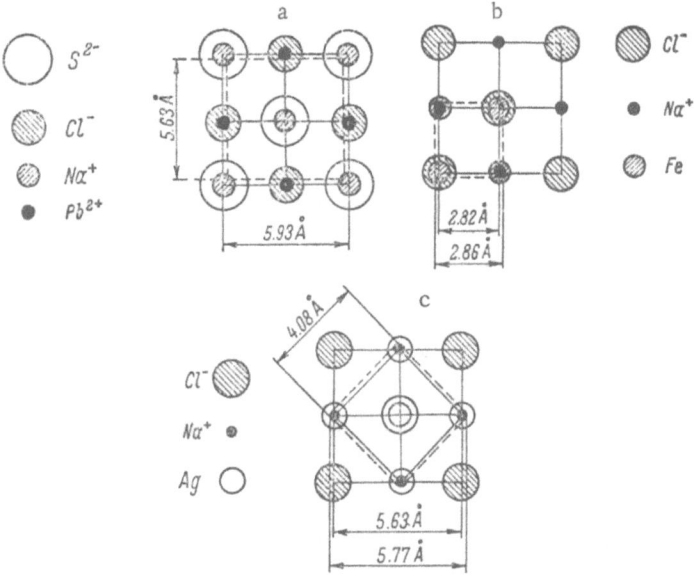

Overgrowth of crystals of one substance by crystals of another. a)
NaCl (a = 5.63 A) on PbS (a = 5.93 A) (direct); b) Fe (a = 2.86 A)
on NaCl (direct); c) Ag (a = 4.08 A) on NaCl (diagonal). Broken
lines indicate the dimensions of undeformed cells of the overgrow-
ing substances.

detect; the angles often approach 90° and it becomes possible to neglect the difference from 90° to regard these
crystals as approximately rhombic.

We investigated the effect of added metals of group A on the crystallization of several binary and ternary
glasses, the primary crystalline phases of which are known from the phase diagrams. Table 1 gives the lattice
parameters of a number of silicates and demonstrates their relation to the parameters of the cubic lattice of Ag
(4.078 A), Pt (3.91 A), Cu (3.61 A), and NaF (4.62 A) [10, 11].

The compositions of the glasses studied and the results of the crystallization studies are given in Table 2.

In addition to silver and copper, a thermal reducing agent (0.2-0.5% $SnO_2$) and an optical sensitizer (0.05%
$CeO_2$) were added to all the glasses (including those without added metals) in order to make it possible to utilize
the photosensitive process for intensifying separation of colloidal copper or silver crystals. Naturally, if crystallization is not preceded by a characteristic colloidal coloration, i.e., if heterogeneous crystallization centers are
not formed, it becomes impossible to determine the influence of crystal-chemical similarity on crystallization.

Glasses of compositions 1, 2, 4-6 were melted in a gas furnace in 100-ml quartz crucibles, and composition 10 in 1-liter crucibles; glasses 3, 7-9 were melted in an electric furnace in platinum crucibles at 1200°. The
melts were poured out onto metal slabs and rolled. Pure pieces showing no signs of crystallization or incomplete
melting were taken for the crystallization studies. Crystallization was investigated in a gradient electric furnace
with temperature recording at 10 points (with the ÉPP-09 instrument) over the 350-1050° range. Ceramic crystallizing vessels with the specimens were heated up to the required temperature during 2 hours together with the
furnace, kept at that temperature for 2 hours, taken out, and cooled in air. Specimens of the same composition
but with different additives were treated simultaneously. We were interested in the differences between the
lower crystallization temperatures of specimens with and without silver or copper, in the nature of crystallization,
and in the primary crystalline phase, which was determined by the x-ray diffraction method.

The results of the investigations (Table 2) agree with the hypotheses which follow from Table 1. Small
amounts of copper and silver have almost no effect on the density, refractive index, and expansion of glasses before crystallization, but have a significant influence on the crystallization process if crystal-chemical similarity
exists between the metals and the primary silicate phase.

TABLE 1. Crystallographic Data for Silica and Certain Silicates [10-12]

| Crystal | Formula | System | Lattice parameters, A | | | Parameters of crystallization centers, A | Deviation, % |
|---|---|---|---|---|---|---|---|
| | | | a | b | c | | |
| α-Cristobalite | $SiO_2$ | Cubic | 7.16 | — | — | $2a_{Cu} = 7.22$ <br> $\sqrt{2}a_{Ag} = 5.77$ <br> $2a_{Ag} = 8.16$ <br> $2a_{Pt} = 7.82$ | + 0.83 <br> −24.1 <br> +12.3 <br> + 8.4 |
| α-Tridymite | $SiO_2$ | Hexagonal | 5.00 | — | 8.26 | $2a_{Ag} = 8.16$ <br> $\sqrt{2}a_{Ag} = 5.77$ <br> $\sqrt{2}a_{Cu} = 5.10$ | for c: − 1.22 <br> for a: +13.4, <br> + 1.96 |
| Lithium metasilicate | $Li_2SiO_3$ | Rhombic | 5.43 | 4.66 | 9.41 | $\sqrt{2}a_{Ag} = 5.77$ <br> $2a_{Ag} = 8.16$ | for a: + 5.9 <br> for b: −14.2 <br> for c: −15.3 |
| Lithium disilicate | $Li_2Si_2O_5$ [12] | Rhombic | 5.82 | 14.46 | 4.79 | $3a_{Ag} = 12.24$ | for a: − 0.87 <br> for b: −19.8 <br> for c: −17.4 |
| Nepheline | β-$Na_2O \cdot Al_2O_3 \cdot 2SiO_2$ | Hexagonal | 10.0 | — | 8.38 | $2a_{Ag} = 8.16$ | for c: − 2.70 |
| α-Carnegieite | α-$Na_2O \cdot Al_2O_3 \cdot 2SiO_2$ | Cubic | 7.37 | — | — | $2a_{Cu} = 7.22$ | − 2.08 |
| Muscovite | $KAl_3Si_3O_{10}(OH, F)_2$ | Monoclinic* | 5.18 | 9.02 | 20.04 | $4a_{NaF} = 18.5$ | for b: + 2.38 <br> for c: − 8.32 |
| Anorthite | $CaO \cdot Al_2O_3 \cdot 2SiO_2$ | Triclinic** | 8.21 | 12.95 | 14.16 | $2a_{Ag} = 8.16$ <br> $3a_{Ag} = 12.24$ <br> $2a_{NaF} = 9.24$ <br> $3a_{NaF} = 13.86$ | for a: − 0.61 <br> for b: − 5.80 <br> for a: +11.1 <br> for b: + 6.57 <br> for c: − 2.16 |

* β = 95°30'.

** α = 93°13', β = 115°56', γ = 91°12'.

23

TABLE 2. Batch Compositions of the Investigated Glasses ($t_s$ — liquidus temperature)

| Specimen No. | Composition, % by wt. | | | | | Relation to phase diagram | Additive | | Color before appearance of crystals | Lowest cryst. temp.,* °C | Nature of initial crystallization |
|---|---|---|---|---|---|---|---|---|---|---|---|
| | $SiO_2$ | $Al_2O_3$ | CaO | $Na_2O$ | $Li_2O$ | | nature | amt, % over 100% | | | |
| 1a | 89 | – | – | 11 | – | $\alpha$-Cristobalite field, $t_s = 1500°$ | – | – | – | 720 | Surface |
| 1b | | | | | | | $Cu_2O$ | 0.5 | Red | 550 | Volume |
| 1c | | | | | | | $AgNO_3$ | 0.2 | Brown | 520 | Volume |
| 2a | 86 | – | – | 14 | – | $\alpha$-Tridymite field, $t_s = 1360°$ | – | – | – | 720 | Surface |
| 2b | | | | | | | $Cu_2O$ | 0.5 | Red | 650 | Volume |
| 2c | | | | | | | $AgNO_3$ | 0.2 | Brown | 630 | Volume |
| 3a | 79 | – | – | 21 | – | $\alpha$-Tridymite field, $t_s = 1100°$ | – | – | – | 750 | Surface |
| 3b | | | | | | | $Cu_2O$ | 0.5 | Red | 750 | Volume |
| 3c | | | | | | | $AgNO_3$ | 0.2 | Yellow | 520 | Volume |
| 4a | 74 | – | 18 | 8 | – | $\alpha$-Tridymite field, $t_s = 1300°$ | – | – | – | 720 | Surface |
| 4b | | | | | | | $Cu_2O$ | 0.2 | – | 720 | Surface |
| 4c | | | | | | | $AgNO_3$ | 0.1 | Yellow | 720 | Surface |
| 5a | 60.4** | 0.4 | 28.7 | 10.5 | – | Near devitrite composition $Na_2O \cdot 3CaO \cdot 6SiO_2$, $\alpha$-$CaSiO_3$ field | – | – | – | 830 | Surface |
| 5b | | | | | | | $Cu_2O$ | 0.5 | Red from 580° | 820 | Surface |
| 5c | | | | | | | $AgNO_3$ | 0.2 | Not irradiated | 820 | Surface |
| | | | | | | | | | Irradiated yellow*** | 680 | Dense crystal. crust down to irrad. depth |
| 6a | 72 | – | 18 | 10 | – | $\beta$-$CaSiO_3$ field near devitrite field, $t_s = 1200°$ | – | – | – | 870 | Surface film |
| 6b | | | | | | | $Cu_2O$ | 0.2 | Red | 870 | Surface |
| 6c | | | | | | | $AgNO_3$ | 0.1 | Yellow | 700 | Volume; intensive from 800° |
| 7a | 72 | – | 10 | 18 | – | Devitrite field, $t_s = 950°$ | – | – | – | 850 | Surface film |
| 7b | | | | | | | $Cu_2O$ | 0.5 | Dark red | 850 | Surface |
| 7c | | | | | | | $AgNO_3$ | 0.2 | Yellow | 800 | Volume |
| 8a | 80.8 | 3.0 | – | – | 16.2 | $Li_2Si_2O_5$ field, $t_s = 1000°$ | – | – | Not irradiated | 630 | Volume |
| 8b | | | | | | | $AgNO_3$ | 0.2 | Irradiated yellow from 480° | 630 / 520 | Volume |

TABLE 2 (continued)

| Specimen No. | Composition, % by wt. | | | | | Relation to phase diagram | Additive | | Color before appearance of crystals | Lowest cryst. temp.,* °C | Nature of initial crystallization |
|---|---|---|---|---|---|---|---|---|---|---|---|
| | $SiO_2$ | $Al_2O_3$ | $CaO$ | $Na_2O$ | $Li_2O$ | | nature | amt., % over 100% | | | |
| 9a | 78.7 | 9.3 | — | — | 12.0 | Field of solid solutions of $\beta$-spodumene, $t_s = 1075°$ | — | — | — | 630 | Volume |
| 9b | | | | | | | $AgNO_3$ | 0.2 | Not irradiated / Irradiated | 630 | |
| 10a | 42.3 | 35.9 | — | 21.8 | — | Nepheline and carnegieite composition | — | — | — | 1000 | |
| 10b | | | | | | | $Cu_2O$ | 0.5 | Red | 900 | Volume |
| 10c | | | | | | | $AgNO_3$ | 0.2 | — | 1000 | |

* Under the experimental conditions (see text).
** Composition by analysis.
*** Irradiated with ultraviolet light (PRK-7 lamp).

In glass compositions 1-4 the primary phases were identified by the x-ray method as tridymite. It is likely that in glass 1b crystobalite is first formed (on the copper crystals) and is then rapidly converted into tridymite in the melt containing alkali oxides [13], whereas, in glass 1c, tridymite is formed at once (on silver crystals). In glasses 2-4, where cristobalite cannot form under equilibrium process conditions (the liquidus temperature is below 1470°), copper crystals produce an intense red color but accelerate crystallization in the case of glass 2 only. The catalytic action of silver, as is to be expected, is retained in glasses 2 and 3 but is for some reason lost in glass 4, despite the appearance of a colloidal color.

Addition of more than 0.1% silver to composition 3 lowers the lower crystallization temperature of the glasses from 700 to 540°. Additions of platinum or palladium (a = 3.86 A) in amounts up to 0.1% to composition 3 have no influence on crystallization, although they cause characteristic gray colorations. It follows that the permissible difference between the parameters of tridymite or cristobalite and of group A additives is less than 10%.

Silver crystals accelerate crystallization of lithium disilicate (compare glasses 8a and 8b). The influence of silver on crystallization of glasses of composition 9 could not be investigated, as even irradiation did not result in colloidal coloration during thermal treatment. Copper crystals accelerate crystallization in the nepheline glass 10. Petrographic investigations suggest that the primary crystalline phase is $\alpha$-carnegieite, similar in structure to $\alpha$-cristobalite (Table 1); x-ray structural investigation is difficult in this case, as the standard x-ray patterns in [10, 11] do not appear to be quite reliable. Silver does not produce a preliminary coloration in glass 10c and has no influence on crystallization. In all the glasses, acceleration of crystallization by metals is accompanied by a change in its nature: it occurs throughout the volume of the specimen.

Glasses in which crystals of the mica type, phlogopite and muscovite, are formed on fluoride crystals by reactions between the latter and silicates have been prepared abroad [5] and by us [14]. This result is satisfactorily explained by the crystal-chemical similarity principle and confirms that fluoride belongs to group A.

These results refute the view that the influence of group A additives is due only to formation of an extensive interface [15]. In cases where there was no similarity between the crystal lattices, the crystallization was independent of the additive, started at the surface, and was macrocrystalline in character. In glasses of composition 4, even similarity between silver and tridymite had no effect on crystallization. Opal glasses in which the presence of fluorides does not initiate silicate crystallization are also known [4]. Apparently, in absence of similarity the surface energy of small group A crystals is insufficient to effect ordering of the cumbersome silicate particles.

We investigated crystallization of three glass compositions in which devitrite is the primary phase (compositions 5-7, Table 2). In determination of the primary phase from the equilibrium diagram for the $Na_2O$-$CaO$-$SiO_2$ system, it must be remembered that in crystallization of glasses rather than melts at temperatures below 1045° (incongruent melting of devitrite), the devitrite region extends at the expense of the wollastonite region and includes the devitrite composition. It may be seen that in all three compositions of the devitrite region the crystallization of devitrite is accelerated in presence of colloidal silver crystals, whereas copper crystals do not influence crystallization. Unfortunately, there is no information in the literature on the lattice parameters of devitrite, and precise interpretation of the x-ray diffraction patterns of crystals in the rhombic system is very difficult [16].

These preliminary results show that catalyzed crystallization of glasses under the influence of group A additives has a crystal-chemical mechanism; similarity between the crystal lattices of the crystallization centers and of the primary silicate phase is necessary to ensure such crystallization.

## LITERATURE CITED

1. R. D. Maurer, J. Appl. Phys. 29(1):1, 1958.
2. R. C. Devries, R. Roy, and E. F. Osborn, Trans. Brit. Ceram. Soc. 53:525, 1954.
3. R. Roy, J. Am. Ceram. Soc. 43(12):670, 1960.
4. S. I. Sil'vestrovich and É. M. Rabinovich, Trudy MKhTI im. D. I. Mendeleeva No. 27, Moscow, 1959; Zhur. Priklad. Khim. 32(8):1690, 1959.
5. St. N. Lungu and D. Popescu-Has, Industria Usoara No. 2, 63, 1958.
6. S. D. Stookey, Glastechn. Ber., Special Volume, V. Intern. Glass Congress 32K(5):1, 1959.
7. J. Brück, Ann. Phys. 26:250, 1936.
8. P. D. Dankov, Zhur. Fiz. Khim. 20(8):853, 1946.
9. V. D. Kuznetsov, Crystals and Crystallization, Gostekhizdat, Moscow, 1953.
10. Cumulative Index of x-Ray Diffraction Data, ASTM, Philadelphia, 1954.
11. V. I. Mikheev, X-Ray Index to Minerals, Gosgeoltekhizdat, Moscow, 1957.
12. F. Liebau, Zs. Naturforsch. 15a(7):467, 1960.
13. I. S. Kainarskii and É. V. Degtyareva, Doklady Akad. Nauk SSR 91(2):355, 1953.
14. S. I. Sil'vestrovich and É. M. Rabinovich, Zhur. VKhO im. D. I. Mendeleeva 5(2):186, 1960.
15. W. A. Weyl, Sprechsaal fur Glas-Email-Keramik 93(6):128, 1960.
16. T. N. Keshishyan and B. G. Varshal, Collection of Scientific Papers on Silicate Chemistry and Technology, Promstroiizdat, Moscow, 1956.

# THE PRECRYSTALLIZATION PERIOD IN GLASS
# AND ITS SIGNIFICANCE

## I. I. Kitaigorodskii and R. Ya. Khodakovskaya

Investigations carried out in the Department of Glass, the D. I. Mendeleev Moscow Institute of Chemical Technology, showed that the so-called p r e c r y s t a l l i z a t i o n  p e r i o d , in which the temperature conditions have a very strong influence on the subsequent crystallization of the glass, must be distinguished in the course of catalyzed crystallization of glass. Elucidation of the nature and character of this effect is not only of theoretical importance, but has great practical significance in view of the fact that formed glassware is annealed in the temperature range of the precrystallization period and the annealing therefore has a considerable influence on the structure and properties of glassceramics.

In studies of the formation of glassceramics we used a glass composition based on cordierite in the $SiO_2$-$Al_2O_3$-MgO system. Oxides of group IV metals ($TiO_2$, $SnO_2$, $ZrO_2$, and PbO) and F were used as the catalysts.

A multiple investigation technique was used, including x-ray phase and differential thermal analyses, electron microscopy, and simultaneous determinations of properties which respond to the slightest changes in the structure of the material. This made it possible to follow the course of catalyzed crystallization and to establish the relationships between the properties, structure, and phase composition of the material and the heat-treatment conditions.

Differential thermal analysis of the glass showed that the first crystalline phase begins to form at about 815° (Fig. 1). The entire temperature range below this is the precrystallization period in the glass.

It must be pointed out that there is no sharp difference between the processes occurring in the glass during the precrystallization period and during the first stage of crystallization. Under certain conditions, crystallization centers may form in the temperature range of the precrystallization period. Moreover, prolonged heat treatment during the precrystallization period may ultimately lead to crystallization of the glass. This applies mainly to the higher temperatures of the precrystallization period (the rising branch of the endothermic effect in Fig. 1). It may be assumed that at these temperatures the glass has a precrystallization period in time.

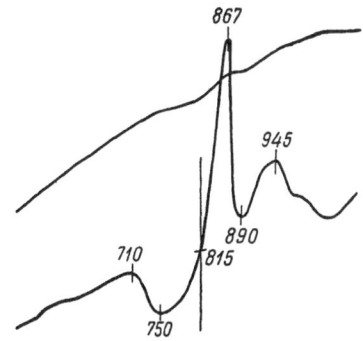

Fig. 1. Thermogram of the original glass.

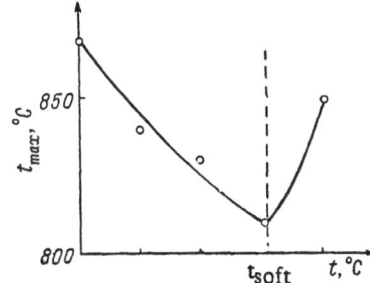

Fig. 2. Dependence of the maximum temperature of the exothermic effect of mullite formation on the heat-treatment temperature of glass in the precrystallization period.

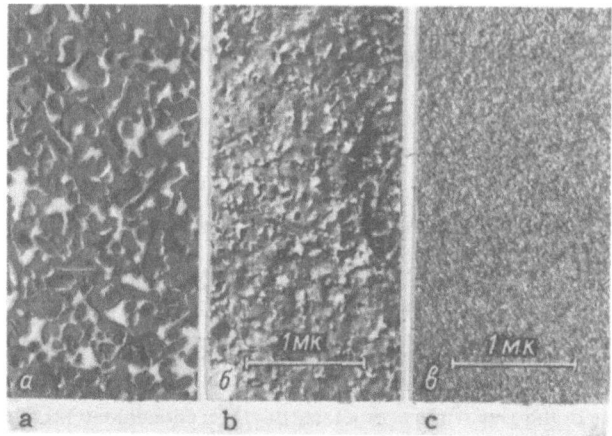

Fig. 3. Structures of crystalline glass materials, magnification 20,000. a, b) Structure of opaque glassceramics obtained without special treatment of the glass in the precrystallization period at 1000 and 900°, respectively; c) structure of transparent crystalline glass material heat-treated at 900° with preliminary treatment for 2 h near the softening temperature.

Crystallization of glass in a gradient furnace in the temperature range between 700 and 1200° without preliminary heat treatment in the precrystallization period showed that with a holding time of 2 h the first crystalline phase detected by x-ray phase analysis appears at 850°. The glass becomes opaque. Further increase of temperature raises the degree of crystallization and makes the structure of the glassceramic more compact; this is accompanied by recrystallization processes and formation of new crystalline phases. As a rule, the first crystalline phase in the crystallization of cordierite compositions is mullite (850°). Higher temperatures lead to the separation of quartz and spinel (900-950°) or quartz and sapphirine (1000°). Finally, at 1200°, the crystalline phase of the glassceramic consists mainly of cordierite. Of titanium minerals, rutile (950-1200°) and possibly geikielite (at lower temperatures) are found. Accordingly, the properties of glassceramics, which are determined by their structure and phase composition, vary over a wide range in accordance with the heat-treatment conditions: the coefficient of expansion from $120 \cdot 10^{-7}$ to $15 \cdot 10^{-7}$ degree$^{-1}$ and the strength from 42 to 8 kg/mm$^2$ (tested with laboratory rods 5 mm in diameter); the dielectric properties vary considerably.

Preliminary heat treatment of the glass in the precrystallization period, in the low-temperature range (in the softening range and considerably below it),has the following effects.

1) The character of the crystallization process is changed sharply — the lower limit of visible crystallization shifts toward higher temperatures (by 150-200°); the glass darkens slightly but remains transparent up to 930° and slightly opalescent up to 950°; the crystallization of glasses without and with preliminary low-temperature heat treatment becomes visually identical only at 1200°.

2) There are radical changes in the phase composition of the glassceramics and in the sequence of deposition of the crystalline phases; the formation temperature of the corresponding titanium minerals is raised and the separation temperature of the first crystalline phase (mullite) is lowered (Fig. 2).

The presence of a minimum on the curve at a temperature close to the softening temperature of the glass is significant. This region of the precrystallization period has the greatest influence on the catalyzed crystallization of glass. With two-hour exposures only this region causes the above anomaly in the course of glass crystallization. The phase composition of the material changes at the same time: mullite and not quartz is the main crystalline phase in glassceramics. Material containing the crystalline phase remains transparent up to 950°. Therefore, the essential nature of the above anomaly is that a wide region of transparent glassceramics is formed (in the stated temperature range, 800-950°) if the glass is subjected to preliminary heat treatment in the precrystallization period.

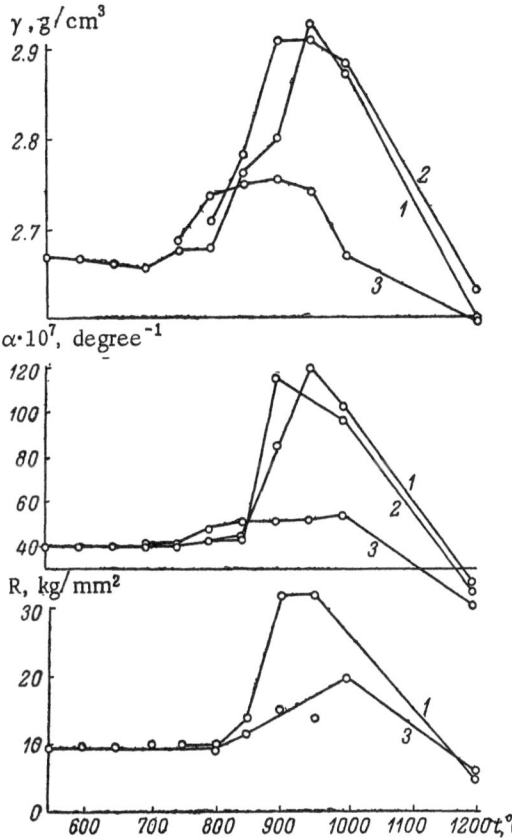

Fig. 4. Variations of the density $\gamma$, the coefficient of thermal expansion $\alpha$, and the strength R with the crystallization temperature with different temperature conditions in the precrystallization period. 1) Material obtained without preliminary heat treatment in the precrystallization period; 2) heat treatment in the precrystallization period at a temperature 50° below the softening point (2 h); 3) heat treatment in the precrystallization period near the softening point (2 h).

The glassceramics are characterized by an extremely disperse and homogeneous structure (the crystal size does not exceed some hundredths of a micron), which differs sharply from the structure of opaque glassceramics formed without preliminary heat treatment in the precrystallization period (Fig. 3).

It is not possible to deal in greater detail with the influence of the preliminary heat-treatment conditions on crystallization. We must merely note that it is most pronounced at the early crystallization stage (at 900°).

Increase of the crystallization temperature leads to a "leveling" of the phase composition and structure of glassceramics regardless of the precrystallization period conditions. At 1200°, the crystalline phase of the glassceramic consists of cordierite and rutile r e g a r d l e s s of the temperature conditions in the precrystallization period, and the structure of the glassceramics is identical. Changes in the structure and phase composition of glassceramics are reflected in the nature of the variations of their properties (Fig. 4).

Figure 4 shows that preliminary heat treatment of the glass for 2 h at a temperature 50° below the softening temperatures does not result in any important changes in a number of properties of the material (curves 1 and 2). On the other hand, treatment of the glass for 2 h near the softening temperature leads to a sharp lowering of these characteristics of the glassceramics over the entire temperature range of crystallization, from 850 to 1000° (curve 3). This is to be expected: the transparent crystalline glass materials formed in the latter case have properties radically different from those of opaque glassceramics.

The coefficient of thermal expansion of the glassceramics is of the order of $(40-50) \cdot 10^{-7}$ degree$^{-1}$, the tangent of the dielectric loss angle at $10^{10}$ cps is approximately seven times the value for opaque glassceramics and double that of the original glass, and the dielectric constant is in the region of 6.5. The strength of transparent glassceramics differs little from that of the original glass, but in some instances reaches 15-20 kg/mm$^2$.

The softening temperature of transparent glassceramics is considerably higher than that of the original glass: the transparent crystalline glass materials gradually become turbid at 950-1000° without softening, and change to opaque glassceramics, the softening point of which is above 1300°.

As already noted, the processes taking place in glass during heat treatment are determined by the aggregate effect of two factors — temperature and time. The question naturally arises whether the same effect might be achieved at other temperatures in the precrystallization period by variation of the treatment time.

Investigation of the influence of the second factor (time) showed that transparent crystalline glass materials can be obtained if the preliminary heat treatment is effected not only in the softening range of the glass, but also considerably below it (Fig. 5). Figure 5 shows that the softening temperature of the glass is a kind of boundary which divides the precrystallization period into two regions, differing sharply in the character of the influence

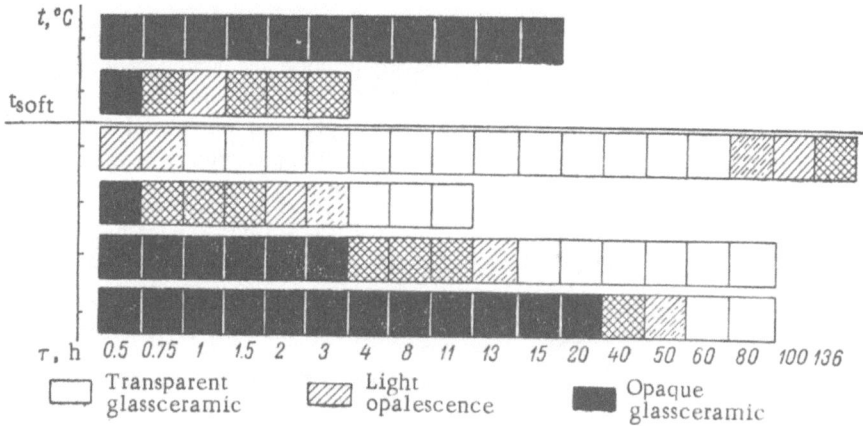

Fig. 5. Visual characteristics of the crystallization of glass in relation to the temperature and duration of the precrystallization period. Darker shading indicates increased opacity.

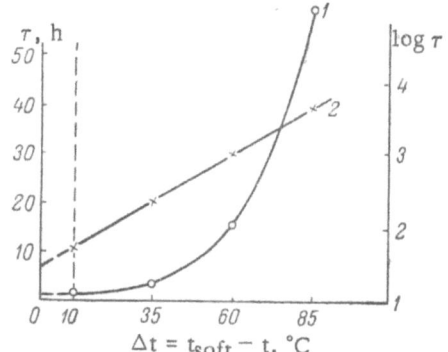

Fig. 6. Effect of the temperature in the precrystallization period on the time of heat-treatment during that period required for subsequent formation of a transparent glassceramic. 1) $\tau = f(\Delta t)$; 2) $\log \tau = f(\Delta t)$.

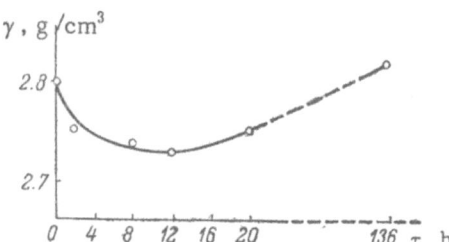

Fig. 7. Dependence of the density of glass-ceramics on the duration of heat treatment in the precrystallization period. Crystallization conditions: $t = 900°$, $\tau = 2$ h.

For the glass under investigation,

on the catalyzed crystallization of glass and the structure of crystalline glass materials: 1) in the temperature range above the softening point, heat-treatment of the glass, whatever its duration, does not lead to subsequent formation of transparent glassceramics, but causes changes in the structure and phase composition of the materials; 2) preliminary heat-treatment of the glass in the temperature range below the softening point for suitable times leads to the formation of transparent glassceramics.

The time of heat treatment necessary for formation of transparent glassceramics increases exponentially with decrease of the temperature in the precrystallization period (Fig. 6). The dependence of this time on the temperature in the precrystallization period is analogous to the dependence of glass viscosity on temperature and can be represented by the equation

$$\log \tau = \log \tau_0 + b\Delta t_0$$

or

$$\tau = \tau_0 \cdot e^{b_1 \Delta t_0},$$

where $\tau_0$ is the time of heat treatment of the glass at the softening point required for the subsequent formation of transparent glassceramics; b and $b_1$ are constants equal to the tangent of the angle formed by the linear plots [$\log \tau = f(\Delta t_0)$ and $\ln \tau = f(\Delta t_0)$] with the temperature axis; their value depends on the rate of change of the viscosity of the glass; $\Delta t_0$ is the difference between the softening temperature of the glass and any given temperature in the precrystallization period, i.e., the degree of undercooling of the glass below the softening point; $\log \tau_0$ is constant and equal to the intercept cut off by the $\log \tau = f(\Delta t_0)$ line along the vertical time axis.

$$\log \tau = 1.425 + 0.025\Delta t_0$$

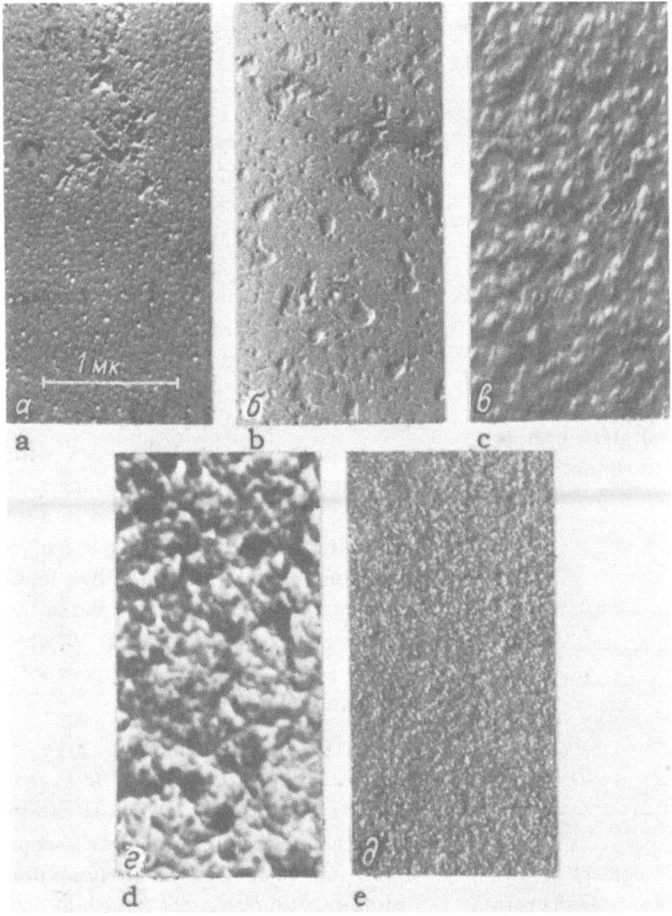

Fig. 8. Variations of the glass structure in relation to the heat-treatment temperature in the precrystallization period. a) Original glass; b) heat-treatment for 2 h at 700°; c) heat-treatment for 2 h near the softening point; d) heat-treatment for 2 h at 800°; e) heat-treatment for 80 h near the softening point.

The above relationships were verified and confirmed for a series of glasses with various catalytic additives, in the $SiO_2$-$Al_2O_3$-BaO and $SiO_2$-$Al_2O_3$-CaO systems, capable of forming transparent or translucent glassceramics.

A fact of particular interest is that excessively long treatment of the glass in the precrystallization period again leads (for the second time) to formation of opaque glassceramics (80 h near the softening point in Fig. 5). This effect is reflected in variations of the properties of the material in relation to the time of heat treatment in the precrystallization period. In particular, the curve for the density of the glassceramic as a function of the time of heat treatment in the precrystallization period (Fig. 7) has a characteristic minimum in the time range corresponding to the formation of transparent crystalline glass materials.

This influence of the conditions in the precrystallization period on the catalyzed crystallization of glass can be explained only if we turn directly to a study of the changes taking place in the glass structure during the precrystallization period. These changes can be noted even at temperatures 150-200° below the softening point and are intensified by rise of temperature while undergoing qualitative alterations.

Electron microscope investigation shows that individual microheterogeneities are already present in the original glass (Fig. 8). Their number and size increase with rise of the temperature of the precrystallization period from 600 to 800°. In the latter case, the heterogeneities become so numerous and large that the glass opalesces appreciably. In the region of the softening point relatively small heterogeneities (of the order of 0.1 to

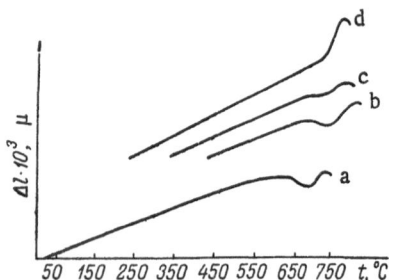

Fig. 9. Expansion curve of the original glass and of the same glass after heat treatment at various temperatures in the precrystallization period ($\tau = 2$ h). a) Original glass; b) heat treatment at 550°; c) heat treatment at 600°; d) heat treatment at 700°.

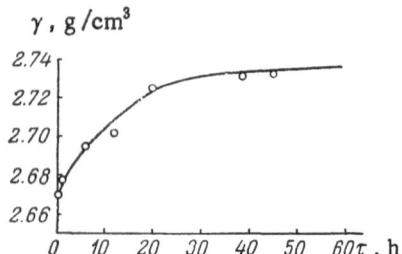

Fig. 10. Dependence of the density of glass on the time of heat treatment near the softening temperature.

0.2 $\mu$) group together and cover the whole cleavage surface (Fig. 8e). The glass darkens sharply but no opalescence is seen.

The absence of a crystalline phase detectable by x-ray phase analysis and the simultaneous presence of well-defined and qualitatively different heterogeneities in the glass lead to the conclusion that they are the result of microscopic immiscibility of the glass. Only certain structural changes, similar to those observed near the softening temperature, lead to the formation of transparent glassceramics during subsequent crystallization of the glass. These changes are accompanied by a deep endothermic effect on the thermogram of the original glass, the magnitude of which diminishes in the course of the structural processes (with increase of the heat-treatment time near the softening temperature to 2 and then to 12 h). They are manifested in a sharp change of the main properties of the material (course of the expansion curve, softening temperature, refractive index, optical density, electrical properties, density, course of the DTA curves). For example, tan $\delta$ of the glass at $10^5$ cps increases from $30 \cdot 10^{-4}$ to $970 \cdot 10^{-4}$ after heat treatment for 2 h near the softening temperature.

The thermal expansion curve of the glass has a characteristic "plateau" (550-650°) followed by a fall with a minimum at 700° (Fig. 9). This indicates that even at 500° structural changes begin in the glass, accompanied by a density increase. Indeed, the anomaly on the expansion curve diminishes with treatment for 2 h at 600° and vanishes completely at higher temperatures (Fig. 9). Heat treatment of the glass for 2 h at 500-550° does not alter the appearance of the $\Delta l$ vs. t curve, but the anomalous region shifts toward higher temperatures and the softening point correspondingly rises by 60-70°.

If the heat-treatment time near the softening temperature is increased to 136 h, x-ray phase analysis does not reveal a crystalline phase in the glass, but after only 40 h the x-ray diagrams show a distinct diffuse maximum in the region corresponding to the most intense mullite peak. Its formation indicates that, although mullite is not yet separated out as an independent crystalline phase, there has already been considerable ordering of the glass structure with formation of regions similar to mullite in structure and preceding the appearance of mullite. The glass is in the precrystalline state and, as electron microscope studies show, its structure is closer to that of transparent crystalline glass material than to the structure of the original glass (Fig. 8). This is also indicated by the sharp rise (by 200°) of the softening temperature of the glass. It is significant, however, that this ordering of the structure in the precrystallization period again leads to subsequent formation of opaque glassceramics (Fig. 5).

The gradual ordering of the glass structure when the time of heat treatment in the precrystallization period is varied proceeds very slowly, and this is reflected in the variations of the properties of the glass. Figure 10 shows that the main structural changes in the glass are completed during the first 20-30 h, after which the rate of change of density falls sharply.

It is significant that the curves representing the dependence of the density of glassceramics (Fig. 7) and the original glass (Fig. 10) on the time of heat treatment in the precrystallization period take opposite courses. This is fully justifiable, as the sharp increase of density in the precrystallization period (Fig. 10) is due to changes in the glass structure which subsequently lead to the formation of transparent glassceramics and, therefore, to a sharp fall of density (Fig. 7).

Finally, we must add that studies of the changes taking place in glass in the precrystallization period and of their influence on the catalyzed crystallization of glass and on the structure and properties of the end product—glassceramics — make it possible to reveal certain general relationships of the heterogeneous crystallization process itself, regardless of the composition of the original glass, and thus enable us to control with greater confidence the course of crystallization of glass by selecting the optimum heat-treatment conditions.

# THE CONNECTION BETWEEN PHASE DIAGRAMS OF SILICATE SYSTEMS AND THE STRUCTURE AND CRYSTALLIZABILITY OF GLASSES

## F. Ya. Galakhov

The formation of a glassceramic is a process of glass crystallization, characterized by the formation of very small crystals of uniform size and uniformly distributed throughout the mass. Therefore, whatever methods are used for studying this process, the results must be interpreted, in particular, in the light of the theory of heterogeneous equilibria with the aid of the appropriate phase diagrams.

Our investigations of various systems have shown that uniform highly disperse crystallization is observed in glasses, the composition of which is either close to the liquation region or in the regions of the phase diagrams exhibiting a tendency to liquation. Roy [1] and Hinz and Kunth [2] put forward the view that one condition for the formation of a good crystalline glass material is submicroscopic immiscibility in the glass, which determines uniform fine crystallization.

These facts and views, important for the theory and practice of glassceramic production, show that it is necessary to establish regions of metastable liquation in the phase diagrams of the systems. These regions can be determined experimentally by the modern methods of electron microscopy and low-angle x-ray diffraction.

However, the theory of heterogeneous equilibria is not always fully utilized in studies of the structure and crystallization of glass. In a study of the structure of sodium borosilicate glasses [3] we demonstrated a close connection between the structures found experimentally and the state of the mixtures as indicated by the phase diagrams. This analysis is of special significance in relation to the formation of glassceramics, as these sodium borosilicate glasses have a pronounced tendency to liquation, and this approach can be extended to other systems, in particular, the $Li_2O$-$SiO_2$ system.

Special studies for establishment of the submicroliquation boundaries in the phase diagrams should, in addition to their significance for the theory of heterogeneous equilibria, assist in the successful development of new kinds of crystalline and porous glass materials.

### LITERATURE CITED

1.  R. Roy, J. Am. Ceram. Soc. 43(12), 1960.
2.  W. Hinz and P. O. Kunth, Glastechn. Ber. 9:431, 1961.
3.  F. Ya. Galakhov, Izv. AN SSSR, Otdel. Khim. Nauk 5:743, 1962.

# ESTABLISHMENT OF PHASE SEPARATION IN SYSTEMS
# COMPOSED OF FLUORINE-CONTAINING SLAGS
# AND RARE-EARTH SILICATES

## I. A. Bondar' and N. A. Toropov

Study of the influence of calcium fluoride on crystallization processes in the ternary system $CaO-Al_2O_3-SiO_2$ is of definite significance for interpretation of the processes of production of crystalline glass materials. Since 1956, we have been investigating equilibria in aluminosilicates with added $CaF_2$ in conjunction with work in the Institute of Metallurgy and Ceramics of the Academy of Sciences of the Chinese People's Republic.

The part of the $CaO-Al_2O_3-SiO_2$ system studied includes the compositions of basic and acidic blast-furnace slags and also relates to the liquation region adjacent to the binary system $CaO-SiO_2$. The amounts of $CaF_2$ added were 1.5, 5, and 10% by weight.

The investigation showed that calcium fluoride is a highly effective mineralizer, lowers the viscosity considerably, and lowers the crystallization temperature of the melts. On addition of 5% $CaF_2$, the viscosity decreased by more than one half and the liquidus temperatures fell by 50-70° in comparison with fluoride-free compositions. Addition to 10% $CaF_2$ to three-component compositions extends the immiscibility region.

In Fig. 1, this region is enclosed by a binodal curve. Investigation with the polarizing microscope reveals drops of the main glass distributed in a "siliceous" glass. The boundaries of coexistence of the two glasses are considerably extended (by 2-5%) by electron microscope investigation as compared with the results given by the ordinary polarizing microscope.

Figure 2 shows electron micrographs obtained by the replica method. It must be pointed out that formerly the existence of liquation regions was regarded as exceptional, but it has now been shown in a number of investigations that the formation of microheterogeneities is a characteristic feature of most artificial and natural silicate melts. The problem of immiscible liquids is very important in relation to the production of new crystalline glass materials. Our subsequent investigations were concerned with detection of liquation fields in systems of simple and complex rare-earth silicates. In particular, studies of lanthanum, samarium, yttrium, and ytterbium silicates demonstrated the existence of immiscible glasses over wide ranges of temperatures and concentrations (35-95% by weight and 1650-2200°). Certain properties of the compositions in the immiscibility regions were studied.

In agreement with Esin's basic views relating to immiscible systems with bivalent cations (Mg, Ca, Ba, Zn, and Fe), we obtained results for trivalent cations of rare-earth elements. It was shown that the smaller the ionic

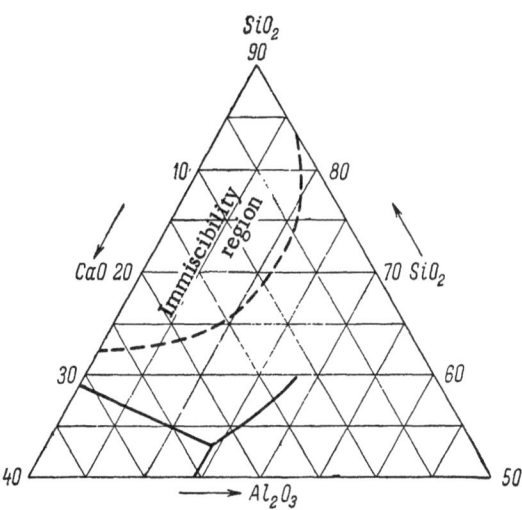

Fig. 1. Immiscibility region in the system $CaO-Al_2O_3-SiO_2$ with addition of 10% $CaF_2$.

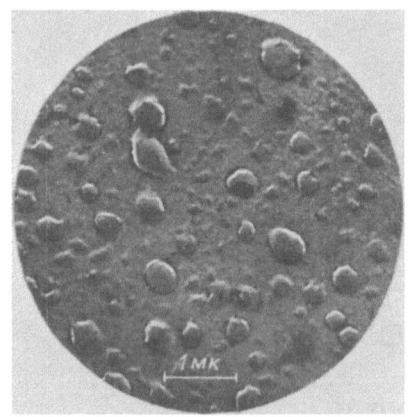

Fig. 2. Formation of two glasses in the system $CaO-Al_2O_3-SiO_2-CaF_2$. Electron microscope, magnification 1200.

Effect of Cation Size on the Immiscibility Limits

| Cation | Cation radius, A | Immiscibility limits, mole% $SiO_2$ |
|---|---|---|
| $Mg^{2+}$.... | 0.78 | 61-98 |
| $Zn^{2+}$.... | 0.83 | 65-98 |
| $Fe^{2+}$.... | 0.83 | 66-98 |
| $Ca^{2+}$.... | 1.06 | 70-98 |
| $Sr^{2+}$.... | 1.27 | 80-98 |
| $Yb^{3+}$.... | 1.00 | 73-98 |
| $Y^{3+}$.... | 1.06 | 75-98 |
| $Sm^{3+}$.... | 1.13 | 75-98 |
| $La^{3+}$.... | 1.22 | 77-98 |

radius of the cation the lower the percentage of $SiO_2$ (the wider the demixing range) at which phase separation begins. The immiscibility region in the series of rare-earth elements increases in the sequence from $La^{3+}$ (r = 1.22 A) to $Yb^{3+}$ (r = 1.00 A). This is also associated with the decreasing basicity of the cations in this series (see table). We used the results of the American workers Levin and Block [1-3] for calculating the limiting compositions of the immiscibility regions, i.e., the compositions of the liquids rich with the modifier oxides. We extended the results to trivalent elements. The oxygen-volume calculation method was used. In this case, it is necessary to know the ionic radii of oxygen and the modifier cation, the type of bonding between the modifier cations and the oxygens of the $(SiO_4)$ tetrahedron, and the volume occupied by an oxygen atom in close packing. The calculated limiting compositions are in good agreement with the experimental data (difference of about 2%).

It was also shown that alumina and niobium oxide have different effects on the position of the immiscibility region when introduced into the binary system yttrium oxide-silica: alumina results in a slight extension of immiscibility into the ternary system; its field occupies a narrow band adjacent to the binary system $Y_2O_3-SiO_2$ (43-95% $SiO_2$ by weight) and extending into the ternary system up to 5% $Al_2O_3$ by weight. With higher $Al_2O_3$ contents (8-10%) opalescent glass is formed, while above 10% $Al_2O_3$ transparent glasses are formed with several compositions. Niobium oxide extends the immiscibility region considerably; it extends from one binary system ($Y_2O_3-SiO_2$) to the other ($Nb_2O_5-SiO_2$).

Immiscibility is a very complex phenomenon which depends on numerous factors: the size and charge of the cation, the nature of the anion-cation bond, the mutual polarization of the anion and cation, the coordination of the metal, etc. For example, Glasser, Warshau, and Roy [4] attribute the absence of stable immiscibility in the $Al_2O_3-SiO_2$ system to the fact that Al and Si are in similar fourfold coordination in the glass. In the light of the theory of solution, mutual immiscibility of components in the liquid state is the consequence of a difference between the intermolecular forces in the two immiscible liquids (expansion and compression coefficients, surface tension, heat of evaporation, etc.). Semenchenko and coworkers showed that addition of a third component which raises the surface tension of one of the liquids lowers the mutual solubility of the components. Addition of alumina probably produces a slight increase in the surface tension of the components and therefore reduces somewhat the mutual solubility of yttrium oxide and silica; niobium oxide causes a large increase of surface tension and decreases the mutual solubility of the components considerably.

It should be pointed out that the third component introduced into a demixing melt determines the boundaries of the stable and metastable immiscibility regions; the compositions near the immiscibility fields (the predemixing regions) are of interest in relation to the production of new materials.

LITERATURE CITED

1. E. M. Levin and S. Block, J. Am. Ceram. Soc. 40(3):95, 1957.
2. S. Block and E. M. Levin, J. Am. Ceram. Soc. 40(4):113, 1957.
3. E. M. Levin and S. Block, J. Am. Ceram. Soc. 1(2):49, 1960.
4. F. P. Glasser, I. Warshau, and R. Roy, Phys. Chem. Glasses 1(2):39, 1960.

# ELECTRON MICROSCOPE INVESTIGATION
# OF MICROCRYSTALLIZATION OF GLASSES

## N. M. Vaisfel'd and V. I. Shelyubskii

The development and applications of microcrystalline materials based on glass [1,2] have made electron microscopy one of the most important methods for investigation of these materials.

The purpose of the present work was a systematic electron microscope investigation of microcrystallization processes in glasses of various types. The investigations were carried out with the ÉM-100 electron microscope with an accelerating voltage of 75 kV and with the Tesla BS-242A instrument at 60 kV. The final magnification was 15,000 in both cases. This may be regarded as the optimum magnification for structure studies of materials with particles of up to 1 $\mu$. The known technique [3] of carbon replicas with preliminary chromium shadowing was used; however, in some cases, chemical interaction between the material and the replica prevented separation of the replicas; in such cases simultaneous oblique coating with platinum and carbon was used. The selection of a suitable method for each particular composition is an essential condition for successful electron microscope studies.

Without going into details of the preparation technique, we merely note that it is necessary to etch the specimen surface before application of the replica. It is etching which makes it possible to reveal the shape and size of the crystalline particles and the nature of their distribution. It is not especially important which of the phases dissolves at the higher rate; the only important factor is selective etching, as this gives rise to the relief necessary for investigations by the replica method.

In investigations of most microcrystalline materials, the structure of the material is characterized most reliably by the structure of the glass surface. However, in a number of cases, such as photosensitive glass where ultraviolet radiation is used to activate crystallization, it is necessary to investigate the specimens from the surface. Here the role of etching is especially important: an electron micrograph of an unetched surface can be used for assessment of the surface finish, but not of the structure. In some instances the polishing process can play the part of mechanical etching owing to the different hardnesses of the glassy and crystalline phases (Fig. 1a), but such micrographs do not give a complete picture of crystallization (compare Fig. 1b, c). Photosensitive glass specimens often become covered with a peculiar crust having a coarse-grained structure in the course of crystallization (Fig. 2a). Deep etching (1-3 $\mu$) makes it possible to reveal the structure of the specimen (Fig. 2b). The etching conditions (concentration of etching agent and treatment time) are chosen experimentally for each particular composition.

In an investigation of the relation between the final structure of the material and the heat-treatment conditions, we studied the crystallization of specimens of one of the glasses at various temperatures and for different times. Figure 3a shows structural changes in the specimens during crystallization at constant temperature (900°) for 0.5-8 h, and Fig. 3b shows crystallization of the same material in the temperature range of 800-1060° for 0.5 h. The specimens were put into a preheated furnace. The illustrations show that the final crystalline particle size increases with rise of the treatment temperature, as growth of particles as the result of coalescence becomes more probable owing to the decreased viscosity. At low temperatures, formation and growth of dendritic aggregates of the particles are characteristic of this material; the dendritic groups become better defined with increase of the crystallization time. A completely different crystallization picture is obtained if the temperature of the specimen is raised continuously from room level (Fig. 3c). At first a similar dendritic structure is formed, but with more numerous and smaller dendrites; at somewhat higher crystallization temperatures a new

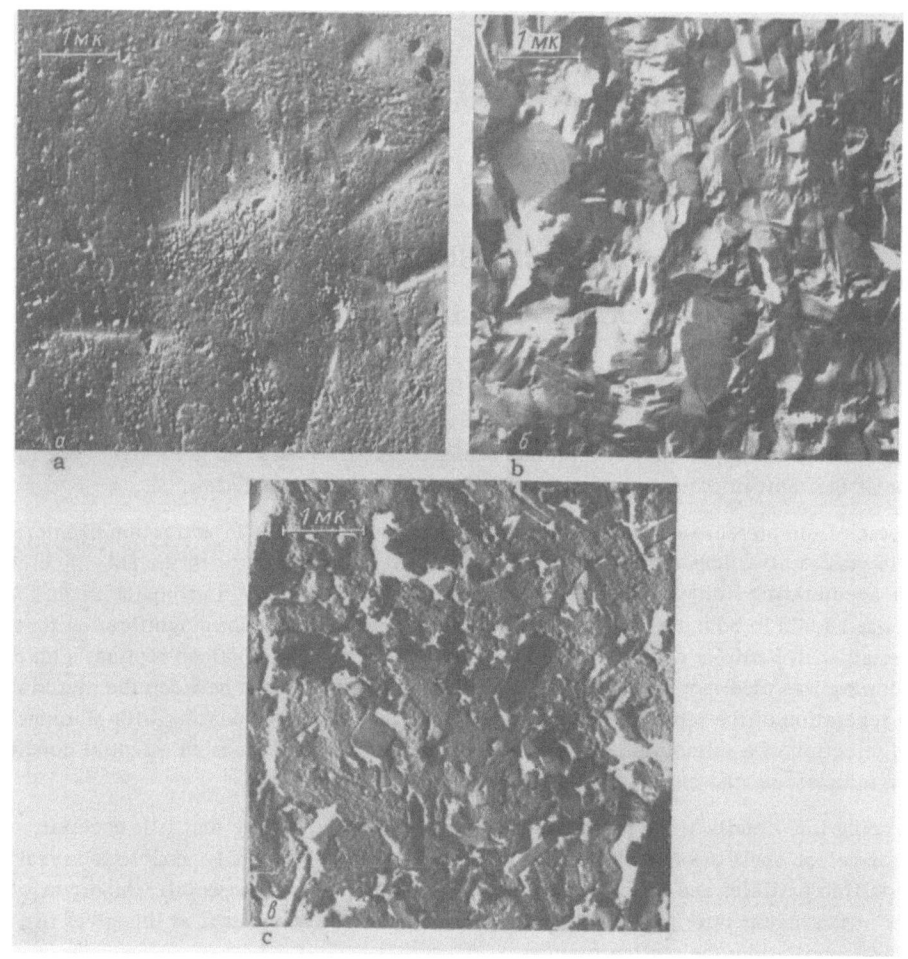

Fig. 1. Electron micrographs of the same microcrystalline glass material. a) From a polished unetched surface; b) from unetched cleavage surface; c) from cleavage surface etched for 10 sec in 10% hydrofluoric acid.

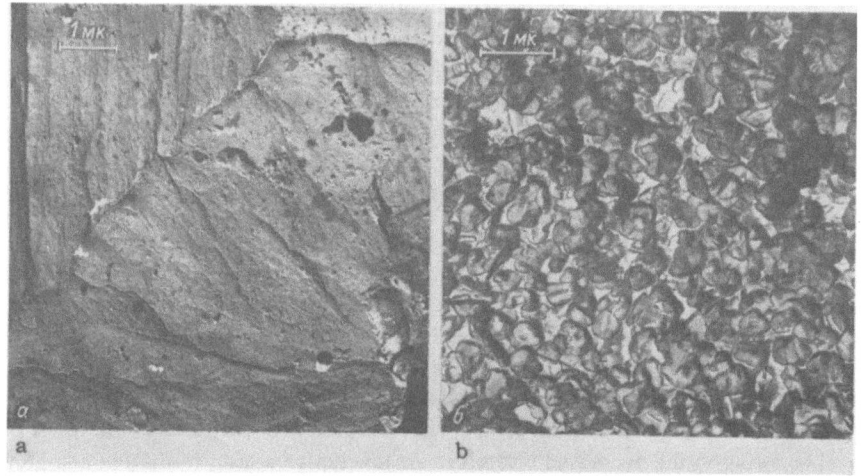

Fig. 2. Structure of crystallized photosensitive glass. a) Slightly etched surface; b) surface of the same specimen etched to a depth of 1-3 $\mu$.

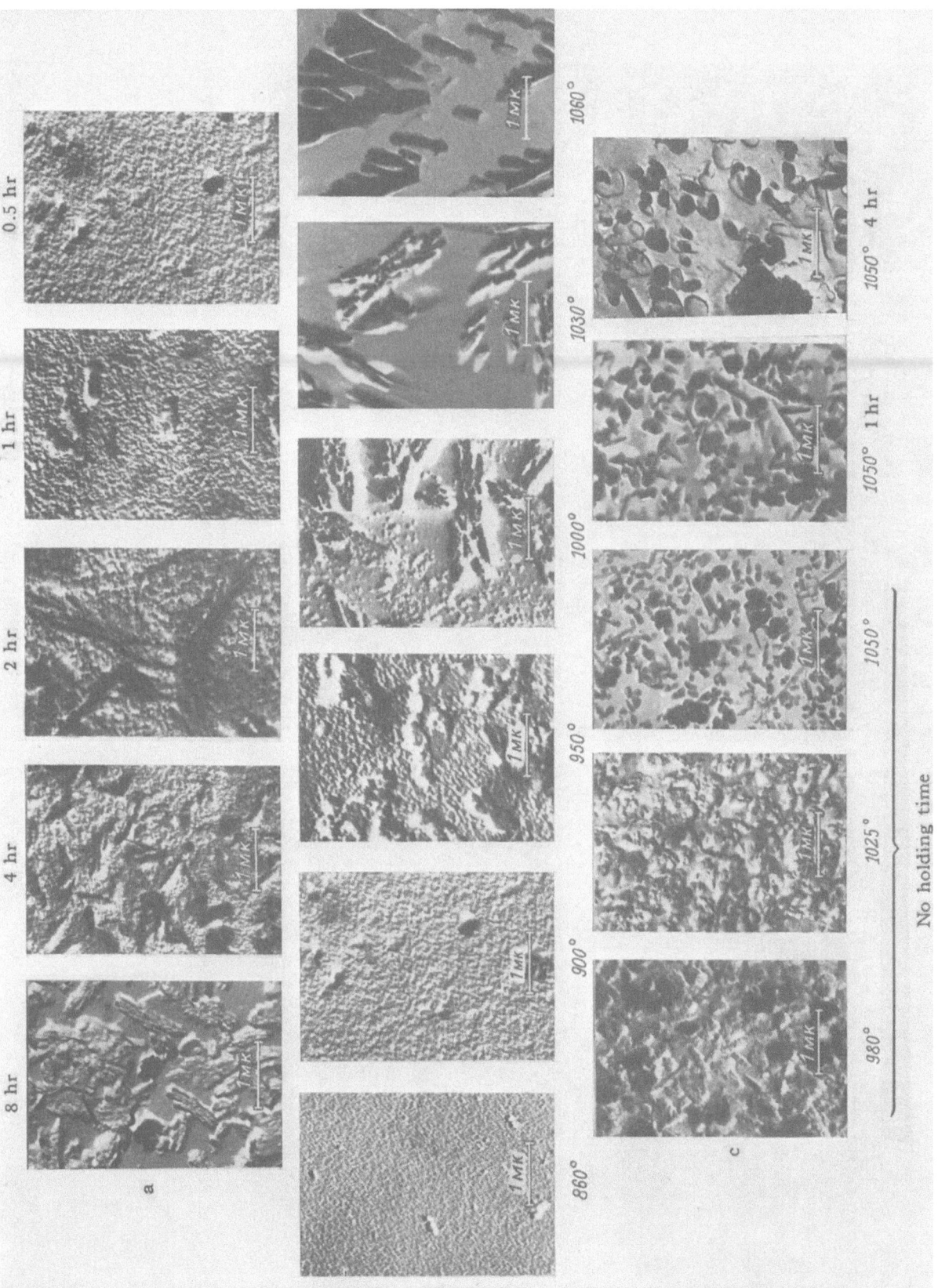

Fig. 3. Structural changes in the course of crystallization of one of the glasses. a) Crystallization for different times at 900°, specimens placed in preheated furnace; b) crystallization for 0,5 h at various temperatures, specimens placed in preheated furnace; c) crystallization with gradual rise of temperature from room level to 1050°, specimens held for different times at that temperature.

39

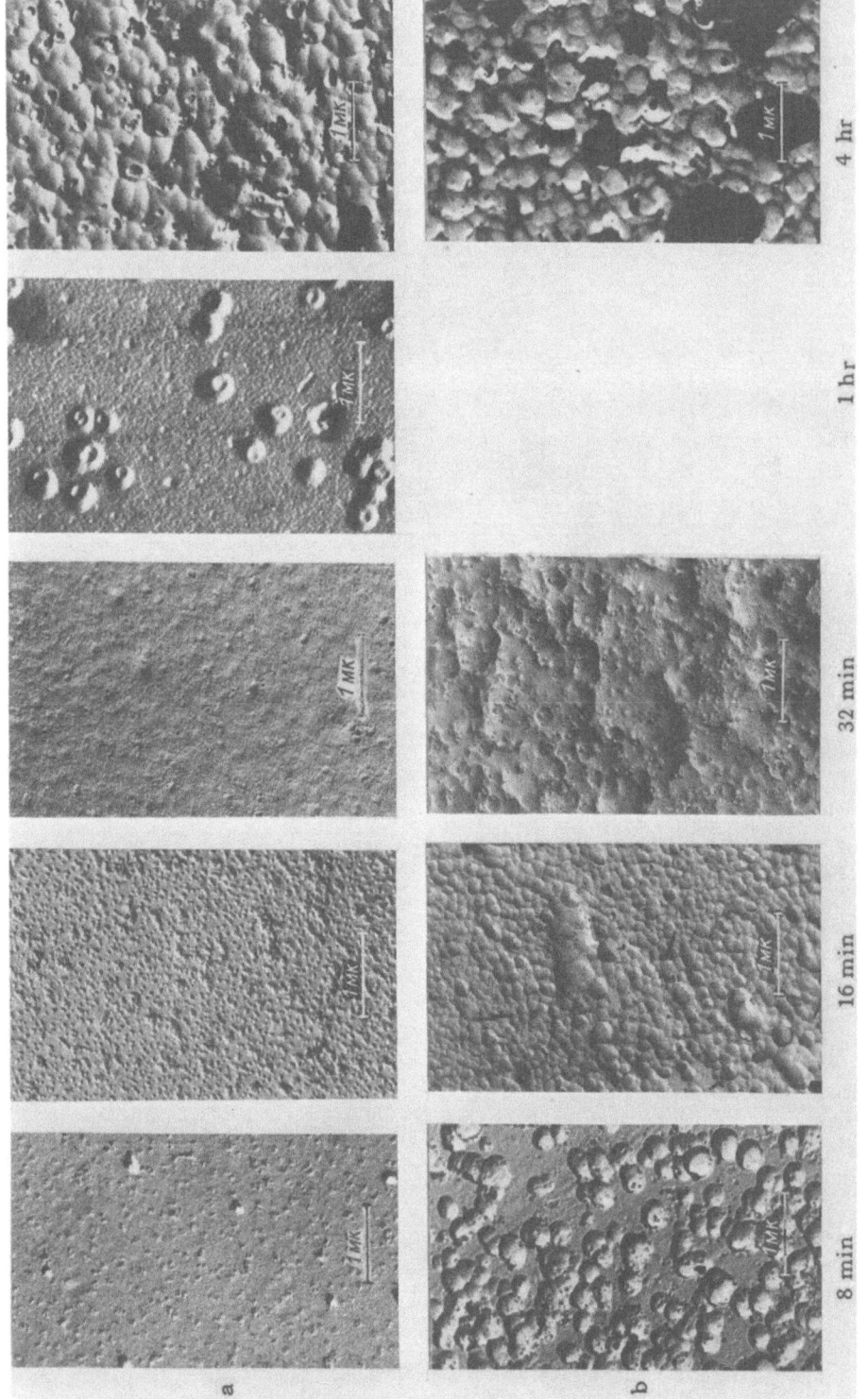

Fig. 4. Electron micrographs of photosensitive glass specimens treated for 3 h at 500° (a) and 550° (b) with different irradiation times.

40

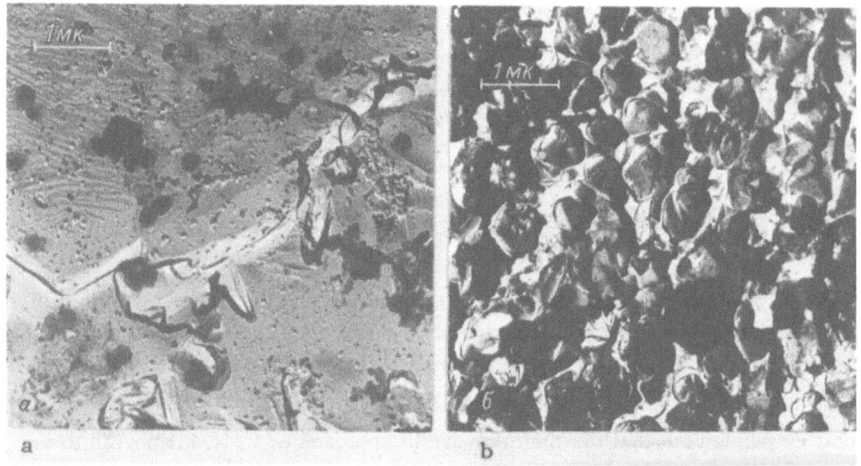

Fig. 5. Structure of two photosensitive glass specimens crystallized by the same procedure, with different irradiation doses. The second specimen (b) is more than twice as strong as the first (a).

crystalline phase appears, not found at the same temperatures in specimens which had not been heated up gradually; this is clear from the sharp difference in the particle morphology. If the specimens are thermostatted at this temperature, the particles grow in size and become less numerous [4]. The electron microscope data were confirmed by x-ray phase analysis.

Electron microscope studies of crystallization processes in photosensitive glass [5] showed that the structure is determined to a considerable extent by the ultraviolet irradiation conditions, as well as by the conditions of heat treatment. At the initial crystallization stages, at 500 and 550° (Fig. 4), variations in the number and size of the particles in accordance with the irradiation dose are clearly seen. The number of crystalline particles, distribution density, and degree of crystallization increase with the irradiation dose. Electron microscope investigation makes it possible to reveal the "precrystalline" state of the material, characterized by the start of intensive growth of the silicate phase on the silver particles. This process is accompanied by the formation of a concentration gradient of the separating phase around the centers, as is indicated by the formation of spherical etching figures. Each such etching pit contains a particle (Fig. 4a, irradiation for 1 and 4 h).

The optimum crystallization conditions can be determined with the aid of electron microscopy. It was found that specimens with a microcrystalline structure, with particles of the order of 0.5 $\mu$ and the minimal intercrystalline layers, were the strongest. For example, the specimen, the structure of which is shown in Fig. 5b, was more than twice as strong as the specimen with the structure in Fig. 5a.

It should be pointed out that in systematic work with the same material it is possible to use the results of electron microscope observations for estimating the strength and also for checking the constancy of the composition and the heat-treatment and irradiation conditions.

## LITERATURE CITED

1. A. I. Berezhnoi, Photosensitive Glasses and Crystalline Glass Materials of the Pyroceram Type, VINITI, Moscow, 1960.
2. V. I. Shelyubskii and N. M. Vaisfel'd, Zhur. Fiz. Khim. 35(11):2652, 1961.
3. V. M. Luk'yanovich, Electron Microscopy in Physicochemical Investigations, Moscow, 1960.
4. N. V. Solomin, V. I. Shelyubskii, and N. M. Vaisfel'd, Doklady Akad. Nauk SSSR 140(5):1087, 1961.
5. V. A. Borgman, V. I. Petrov, and V. G. Chistoserdov, Zhur. Fiz. Khim. 35(6):1383, 1961.

# DISCUSSION

N. A. Toropov, answering questions, emphasized that separation of a crystalline phase which does not correspond to the crystallization field in which the composition of the glass lies is possible during the crystallization of glasses containing nuclei introduced from outside, especially under conditions of considerable undercooling. Several phases may separate out, especially if their lattice parameters are close to those of the nuclei.

Yu. N. Kondrat'ev pointed out that the thermodynamic approach of V. N. Filipovich to the initial stages of the crystallization process in glasses does not describe the kinetics of the process, as none of the equations given in the paper contains a time parameter.

V. N. Filipovich replied that nearly all the parameters in the equations depend on time. The composition of the glassy phase varies in the course of crystallization, especially in the case of multicomponent glasses, and the magnitude of the potential barrier which the nucleus must overcome in order to grow varies accordingly.

R. Ya. Khodakovskaya, replying to a question concerning heat treatment in the precrystallization period, said that she based her determination of this period on DTA curves and on absence of any signs of crystallization during heating even for 200 h; in the experiments the usual times of the order of several hours were used.

E. A. Porai-Koshits, in his comments on V. N. Filipovich's paper and statements during the discussion,* did not agree that uniform fine crystallization is possible without preliminary formation of a chemically heterogeneous glass structure, as chemical differentiation occurs even in the melt itself (as Filipovich himself said). Porai-Koshits considered that crystallization is always preceded by formation of regions with the stoichiometric composition of the future crystals, although formation of chemically heterogeneous regions (submicroscopic immiscibility) does not always cause the glass to crystallize. This is proved by the disappearance of such regions and of the opalescence caused by them when the temperature is raised above the clearing temperature [in connection with increase of the entropy term in Eq. (4) of Filipovich's paper even before crystallization]. However, if crystallization begins after this, it is preceded by formation of new nuclei in the form of disordered heterogeneity regions of the corresponding composition. The detection of such regions (which may possibly exist for only a very short time) requires the development of new sensitive methods or refinement of known structure techniques.

---

*V. N. Filipovich's statements in the discussion were included by him in the paper published in the first section.

*Two-Component Systems*

# DETECTION AND STUDY OF VERY SMALL HETEROGENEITIES
# WITH THE AID OF A NEW LOW-ANGLE X-RAY APPARATUS

## D. A. Goganov, E. A. Porai-Koshits, and Yu. G. Sokolov

For study of the initial stage of the crystallization of glasses it is important to know their submicroscopic structure. This can be achieved by studying the scattering of x-rays at low angles; detection of such scattering provides direct proof of heterogeneous structure in glass. After a number of unsuccessful attempts [1], low-angle scattering free from parasitic effects was detected for the first time in leachable sodium borosilicate glasses, in which the scattering heterogeneity regions are relatively large ($R \geq 100$ A) [2]. The difference between the electron densities of regions differing in chemical composition proved to be so small (about $0.02$ el./$A^3$) that a new apparatus, capable of recording very weak intensities, had to be designed in order to extend these investigations to glasses with smaller heterogeneity regions.

Accordingly, we designed an apparatus with the same Kratky collimation as in [2], but with recording of the scattered radiation by means of a proportional quantum counter and with a pulse-height analyzer. The counter, filled with a mixture of xenon (290 mm Hg), had an ethane (45 mm Hg) had an efficiency of about 80% (experimental value) and energy resolution of about 14% for the $CuK_{\alpha}$ line energy. The background of the apparatus itself in recording of copper radiation did not exceed 5-7 pulses/min. Owing to the good energy resolution and the fact that the counter efficiency falls rapidly with decreasing wavelength, it is possible to work with nonmonochromatic radiation by tuning the analyzer channel to recording of the characteristic radiation and suppressing the $K_{\beta}$ line with a fine filter. The advantages of the new apparatus are illustrated by the following examples.*

Figure 1 gives the intensity curves of three specimens of a sodium borosilicate glass containing 7% $Na_2O$, 23% $B_2O_3$, and 70% $SiO_2$ (mole%), two of which were obtained in [2] with photographic recording. Curves 1 and 2 refer to specimens heated at temperatures in the opalescence zone. The intensity — angle relationship and the radii of the heterogeneity regions (190 A for specimen 1 and 110 A for specimen 2) were close to the values found in [2]; however, with photographic recording, the exposure times needed for determination of the curves were 30 h for curve 1 and 100 h for curve 2, whereas with the new apparatus each of the two curves was plotted from the points in 4-6 h.

In the earlier work [2] it was not possible to record scattering by a specimen of the same glass heated at 750° for 5 min (at a temperature outside the opalescence zone), as this would have required exposures of the order of 300-400 h. Such heating makes the glass optically transparent and the glass leaches out to form pores about 50 A in radius. Curve 3 in Fig. 1 relates to this specimen, and was obtained with the new apparatus in 10-15 h; the size of the heterogeneity regions was found to be 55 A.

As an example of the measurement of even lower intensities, Fig. 2 shows the intensity curves for two lithium silicate glasses containing 23.5% $Li_2O$, 76.5% $SiO_2$ (curve 1) and 25.5% $Li_2O$, 74.5% $SiO_2$ (curve 2). The radii of the heterogeneity regions in these glasses are 200 and 160 A, respectively. The low-angle scattering intensity in this last case is the lowest of all the examples given; even in the range of scattering angles between 6 and 12', it does not exceed 8 pulses/min,while the last point (at a scattering angle of about 27') corresponds to an intensity of less than 1 pulse/min. It is quite evident that such low intensity is associated with an even smaller difference between the electron densities of the chemically heterogeneous regions.

---

*A detailed account of the apparatus is published separately.

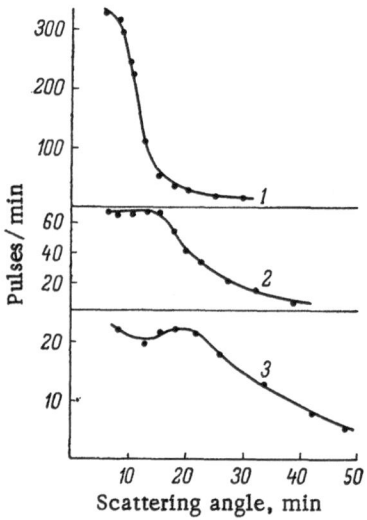

Fig. 1. Intensity curves of three specimens of sodium borosilicate glass.
1) Heated for 8 h at 600°; 2) for 5 h at 530°; 3) for 5 min at 750°.

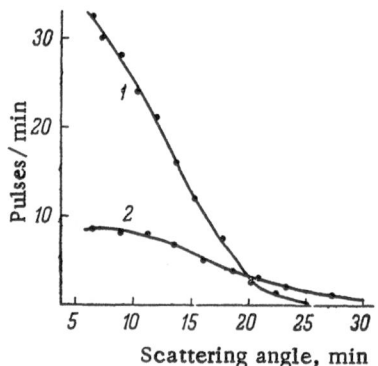

Fig. 2. Intensity curves of two lithium glasses of similar composition; melt chilled to different degrees.

We have already stated that scattering of x rays at low angles is proof of a chemically heterogeneous structure in any glass. The new apparatus offers considerably wider experimental possibilities in this respect, as it can be used for obtaining such proof for glasses with considerably smaller heterogeneity regions and with smaller electron-density differences than could be achieved with the apparatus known earlier. In addition, the absolute value of the scattering intensity can be easily determined by comparison of the low-angle scattering intensity with the intensity of the incident beam; the result can then be used for determining the r.m.s. difference between the electron densities of the heterogeneity regions, which is an objective quantitative measure of the degree of chemical heterogeneity of the glass. Both the results make it possible to study the precrystallization structure of glass and its influence on the initial stage of crystallization. Examples of this application are given in [3].

LITERATURE CITED

1.  L. C. Hoffman and W. O. Statton, Nature 176:561, 1955; L. Kartz, Nature 180:1115, 1957.
2.  N. S. Andreev and E. A. Porai-Koshits, Doklady Akad. Nauk SSSR 118:535, 1958; E. A. Porai-Koshits and N. S. Andreyev, Nature 182:535, 1958; J. Soc. Glass Technol. 43:213, 1959.
3.  N. S. Andreev, D. A. Goganov, E. A. Porai-Koshits, and Yu. G. Sokolov, this collection, p. 47.

# CHEMICALLY HETEROGENEOUS STRUCTURE
# OF TWO-COMPONENT SODIUM AND LITHIUM SILICATE GLASSES

## N. S. Andreev, D. A. Goganov, E. A. Porai-Koshits, and Yu. G. Sokolov

Direct experimental evidence in favor of the hypothesis of a chemically heterogeneous structure in complex glasses has been obtained by means of x-ray diffraction [1,2] and electron microscope [3-7] studies of multicomponent glasses carried out in recent years. This structure should have a significant influence on the initial stage of glass crystallization during formation of glassceramics. However, some comments should be made on these investigations.

First, nearly all of them give obviously incomplete information on the submicroheterogeneous structure of the glasses; the information relates almost entirely to the size and shape of the heterogeneity regions and not to their compositions or the ratios of the total volumes occupied by regions of different composition.

Second, three-component or even more complex glasses were (with rare exceptions) investigated in these studies; this makes it difficult to reveal the general laws governing the formation of chemically heterogeneous regions in them.

Third, the most convincing and complete results were obtained for sodium borosilicate glasses, which have numerous anomalous properties; accordingly, certain investigators [8,9] regarded chemically heterogeneous structure as a peculiar structural anomaly.

We therefore chose for the present investigation simple silicate glasses of the binary systems $Na_2O-SiO_2$ and $Li_2O-SiO_2$, which are components of many real glasses. For fuller characterization of the submicroscopic structure of the glasses, the r.m.s. (root mean square) differences between their electron densities were determined, in addition to the dimensions of the heterogeneity regions.

## Root Mean Square Difference of the Electron Densities $\overline{(\Delta\rho)^2}$ – Measure of Glass Heterogeneity

According to [10], the intensity of x-ray scattering at small angles by an isotropic body may be written as

$$I = 4\pi \int_0^\infty \varphi(\mathbf{r})\, \mathbf{r} \frac{\sin sr}{s}\, dr, \tag{1}$$

where $|s| = \frac{4\pi \sin \vartheta}{\lambda}$; $\lambda$ is the wavelength; $\vartheta$ is half the scattering angle. The function $\varphi(\mathbf{r})$ depends on the electron density fluctuations in the following manner:

$$\varphi(\mathbf{r}) = \int \Delta\rho(\mathbf{r}')\, \Delta\rho(\mathbf{r}' + \mathbf{r})\, dv', \tag{2}$$

where r is the distance from the running point, determined by the vector $\mathbf{r}'$; $\Delta\rho(\mathbf{r}) = \rho(\mathbf{r}) - \rho_0$ is the deviation of the electron density from its mean value. It follows from the definition of $\varphi(\mathbf{r})$ that $\varphi(0) = \overline{(\Delta\rho)^2}$.

By the Fourier theorem, we have from (1),

$$\varphi(r) = \frac{1}{2\pi^2} \int_0^\infty sI(s) \frac{\sin sr}{r}\, ds. \tag{3}$$

Composition and Heat Treatment of the Glasses Studied

| Contents, mole% | | | Clearing temperature $T_c$, °C | Heating temperature, °C | Heating time, min | Radius of heterogeneity regions, A | $\overline{(\Delta\rho)^2}$, rel. units |
|---|---|---|---|---|---|---|---|
| SiO$_2$ | Na$_2$O | LiO$_2$ | | | | | |
| 95 | 5 | – | 855 | 850 | 10 | 380 | 0.38 |
| 92.5 | 7.5 | – | 860 | 860 | 10 | 115 | 0.56 |
| 90 | 10 | – | 855 | 860 | 10 | 155 | 0.83 |
| 88.5 | 11.5 | – | 840 | 850 | 10 | 125 | 1.00 |
| 87.5 | 12.5 | – | 825 | – | – | – | – |
| 86.5 | 13.5 | – | 805 | 820 | 10 | 90 | 0.63 |
| 86 | 14 | – | 785 | – | – | – | – |
| 85 | 15 | – | 760 | – | – | – | – |
| 84 | 16 | – | 725 | 700 | 10 | 200 | 0.30 |
| 82.5 | 17.5 | – | 690 | | | | |
| 81.5 | 18.5 | – | 655 | | | | |
| 76.5 | – | 23.5 | 860 | * | * | 175 | 0.41 |
| 74.5 | – | 25.5 | – | * | * | 160 | 0.26 |
| 72.5 | – | 27.5 | – | ** | ** | 120 | 0.16 |
| 71.5 | – | 28.5 | 700 | – | – | – | – |
| 70 | – | 30 | 620 | – | – | – | – |

*Glasses chilled from 1550° on a metal support.

**Glass chilled from 1500° on a metal support.

Therefore, a relation of the following form exists between the r.m.s. difference of the electron densities and the small-angle scattering intensity:

$$\overline{(\Delta\rho)^2} = \frac{1}{2\pi^2} \int s^2 I(s)\, ds. \qquad (4)$$

Expression (4) is applicable to determination of $\overline{(\Delta\rho)^2}$ only when apparatus in which collimation effects may be disregarded (apparatus with point slits) is used. If collimation effects are taken into account [11] for apparatus with finite slits, the following expression is obtained:

$$\overline{(\Delta\rho)^2} = c \int s I(s)\, ds, \qquad (5)$$

where c is a constant for the apparatus, which can be calculated or measured with the aid of a standard specimen.

It is essential in the use of Eq. (5) to correct the experimental intensities for absorption, to calculate them per unit volume of the scattering specimen, and to express them in absolute electron units.

It is easy to show that in the case of a body containing only two types of heterogeneities the r.m.s. difference between the electron densities must be

$$\overline{(\Delta\rho)^2} = (\rho_1 - \rho_2)^2\, w_1 w_2, \qquad (6)$$

where $\rho_1$ and $\rho_2$ are the electron densities of the respective "phases," $w_1 = v_1/v$, and $w_2 = v_2/v$, where $v_1$ and $v_2$ are the volumes occupied by the first and second "phases," and v is the total scattering volume of the body, so that $v_1 + v_2 = v$ and $w_1 + w_2 = 1$.

Thus, the magnitude of the r.m.s. difference between the electron densities is determined simultaneously by the differences in the structure and composition of the "phases" present in the heterogeneous body and by the ratio of their volumes. Therefore, the r.m.s. difference between the electron densities (like the r.m.s. difference between the refractive indices, determined from the scattering of visible light) should be regarded as a most general characteristic of any submicroheterogeneous body, including glass.

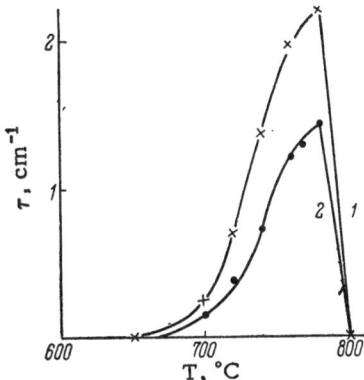

Fig. 1. Effect of temperature on the turbidity of glass containing 14% $Na_2O$. 1) $\lambda = 546$ m$\mu$; 2) $\lambda = 492$ m$\mu$.

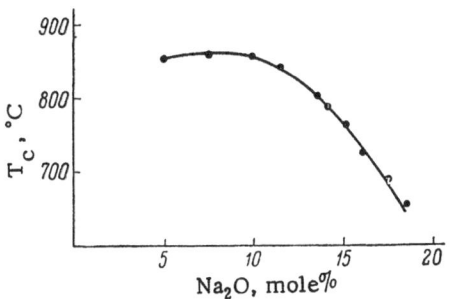

Fig. 2. Effect of the composition of sodium silicate glasses on the clearing temperature.

Equation (6) shows that at a constant value of the product $w_1 w_2$, the r.m.s. difference does not depend on the size of the heterogeneity regions. Therefore, $\overline{(\Delta\rho)^2}$ should remain unchanged when processes involving changes in the size of the heterogeneity regions without alteration of their composition, internal structure, or total volume take place in the glass.

It should be noted that it is possible to determine $\overline{(\Delta\rho)^2}$ from low-angle x-ray scattering only for bodies in which the linear dimensions of the heterogeneity regions exceed 20-25 A. Electron-density fluctuations at distances closer to interatomic are revealed by the scattering of x rays at large angles.

Experimental Results

The compositions of the glasses studied, based on chemical analysis, are given in the table. Sodium silicate glasses containing from 11.5 to 18.5 mole% of sodium oxide were melted in 3-liter quartz pots and annealed at 480° for 10 h.* The subsequent heat treatment is indicated in the table. All the lithium silicate glasses were obtained by sharp chilling from melts onto metal supports, and these glasses were not subjected to additional heat treatment.

A characteristic feature of all the glasses given in the table is their ability to exhibit opalescence provided that they have been subjected to definite heat treatment, characteristic for each composition.

As an example, Fig. 1 shows variations of the turbidity $\tau$ with temperature for a glass containing 14% $Na_2O$. The determinations were performed with the same specimen at room temperature. The specimen was held for 20 min at the required temperature, chilled in air, and after determination of turbidity in the spectrophotometer it was heated again at the higher temperature. At about 785° the opalescence vanished and the glass became transparent again. If this was accompanied by surface crystallization which spread gradually inward, the crystallized layer was ground off before the measurements, so that all the relationships described refer only to the internal volume of the glass; the x-ray diagrams were free from the slightest traces of discrete lines regardless of the degree of opalescence. Thus, heterogeneity regions in alkali silicate glasses have only qualitative and not geometrical order.

It was shown earlier by one of the present authors [12] that the structure of sodium silicate glasses remains heterogeneous even after opalescence has disappeared, but the heterogeneity regions decrease considerably in size.

Complete disappearance of opalescence at the "clearing temperature" $T_c$ is preceded by a gradual fall of light-scattering intensity in a narrow temperature region which begins 30-35° below $T_c$.

Figure 2 shows the dependence of the clearing temperature on the alkali oxide content of glasses in the $Na_2O$-$SiO_2$ system. The corresponding numerical data are given in the table, which also contains $T_c$ values for certain lithium silicate glasses. The precision in determination of the clearing temperature does not exceed ±5°. It follows from Fig. 2 and the table that the clearing temperature falls with increasing alkali content in the glass.

The intensity of low-angle x-ray scattering was measured in an apparatus with a proportional counter as receiver [13], at 30-kV tube voltage and 10 mA. The experimental low-angle scattering intensity curves, uncorrected for absorption, are given in Figs. 3 and 4.

*Glasses containing from 5 to 10 mole% $Na_2O$ were made by E. V. Podushko in a high-frequency furnace. The authors offer E. V. Podushko their sincere thanks.

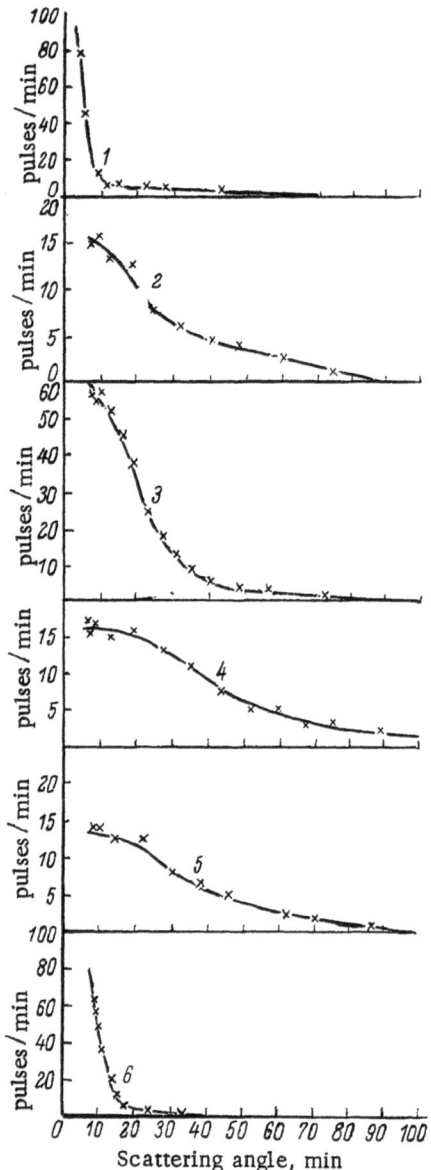

Fig. 3. Low-angle scattering curves for sodium silicate glasses. 1) 5% $Na_2O$, 95% $SiO_2$; 2) 7.5% $Na_2O$, 92.5% $SiO_2$; 3) 10% $Na_2O$, 90% $SiO_2$; 4) 11.5% $Na_2O$, 88.5% $SiO_2$; 5) 13.5% $Na_2O$, 86.5% $SiO_2$; 6) 16% $Na_2O$, 84% $SiO_2$.

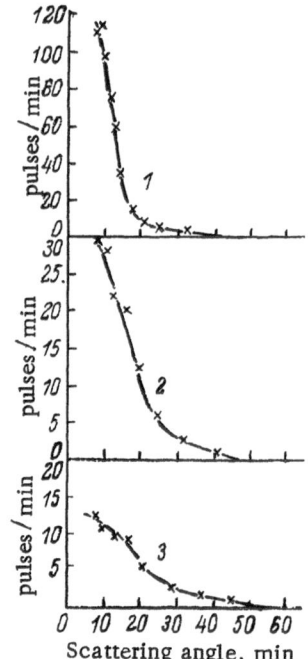

Fig. 4. Low-angle scattering curves for lithium silicate glasses. 1) 23.5% $Li_2O$, 76.5% $SiO_2$; 2) 25.5% $Li_2O$, 74.5% $SiO_2$; 3) 27.5% $Li_2O$, 72.5% $SiO_2$.

The dimensions of the heterogeneity regions were determined by the method of tangents for glasses containing 5 and 16% $Na_2O$ and 23.5% $Li_2O$. For the intensity curves of the other glasses a collimation correction for slit height was applied and the size of the heterogeneity regions was found from the angular position of the interference maximum thus revealed [10, 14]. The results of the calculations are given in the table.

The experimental curves used for calculation of the size of the heterogeneity regions were then corrected for absorption, the scattering intensities were calculated per unit scattering volume, and the curves were used to calculate the r.m.s. differences $\overline{(\Delta\rho)^2}$ between the electron densities of the "phases." The integral in Eq. (5) was found graphically for this purpose. The dependence of $\overline{(\Delta\rho)^2}$ on composition is plotted in Fig. 5 for sodium silicate glasses and in Fig. 6 for lithium silicate glasses. The corresponding numerical values are given in the table. The value of $\overline{(\Delta\rho)^2}$ for a specimen with 11.5% $Na_2O$ is taken as unity in each case.

### Discussion of Results

Although there have been numerous studies of the structure of alkali silicate glasses, especially sodium silicates, there are no generally accepted views on their structure.

Warren and co-workers used Fourier analysis of the diffraction patterns of sodium silicate glasses and concluded that their structure can be represented by a random but completely uniform spatial network [15]. Valenkov

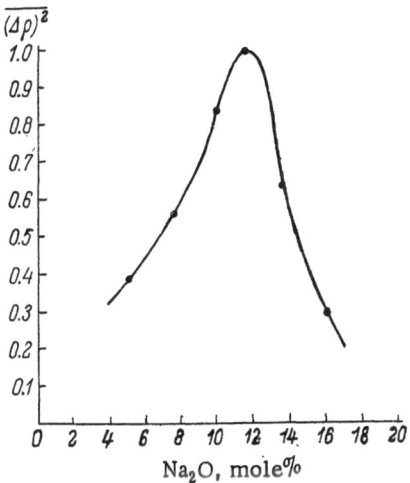

Fig. 5. Effect of composition of sodium silicate glasses on the r.m.s. difference between electron densities.

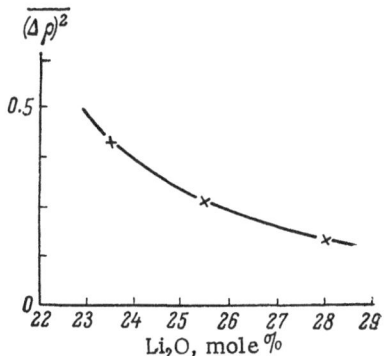

Fig. 6. Effect of composition of lithium silicate glasses on the r.m.s. difference between electron densities.

and Porai-Koshits found that x rays are scattered by glassy silica and sodium metasilicate independently in all sodium silicate glasses and concluded that the structure of such glasses is chemically heterogeneous [16]. In the interpretation of their results on Raman spectra of sodium and potassium silicate glasses, Kolesova and Gross consider that they "must be regarded as being completely homogeneous chemically . . . or as having very small regions of chemical heterogeneity − so small that atoms of one heterogeneity region may influence the natural vibrations of the atoms in another region" [17]. On the basis of their own Raman spectrum data, Bobovich and Tulub concluded that sodium silicate glasses are structurally heterogeneous, containing both indefinite and definite chemical compounds [18]. Florinskaya, who studied sodium silicate glasses by infrared spectroscopy, concluded that only definite chemical compounds exist in them [19].

Obviously, the results of the present investigation likewise cannot give a detailed picture of the structure of sodium and lithium silicate glasses. However, the detection of low-angle x-ray scattering by glasses is in itself irrefutable proof of their heterogeneous structure, and the relationships found between this scattering and the composition and heat treatment show that the following facts are certain.

In sodium and lithium silicate glasses, submicroscopic regions enriched with alkali-metal ions form even at low contents of the alkali oxide. The number of these regions and their total volume increase with increasing alkali content in the glass, corresponding to increase of the r.m.s. difference between the electron densities of the "phases" formed. With 11.5% $Na_2O$, sodium silicate glass reaches the maximum degree of heterogeneity, which corresponds to maximum $\overline{(\Delta\rho)^2}$.

If the composition of the "phases" is unchanged, $(\rho_1 - \rho_2)$ is constant, and the maximum r.m.s. difference of electron densities $\overline{(\Delta\rho)^2} = (\rho_1 - \rho_2)^2 w_1 \cdot w_2$ is attained at $w_1 = w_2 = 0.5$. If the composition of the heterogeneity regions really remains unchanged, at 11.5% $Na_2O$ half the volume of the glass should be occupied by regions enriched with sodium ions, and half by regions enriched with silica. If, on the other hand, the composition of the heterogeneity regions varies somewhat, the volumes become equal when the amount of $Na_2O$ is somewhat different from 11.5%.

With further increase of the amount of sodium, $\overline{(\Delta\rho)^2}$ decreases. This means that the ratio of the total volume of the silica-rich regions to the total volume of the sodium-rich regions also decreases; the glass becomes progressively more homogeneous. However, it is not possible to determine from these experimental results when complete homogeneity is reached, or whether it is reached at all, although a tendency to increasing homogeneity is quite obvious.

Because of the difficulty in melting lithium silicate glasses with low lithium oxide contents, it was possible to obtain only the right branch of the curve representing the dependence of $\overline{(\Delta\rho)^2}$ on composition. The position of the maximum $\overline{(\Delta\rho)^2}$ is still unknown. On the other hand, it is obvious that the structure of lithium silicate glasses remains heterogeneous even at compositions close to lithium disilicate.

The laws governing changes in the size of the heterogeneity regions remain to be investigated. However, even now the existence of a certain critical temperature $T_{cr}$ seems certain. At $T < T_{cr}$, the size of the heterogeneity regions increases both with rise of temperature and with increase of the heating time at a given temperature, whereas at $T > T_{cr}$ the size of the regions decreases rapidly with rise of temperature, down to values characteristic of completely clear glasses. As was already noted, this critical temperatures is 30-35° lower than the clearing temperature $T_c$ (see table).

The objects of the present work did not include study of the relationship between the results and the crystallizability of glasses. However, it seems certain to us that there should be correlation between variations of the r.m.s. difference $\overline{(\Delta\rho)^2}$ of the electron densities of the two "phases" in the glass, its crystallizability, and the amounts of resultant crystals of different compositions.

## LITERATURE CITED

1. E. A. Porai-Koshits and N. S. Andreyev, J. Soc. Glass Technol. 43(213):235, 1959.
2. N. S. Andreev, E. A. Porai-Koshits, and Yu. G. Sokolov, Izv. AN SSSR, Otdel. Khim. Nauk No. 4, 636, 1960.
3. W. Vogel, "The cellular structure of glass," in The Structure of Glass, Vol. 2, Consultants Bureau, New York, 1960, p. 17.
4. K. T. Bondarev and V. A. Minakov, Steklo i Keramika No. 12, 22, 1960.
5. F. Oberlies, Naturwissenschaften 43(10):224, 1956.
6. J. Warschaw, J. Am. Ceram. Soc. 43(1):4, 1960.
7. F. K. Aleinikov, V. A. Sliskis, R. B. Paulavichus, and P. V. Dundzis, Doklady Akad. Nauk SSSR 141(3):674, 1961.
8. A. A. Appen, "The coordination principle of ion distribution in silicate glasses," in The Structure of Glass, Vol. 1, Consultants Bureau, New York, 1958, p. 75.
9. M. V. Vol'kenshtein, "Structural and kinetic characteristics of the glassy state," in The Structure of Glass, Vol. 2, Consultants Bureau, New York, 1960, p. 111.
10. V. N. Filipovich, Zhur. Tekh. Fiz. 26(2):398, 1956.
11. D. Heirens, J. Polymer Sci. 35:139, 1959.
12. N. S. Andreev, Author's abstract of candidate's disseration, Leningrad, 1961.
13. D. A. Goganov, Yu. G. Sokolov, and E. A. Porai-Koshits, this collection, p.45.
14. V. N. Filipovich, Zhur. Tekh. Fiz. 27(5):1029, 1957.
15. B. E. Warren and J. Biscoe, J. Am. Ceram. Soc. 21(7):259, 1938.
16. N. Valenkov and E. A. Porai-Koshits, Zs. Krist. 95:195, 1936.
17. E. F. Gross and V. A. Kolesova, "Raman spectra and structure of glassy substances," in The Structure of Glass, Vol. 1, Consultants Bureau, New York, 1958, p. 45.
18. Ya. S. Bobovich and T. P. Tulub, "Raman spectra and structure of certain silicate glasses," in The Structure of Glass, Vol. 2, Consultants Bureau, New York, 1960, p. 173.
19. V. A. Florinskaya, "Infrared reflection spectra of sodium silicate glasses and their relationship to structure," in The Structure of Glass, Vol. 2, Consultants Bureau, New York, 1960, p. 154.

# CRYSTALLIZATION PRODUCTS OF LITHIUM SILICATE GLASSES

A. M. Kalinina, V. N. Filipovich, V. A. Kolesova, and I. A. Bondar'

The increasing practical utilization of crystallized glasses makes necessary a precise knowledge of the course of their crystallization. There are few publications dealing with studies of the spontaneous crystallization of glasses over a wide temperature range (up to the softening temperature), with identification of the devitrification products by simultaneous use of x-ray phase analysis, microscopy, and molecular spectroscopy. To the best of our knowledge, no work of this kind has been published on the $Li_2O$-$SiO_2$ system, which is of special interest as a part of the $Li_2O$-$Al_2O_3$-$SiO_2$ system, which is one of practical importance.

According to the phase diagram of the $Li_2O$-$SiO_2$ system [1], three chemical compounds can crystallize in this system: lithium ortho-, meta-, and disilicate ($2Li_2O \cdot SiO_2$, $Li_2O \cdot SiO_2$, and $Li_2O \cdot 2SiO_2$). It must be pointed out that there are no complete data in the literature on the interplanar spacings d of lithium di- and metasilicate. In fact, a comparison of Austin's tables [2] for d of lithium ortho-, meta-, and disilicates showed that many metasilicate lines are present in the disilicate diffraction pattern, and lithium meta- and disilicate lines occur in the orthosilicate diffraction pattern. This suggests that the crystalline substances regarded by Austin as standards (meta-, di-, or orthosilicate) were in fact mixtures of two or three phases. The probable explanation is that, because of the high volatility of lithium oxide, it is difficult to obtain compounds of predetermined composition in the preparation of lithium silicate glasses. This was also noted by Roy and Osborn [3].

The existence of two low-temperature forms of lithium disilicate $Li_2O \cdot 2SiO_2$ is reported in the literature. The first form has the following refractive indices: $n_g = 1.560 \pm 0.002$, $n_p = 1.549 \pm 0.002$ (according to Matveev and Velya [4]); $n_g = 1.558$ and $n_p = 1.547$ (according to Kracek [1]). The first form has a lower refractive index: $n_g = 1.531 \pm 0.003$ and $n_p = 1.528 \pm 0.003$ [4]. Jaeger and Klooster [5] and Matveev and Velya [4] regard the second form as a solid solution of $SiO_2$ in lithium disilicate. They detected this form only in a crystallized glass of the composition $Li_2O \cdot 3SiO_2$; it was not found for the composition $Li_2O \cdot 2SiO_2$. In addition, Liebau [6] reported enantiomorphic transition of lithium disilicate at 936° into a high-temperature form. However, to the best of our knowledge there are no microscope or x-ray diffraction data on the high-temperature form in the literature.

In the case of lithium metasilicate $Li_2O \cdot SiO_2$, only one crystalline form is known, with refractive indices $n_g = 1.609$ and $n_p = 1.584$ [7].

As regards the infrared spectra of lithium silicates, only the absorption spectrum of lithium metasilicate [8] and the spectra of different silica modifications [9,10] are known in the literature.

Thus, in studies of the crystallization of glasses in the $Li_2O$-$SiO_2$ system, attention should be focused mainly on the following aspects: 1) identification and sequence of formation of crystalline phases in relation to the composition and heat treatment of the crystallizing glass; 2) precise determination of the nature of the heat effects on the DTA curves; 3) detailed study of the existence of solid solutions of silica in lithium disilicate in the crystallization products of high-silica glasses.

## Experimental

The compositions of the lithium silicate glasses studied are given in the table. The x-ray diffraction method was used (with recording of the scattered radiation by the ionization and photographic methods) in conjunction with infrared absorption spectra (in the 420-1300 $cm^{-1}$ region) and thermographic and microscopic investigations. The starting materials for the glasses were amorphous spectrally pure $SiO_2$ and lithium carbonate of

Compositions of Glasses in the $Li_2O$-$SiO_2$ System

| Oxide ratio $SiO_2 : Li_2O$ | Composition by analysis | | | | Notes |
|---|---|---|---|---|---|
| | mole% | | weight % | | |
| | $Li_2O$ | $SiO_2$ | $Li_2O$ | $SiO_2$ | |
| 4 | 20.0 | 80.0 | 11.8 | 88.2 | Opalescent glass |
| 3.27 | 23.4 | 76.6 | 13.1 | 86.9 | |
| 3.27 | 23.4 | 76.6 | 13.1 | 86.9 | Clear transparent glass |
| 2.84 | 26.0 | 74.0 | 14.9 | 85.1 | Opalescent glass |
| 2.64 | 27.5 | 72.5 | 15.8 | 84.2 | Transparent glass |
| 2.40 | 29.6 | 70.6 | 17.1 | 82.9 | |
| 2.30 | 30.3 | 69.7 | 17.8 | 82.2 | |
| 1.93 | 34.2 | 65.8 | 20.5 | 79.5 | |
| 1.78 | 36.0 | 64.0 | 22.0 | 78.0 | |
| 1.70 | 37.1 | 62.9 | 22.7 | 77.3 | |
| 1.40 | 41.6 | 58.4 | 26.3 | 73.7 | |
| 1.29 | 43.7 | 56.3 | 27.9 | 72.1 | |
| 1.08 | 48.0 | 52.0 | 31.5 | 68.5 | Glass obtained in thin layers only, by chilling of melt between two metal plates |

analytical grade. The glasses were melted in a platinum crucible at 1250-1500° and chilled by pouring onto a metal slab. If special precautions are not taken, glasses containing from 20 to 29 mole% $Li_2O$ are opalescent, and the opalescence intensifies with decrease of the lithium oxide content.

It was shown by x-ray phase analysis and infrared absorption spectroscopy that, despite the results in [11], opalescence of these glasses is not associated with deposition of a crystalline phase. Transparent specimens, even of glasses with low $Li_2O$ contents, can be obtained by sharp chilling of thin glass layers cast from melts.

The glass compositions taken are indicated in Fig. 1 by crosses. They lie both in the region from the eutectic composition (30.3 mole% $Li_2O$) to pure $SiO_2$ (field I) and in the composition region from the metasilicate to the disilicate (field II). According to the phase diagram, the crystalline phases lithium disilicate and $SiO_2$ should be deposited from the glass in field I, and lithium meta- and disilicate in field II.

The temperatures at which the glasses were heated for crystallization were taken in accordance with the thermrograms (Fig. 2): crystallization was effected mainly at 430, 480, 630, and 900-960° with heating time from 1 to 100 h at the chosen temperatures. The temperatures of 430 and 480° are in the region of the initial crystallization stage. The temperature of 630° corresponds to the region of an intense exothermic effect due to main crystallization of the glass. The temperatures of 900-960° are in the region of high-temperature thermal effects corresponding to additional crystallization – secondary processes. Some glasses were treated in the 430-960° range at intervals of 50°.

Crystallization in Field I (20-30.3 mole % $Li_2O$)

Primary deposition of lithium disilicate at 480-630° is characteristic for all compositions in field I. At 900-960° cristobalite lines and one strong tridymite line appear. The lines are identified and the interplanar spacings are given in Fig. 3. The lines were identified by study and comparison of numerous diffraction patterns for all compositions in fields I and II. The main difficulty was in the separation of the disilicate from the metasilicate lines, as there is a small amount of nonequilibrium deposition of metasilicate in field I also, although, according to the equilibrium diagram (Fig. 1) it should not occur here.* The solution of the problem was also

---

*Analogously, as will be shown later, in field II glass of almost disilicate composition (Fig. 4c, d) gives appreciable amounts of lithium metasilicate at 630°.

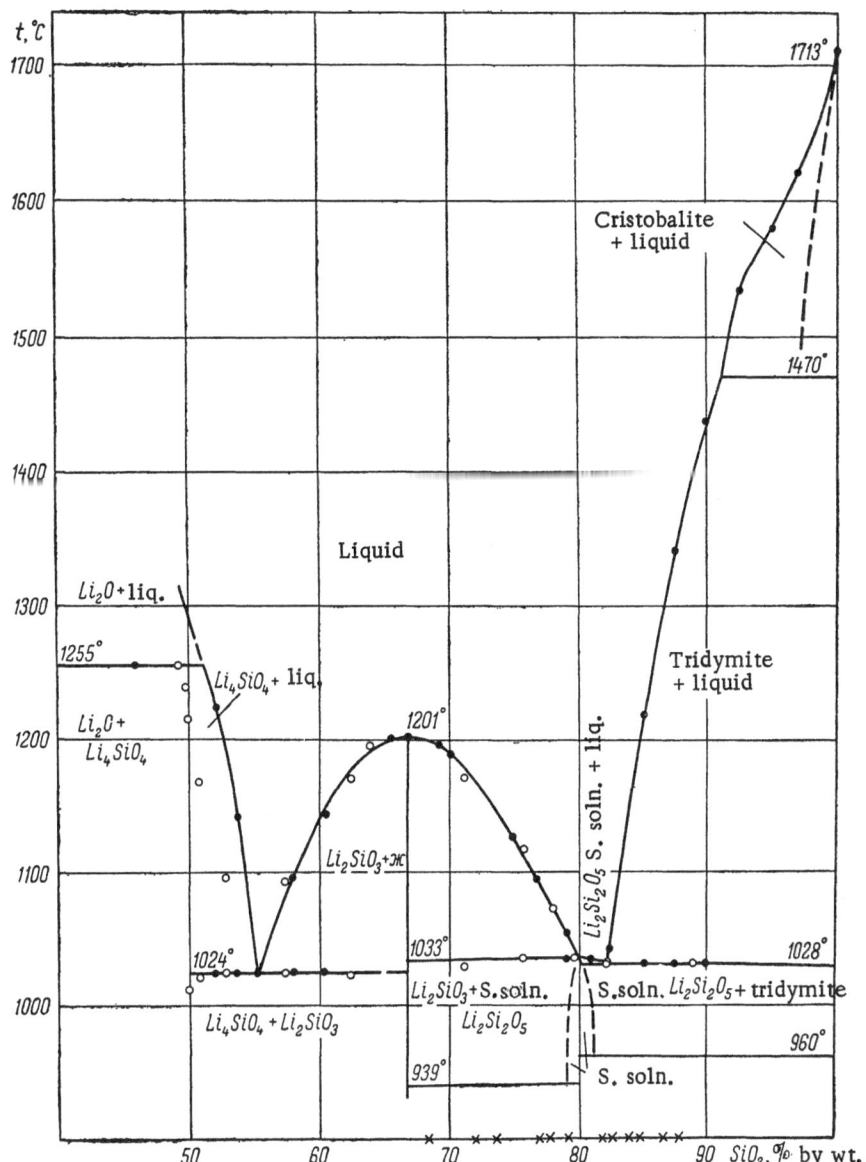

Fig. 1. Phase diagram of the $Li_2O$-$SiO_2$ system, after Kracek [1]. Crosses indicate the glass compositions studied.

assisted by study of the infrared absorption spectra of the same specimens. As a result, standard x-ray diffraction diagrams (Fig. 4e, Fig. 5b) and spectra (Fig. 7a,b) of lithium disilicate and metasilicate were obtained.

The course of crystallization in field I was studied most fully for the glass with 23.4 mole% $Li_2O$. Treatment for 100 h at 430° did not lead to crystallization. At 480°, with increase of the treatment time from 24 to 96 h, mainly lithium disilicate crystals are deposited. At the same time, the specimens exhibit blue opalescence, weak at first and then increasing, but remain transparent. On prolonged heating at 480° and also in the 630-900° range, the diffraction picture becomes more distinct and a line with d = 3.33 A, corresponding to lithium metasilicate, also appears. Cristobalite lines appear after 24 h in the 800-900° range. The appearance of considerable amounts of cristobalite in the crystallization products of the glass containing 23.4 mole% $Li_2O$ is also reflected in the infrared spectrum (Fig. 7f). It may be noted that the strongest cristobalite and lithium disilicate bands (Fig. 7b,c) coincide in frequency, and the presence of cristobalite in the specimens can be determined only from changes in the band contours in the 640 and 780 cm$^{-1}$ regions.

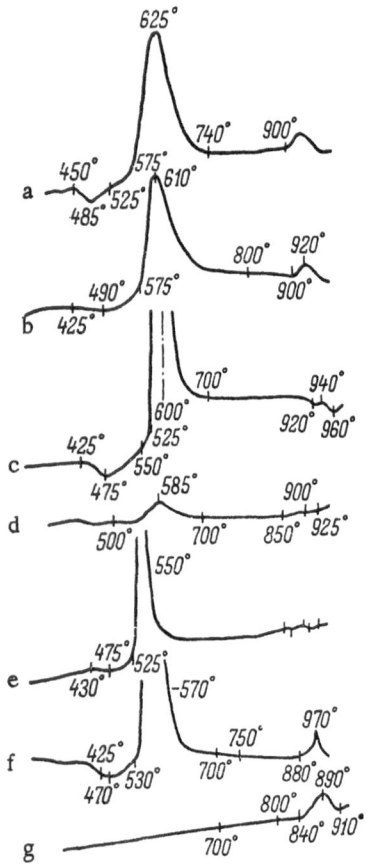

Fig. 2. Thermograms of the glasses.
a,b) 23.4 mole% Li₂O, opalescent and
cleared, respectively; c,d) 34.2 mole%
Li₂O, original and heated for 24 h at
480°; e) 41.6 mole% Li₂O; f,g) 43.7
mole% Li₂O, original and heated for
24 h at 630°.

After prolonged treatment at 960°, the strongest tridymite line (d = 4.33 A) appears, together with the cristobalite lines, while the line with d = 3.33 A, corresponding to metasilicate, disappears. Its disappearance is associated with the establishment of equilibrium in accordance with the phase diagram. From a comparison of the results of phase analysis with the thermograms (Fig. 2a,b) of glasses in field I, it follows that the first maximum at 610-630° should be assigned mainly to crystallization of lithium disilicate, and the second, at 920-930°, to crystallization of cristobalite. It should be noted in this connection that Matveev and Velya [4] observed that the second exothermic maximum on the DTA curves increases in the transition to compositions of higher silica content, Li₂O · nSiO₂ (n = 2-6), while the intensity of the first exothermic maximum (630°) decreases. This, of course, is due to an increase of the amount of crystallizing cristobalite and a decrease of the amount of disilicate. It must be pointed out that these workers tentatively assigned the high-temperature effect (920-930°) to additional secondary crystallization of the glass, without indicating the nature of the effect. Our data suggest the crystallization of cristobalite here.

Microscope investigations confirm the picture described above: the glass with 23.4 mole% Li₂O crystallized at 480° was found to contain spherulite formations with refractive indices $n_g$ = 1.527 and $n_p$ = 1.521 (birefringence 0.006). These values agree with Matveev and Velya's data for the composition Li₂O · 3SiO₂. The fact that the refractive index of the crystals in the glass with 23.4 mole% Li₂O is lower than that of crystals deposited from glass of the stoichiometric composition Li₂O · 2SiO₂, is probably attributable to the presence of a relatively considerable amount of glass, which fills the interstices between the crystals and thereby lowers the effective value of the refractive index of the crystals. It is also possible to attempt to attribute this fact to the formation of solid solutions of SiO₂ in Li₂O · 2SiO₂, as was done by Matveev and Velya [4], but our results do not confirm this.

## Crystallization in Field II (36-48 mole % Li₂O)

Glasses close in composition to lithium metasilicate (for example with 48 mole% Li₂O) crystallize extremely easily. They can be obtained only in the form of very thin plates. The crystallizability of the glasses decreases as the composition approaches the disilicate. Glasses containing 37.1 and 36 mole% Li₂O are relatively easy to obtain. The heat-treatment temperatures were the same as for field I (480, 630, 900-960°). This choice is also justified, in general, in relation to the DTA curves (Fig. 2f, g; the temperatures of the effects are 550-600 and 850-920°).

a) Let us examine the course of crystallization of glasses with 43.7 and 41.6 mole% Li₂O. Glass with 43.7 mole% Li₂O was subjected to consecutive heat treatments of 24 h at 480°, 16 h at 630°, and 1 h at 960°, and also 24 h at 630° and 16 h at 900°. The x-ray diffraction diagrams of the crystallization products of this glass (Fig. 5) show that lithium metasilicate is deposited first, and the amount of it increases with rise of temperature. At 630° the crystallization of lithium metasilicate is virtually completed in 24 h, as is indicated by the duplicate thermogram of the crystallized glass, which does not contain the first maximum (Fig. 2g). A similar result was obtained from crystallization for 16 h at 630° after preliminary treatment for 24 h at 480°. Despite the fact that, according to the phase diagram (Fig. 1), lithium disilicate should also crystallize in field II, it does not appear in this temperature range (Fig. 5b, c). A large amount of lithium disilicate appears in specimens previously treated for 24 h at 630° or 24 h at 480 and 630°, after additional heat treatment for 16 h at 900° or 1 h at 960°. Comparison of the spectrum of a specimen subjected to two-step heat treatment at 630 and 900° (Fig. 7d) with

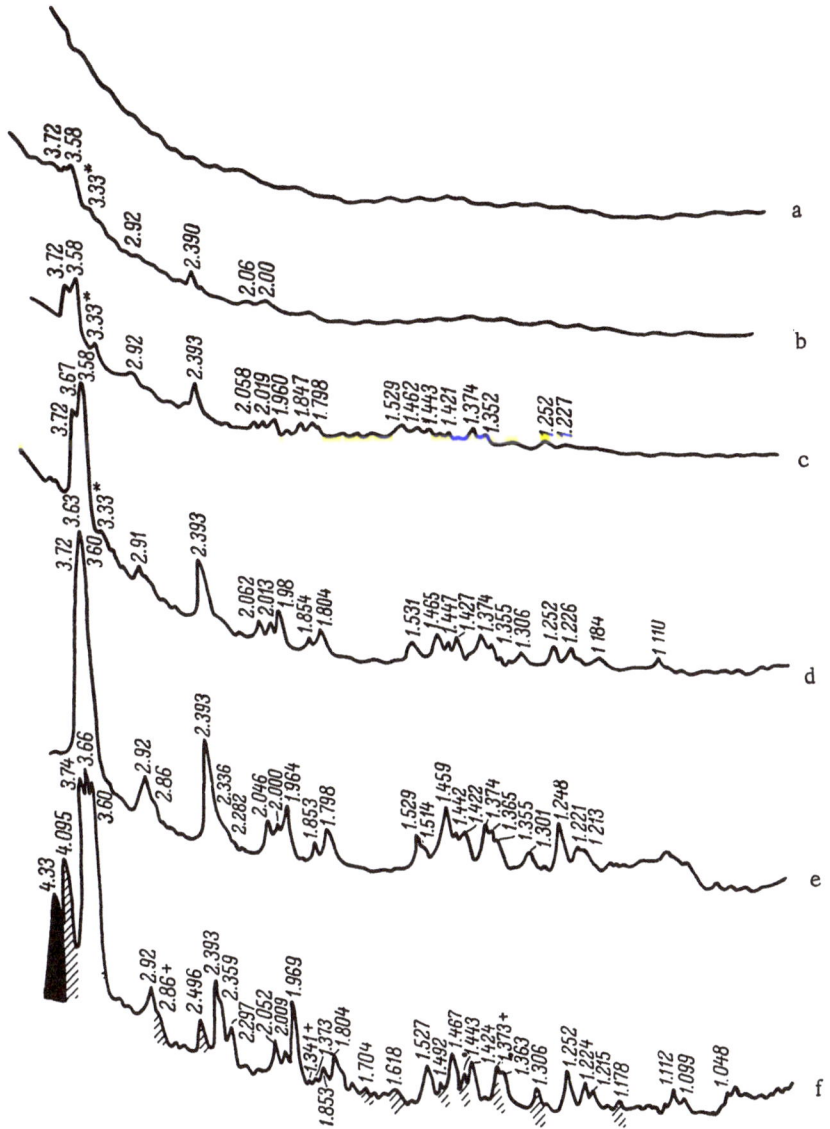

Fig. 3. X-ray diffraction intensity curves of cleared glass with 23.4 mole% $Li_2O$ after heat treatment. a,b,c,d) At 480° for 24, 48, 72, and 96 h; e) at 480 and 630° (24 h at each temperature); f) at 480, 630, and 960° (24 h at each temperature). The cristobalite lines are indicated by hatching, the tridymite lines are darkened, the metasilicate lines (M) are indicated by asterisks. The remaining lines correspond to lithium disilicate (D). The numbers represent the interplanar spacings (d) in angstroms. A plus sign indicates superposition of lines of two phases.

the spectra of lithium metasilicate ("analytical" bands at 854, 737, 613 $cm^{-1}$, Fig. 7a) and disilicate ("analytical" bands at 1218, 787, 761, 528 $cm^{-1}$, Fig. 7b) shows that the specimen contains roughly equal amounts of lithium meta- and disilicate. Because of all this, we can associate the high-temperature maximum on the thermogram (910°, Fig. 2g) with crystallization of lithium disilicate. It should be noted that in this case the crystallization of lithium disilicate is retarded considerably in comparison with field I, where it is readily apparent even at 630°.

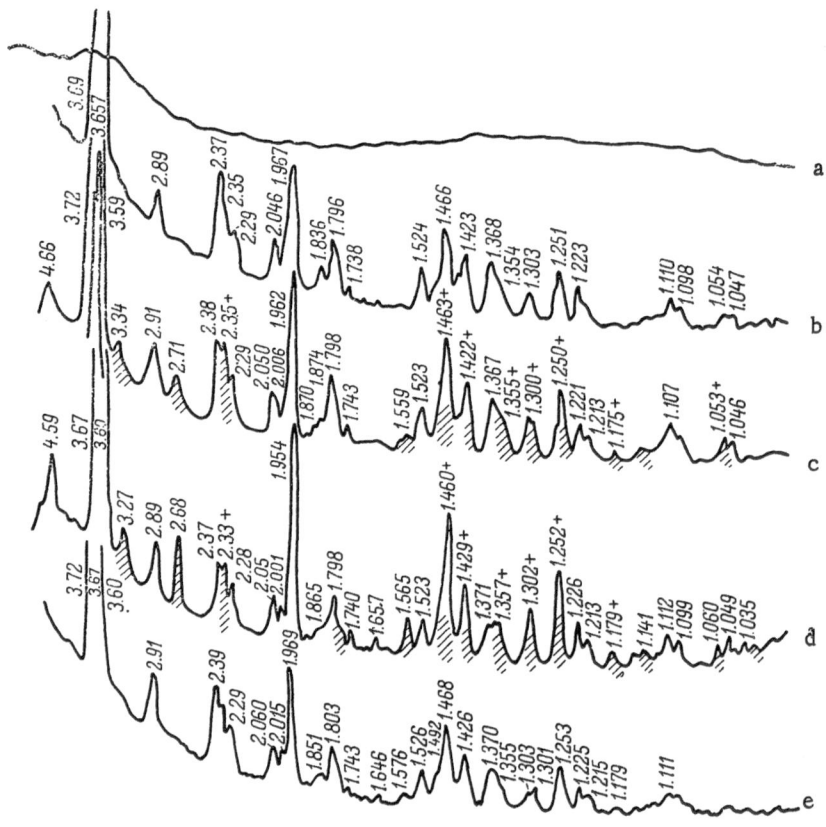

Fig. 4. X-ray diffraction intensity curves for glass with 34.2 mole% Li₂O. a) Original; b) crystallized for 24 h at 480°; c) for 24 h at 630°; d) for 24 h at 630° + 24 h at 960° (specimens b-d were cooled in air); e) for 90 h at 480° (cooled in the furnace). Metasilicate lines are indicated by hatching; plus signs indicate superposition of di- and metasilicate lines.

In glass with 41.6 mole% Li₂O, metasilicate and disilicate crystallize simultaneously (treatment for 24 h at 480°, Fig. 6b), and there is apparently more of the former than of the latter. As the temperature is raised to 630 and 960°, the disilicate lines increase more in intensity than the metasilicate lines, and the entire diffraction pattern becomes sharper (Fig. 6c, d). Separation of these phases was also observed under the microscope.

b) If we now turn to glasses close to lithium disilicate in composition (37.1, 36, and 34.2 mole% Li₂O), we find that the course of crystallization and the sequence of phase formation are the same for all: heat treatment at 480° leads mainly to the formation of lithium disilicate only, whereas treatment at higher temperatures (630 and 940-960°) also induces crystallization of lithium metasilicate (Fig. 6e-g, Fig. 4b-d; Fig. 7e).†

Special attention was devoted to crystallization of glass containing 34.2 mole% Li₂O (composition close to lithium disilicate, 33.3 mole% Li₂O). Specimens of this glass were subjected to the following heat treatment: 1) heating from room temperature to 480, 630, and 940° followed by exposure for 24 h at each temperature; 2)

†Glasses with 41.6 and 37.1% Li₂O yield products with very fine crystals as the result of stepwise crystallization; instead of spherulites (as in field I), uniform crystallization of the acicular-lamellar type occurs here (fuller details are given in [12]). These glasses are examples of hard and strong crystalline glass materials (superior in quality to the spherulite-crystallized materials with compositions of field I, where metastable phase separation occurs) formed by spontaneous volume crystallization without observed phase separation and in absence of catalyzing admixtures.

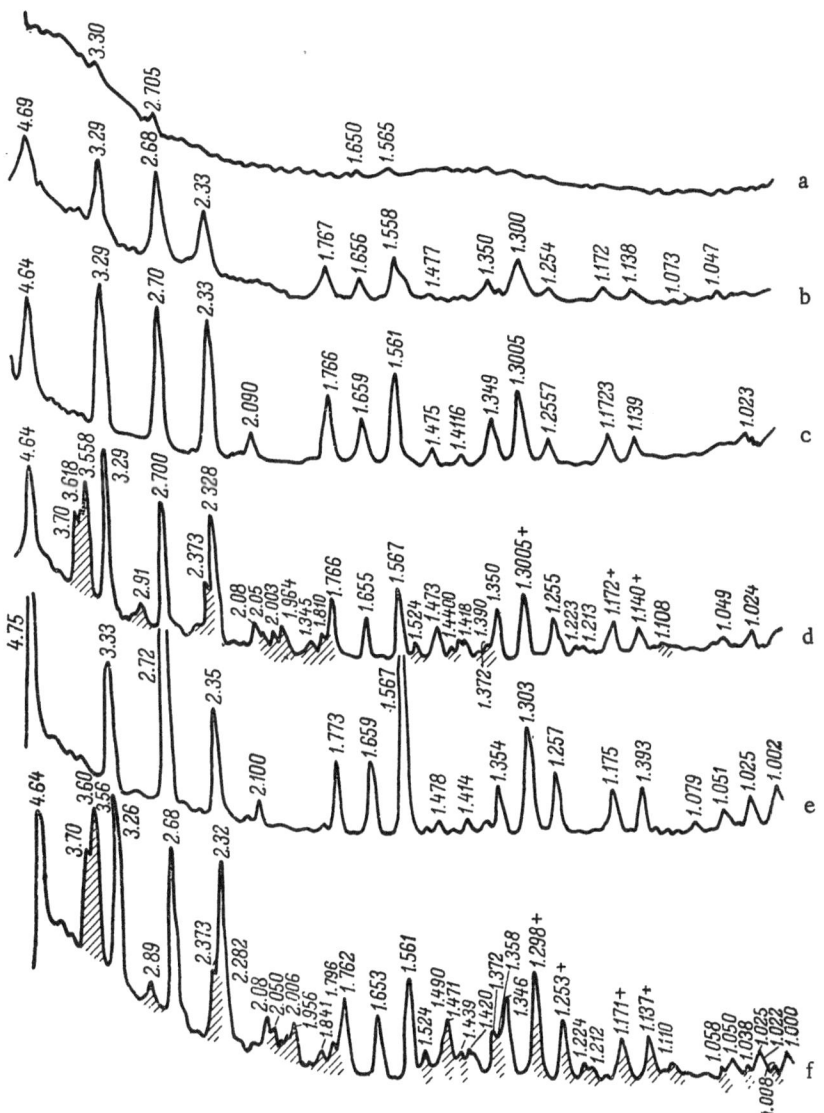

Fig. 5. X-ray diffraction intensity curves for glass with 43.7 mole% Li₂O after heat treatment. a) Original glass; b) treated for 24 h at 480°; c) 24 h at 480° + 16 h at 630°; d) 24 h at 480° + 16 h at 630° + 1 h at 960°; e) crystallization from melt; f) 24 h at 630° + 16 h at 960°. Specimens b-d were chilled in water; specimen f was cooled in air. The disilicate lines are indicated by hatching.

consecutive treatments of 24 h each at 480, 630, and 940°; 3) heating to 1000° followed by exposure at 940, 630, and 480° for 24 h. All these treatments, with the exception of heating from room temperature to 480° and subsequent holding at that temperature, lead to crystallization of lithium metasilicate as well as the disilicate. Heating of the glass from room temperature to 480° leads to formation of lithium disilicate only, and the situation is not changed substantially if the treatment time at this temperature is increased (Fig. 4b, e). The thermogram of the specimen heat-treated at 480° has a diffuse exothermic maximum at 585° (Fig. 2d), which indicates that the specimen contains a certain amount of glass which did not have time to crystallize at 480°. Indeed, this maximum is no longer present in the thermogram of a specimen previously heat-treated at 630°. However, it cannot be asserted that the residual maximum at 585° in this case corresponds only to further crystallization of the disilicate, as phase analysis indicates that the glass also contains lithium metasilicate after heat treatment at 630°.

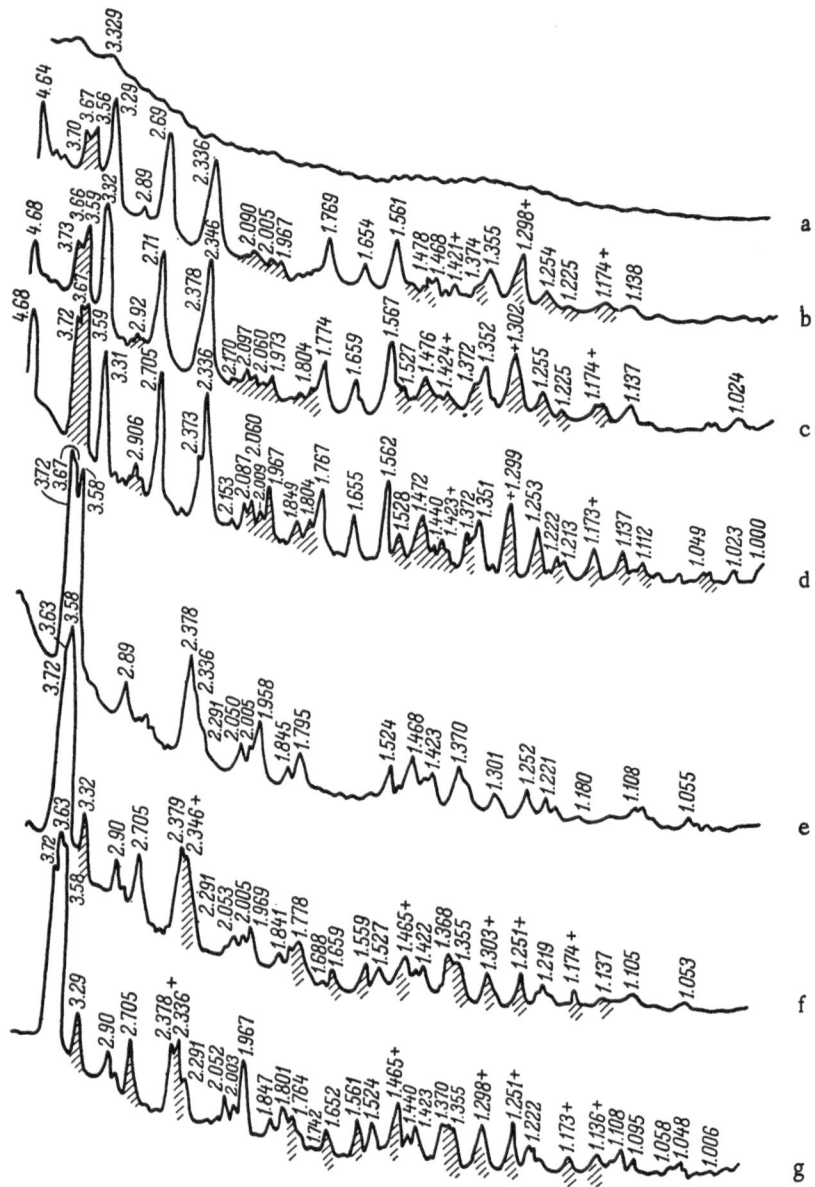

Fig. 6. X-ray diffraction intensity curves for glasses with 41.6 mole% $Li_2O$ (a-d) and 37.1 mole% $Li_2O$ (e-g) after heat treatment. a) Original glass; b) treated for 24 h at 480°; c) 24 h at 480° + 24 h at 630°; d) 24 h at 480° + 24 h at 630° + 24 h at 960° (the disilicate lines on curves b-d are indicated by hatching); e) 24 h at 480°; f) 24 h at 480° + 24 h at 630°; g) 24 h at 480° + 24 h at 630° + 24 h at 960° (the metasilicate lines on curves e-g are indicated by hatching).

Thus, a glass corresponding to lithium disilicate in composition yields crystalline disilicate only during low-temperature heat treatment (480°), while at 580-630° lithium metasilicate also crystallizes. It will be remembered that traces of metasilicate appear even in field I.

Comparison of the results of phase analysis and thermograms of glasses in field II, close to disilicate in composition, leads to the conclusion that the strong exothermic maximum at 580-600° is associated with

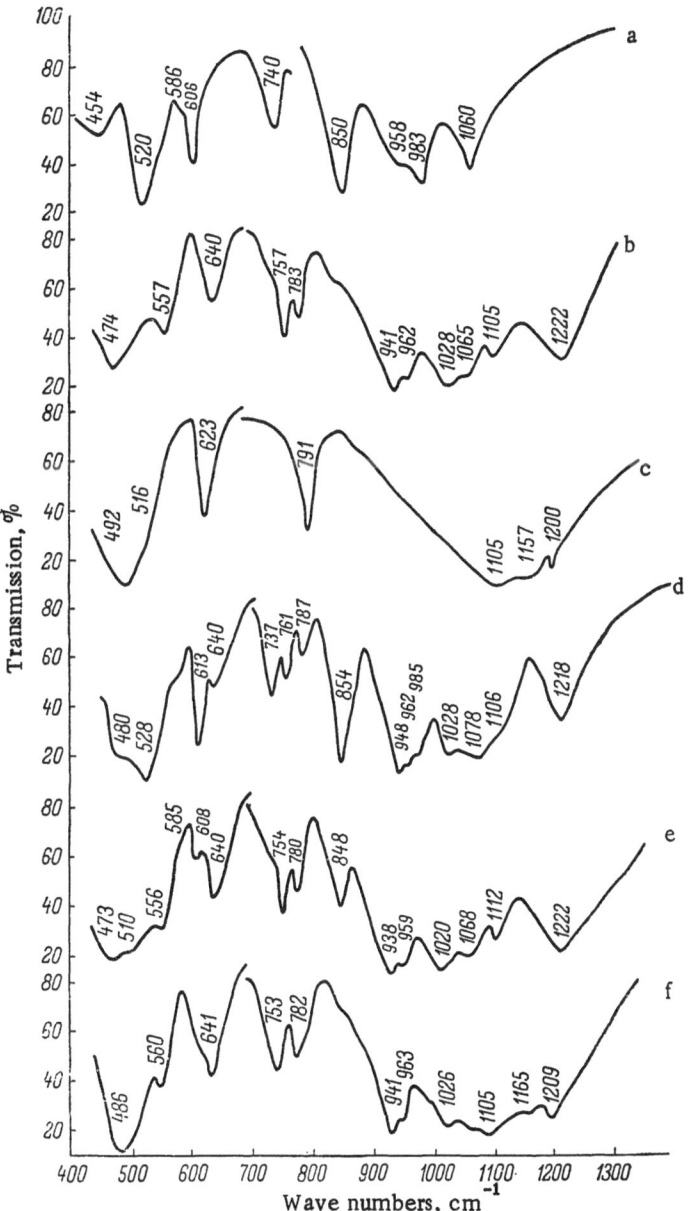

Fig. 7. Infrared absorption spectra. a) Lithium metasilicate obtained by crystallization from a melt; b) lithium disilicate formed by crystallization of glass with 34.2 mole% $Li_2O$ for 96 h at 480°; c) cristobalite; d) crystallization product of glass with 44 mole% $Li_2O$, heat-treated for 24 h at 630° and 16 h at 900°; e) crystallization product of glass with 34 mole% $Li_2O$, heat-treated for 24 h at 940°; f) crystallization product of glass with 23.4 mole% $Li_2O$, heat-treated for 48 h at 960°. Specimens in the form of disks pressed from mixtures of the substances with KBr.

simultaneous crystallization of di- and metasilicate; lithium disilicate makes the principal contribution to this. As regards the high-temperature heat effects on the thermograms of these glasses, an attempt at their interpretation will be made later.

## Discussion of Results

First, it must be emphasized that crystallization of the glasses studied proceeds mainly in accordance with the phase diagram.

One interesting feature is the sequence of phase deposition. In field I, at relatively low temperatures (480-800°), lithium disilicate crystallizes first. With regard to $SiO_2$, because of the high viscosity (high activation energy $\Delta \Phi_A^*$ of growth of the critical nucleus [13]) it is formed only as a glass. This metastable glassy phase begins to crystallize in the form of cristobalite only at high temperatures (800-960°). This is a manifestation of the Ostwald step rule. Separation into phases proceeds in such a manner that the thermodynamic potential barrier in formation of nuclei of the critical size ($\Delta \Phi^* + \Delta \Phi_A^*$) has the minimum value. Therefore, disilicate crystal nuclei appear first (the value of $\Delta \Phi_A^*$ for them is less than for $SiO_2$). These nuclei have the highest degree of order in the center, while the boundary layer gradually merges in composition and structure with the surrounding metastable glassy medium. This ensures the minimum surface energy $\sigma$ at the interphase boundary, and $\Delta \Phi^*$, which is proportional to $\sigma^3$, depends strongly on this [13]. It is also significant that at the next stage, in the crystallization of $SiO_2$, cristobalite nuclei are formed; cristobalite is also metastable at the crystallization temperature (800-960°).

Cristobalite is a high-temperature, and the most symmetrical cubic, form of silica, and its deposition as the first phase from isotropic (spherically symmetrical) glass is most probable, as the surface energy is then minimal. Subsequently cristobalite should pass into tridymite, stable at 870-1470°; small amounts of tridymite are in fact noted in the x-ray patterns (Fig. 3).

Glasses in field I (20-29 mole% $Li_2O$) are characterized by opalescence (up to opacification) due to metastable phase separation. In the same temperature and composition region where metastable phase separation is possible, the separation begins, in accordance with the Ostwald rule, with the formation of amorphous nuclei of metastable glassy phases, as $\sigma$ is evidently least at the boundary of amorphous phases. Subsequently crystal nuclei appear in these metastable phases, the compositions of which are shifted toward lithium disilicate or $SiO_2$. These matters are discussed more fully in [13].

Glasses with 23.4 mole% $Li_2O$ begin to opalesce when heated at 480° (after only 48 h), and the visible opalescence increases with the time of heating. Appearance of the blue opalescence is accompanied by the appearance of the diffraction pattern of lithium disilicate, which becomes progressively sharper (Fig. 3). Here, apparently, phase separation proceeds simultaneously with crystallization.

Distinct azure opalescence without signs of crystallization can be obtained in glasses with 29 mole% $Li_2O$ (composition close to the eutectic) by heat treatment for 24 h at 480°. Rapid heating of the opalescent glass to higher temperatures (620°) leads to disappearance of the opalescence, without appreciable crystallization (the glass becomes colorless and transparent). This indicates that metastable phase separation is an independent effect, inherent in glass as an undercooled liquid, and not purely a precrystallization effect. A fuller account of opalescence in lithium silicate glasses is given in [14].

Let us now consider the appearance of small amounts of lithium metasilicate in field I, where, according to the phase diagram, it should not exist. The possible explanation is that, because of ineffective mixing of the melt during formation of the glass, regions with compositions lying in field II are formed, and these can give rise to metasilicate together with disilicate during crystallization.

Moreover, in phase separation of the original glass, the appearance of regions of glassy $SiO_2$ is associated, in particular, with loss of lithium atoms from these regions. This may lead to accumulation of lithium atoms at the boundaries of the pure $SiO_2$ regions. The composition of these boundary regions lies in field II, and metasilicate can then form before this nonequilibrium accumulation has time to disperse.

We shall now discuss the crystallization sequence in field II (from lithium metasilicate to disilicate). Here glasses with the composition 34-37 mole% $Li_2O$ give lithium disilicate as the first phase (Fig. 4f, Fig. 6e), deposited even at a low temperature (480°). The second phase, with a composition shifted toward metasilicate, remains glassy. This may be attributed to the relatively high fluctuation frequency of the composition in individual regions of the glass; as a result, the stoichiometric disilicate composition becomes established in these regions and crystallization follows. In this case, nucleus regions of the metasilicate composition are formed much less frequently, as their composition differs more from the average glass composition. Moreover, the thermodynamic potential barrier $\Delta \Phi^*$ for formation of the critical nucleus of any crystalline phase in general decreases as the composition of the glass approaches that of the given phase [13]; therefore, crystallization of metasilicate lags behind that of disilicate and large amounts of metasilicate appear only at temperatures of 630° and higher.

Conversely, in compositions closer to metasilicate, the latter appears first at 480° and there is hardly any disilicate. It is significant that in this case the signs of disilicate are in general very weak up to 900°, and appreciable amounts appear (Fig. 5d, f) only above that temperature (910° maximum on the thermogram, Fig. 2f). At the same time, as already noted, disilicate can crystallize even at 480-630° in field II (as in field I) in compositions with 34-37 mole% $Li_2O$. The question arises: what is the cause of this retardation in the crystallization of disilicate?

One explanation may be that, since metasilicate crystallizes very easily, a nonequilibrium excess may appear. Numerous metasilicate nuclei are formed; these grow actively and capture lithium atoms from the surrounding medium, bringing its composition closer to $SiO_2$, before disilicate crystal nuclei can appear in appreciable amounts. However, at least at high temperatures, establishment of equilibrium must begin, with redissolution of metasilicate in silica and formation of disilicate. A temperature in the region of 900° is apparently the temperature at which significant structural changes in the glass become possible, including the conversion of nonequilibrium metasilicate into disilicate. In fact, the thermograms in Fig. 2 show that the 900-960° region is the region in which such hindered processes as the crystallization of cristobalite occur. The temperature of enantiomorphic transition of lithium disilicate also lies in this region.

In compositions intermediate between lithium meta- and disilicate, crystals of these two phases appear simultaneously (Fig. 6b-d), although, of course, nonequilibrium states are also possible here, with deviations of the amounts of crystalline substances deposited from the lever rule.

The formation of metasilicate even in the composition with 34.2 mole% $Li_2O$, directly adjacent to the disilicate composition, is probably associated with incongruent melting of the disilicate and formation of metasilicate regions during cooling of the melt (Fig. 1; Fig. 4c, d).

We now examine the origin of certain details, not yet discussed, in the thermograms of Fig. 2. The minimum on each DTA curve before the maximum corresponding to primary crystallization of the glass is apparently associated with precrystallizational increase of the heat capacity of the glass due to increased mobility of the structural elements in the glass and to intensified fluctuations (generally reversible) of concentration and structure, preceding the formation of crystal nuclei of critical size, and also due to relaxation processes in establishment of the metaequilibrium "liquid" structure of the glass, corresponding to the temperature of this minimum [15]. In the case of opalescent original glasses (Fig. 2a) this minimum has a complex structure, because an opalescent glass really comprises two glasses in which the precrystallization effects are different, so that the thermograms become complicated.

The minimum at 940-960° might be attributed to enantiomorphic conversion of disilicate [6] from the low-temperature into the high-temperature form, which should be accompanied by absorption of heat. Special x-ray studies, the results of which will be published later, confirm this polymorphic conversion. This minimum is not found in Fig. 2d for a lithium disilicate sample after previous crystallization. On the contrary, a maximum is found at 850-920°. This may be attributed to superposition of the maximum corresponding to partial recrystallization of nonequilibrium metasilicate (analogously to Fig. 2e, f, g) on the minimum. It is also possible that the more perfect and larger crystals in the case of Fig. 2d are more resistant to the polymorphic transition than the less perfect crystals in the specimen of Fig. 2c, not subjected to previous crystallization at 480°. This should level out the minimum. Additional investigations are needed for more definite conclusions.

Finally, we must point out that explanations other than those put forward here may account for some of these characteristics in the crystallization of lithium silicate glasses, but, in any event, they must be associated with hindered establishment of thermodynamic equilibrium and with different rates of this process.

## LITERATURE CITED

1. F. C. Kracek, J. Phys. Chem. 34:2641, 1930.
2. A. E. Austin, J. Am. Ceram. Soc. 30:218, 1947.
3. R. Roy and E. Osborn, J. Am. Chem. Soc. 71:2086, 1949.
4. M. A. Matveev and V. V. Velya, Steklo i Keramika 10:14, 1959.
5. T. Jaeger and H. S. Van Klooster, Proc. Amst. Acad. Sci. 16:857, 1914.
6. F. Liebau, Acta Cryst. 14:389, 1961.
7. G. Donney and J. D. H. Donney, Am. Miner. 38:163, 1953.
8. A. N. Lazarev and T. F. Tenisheva, Optika i Spektroskopiya, 10:79, 1961.
9. N. A. Savchenko and V. A. Florinskaya, Doklady Akad. Nauk SSSR 109:1115, 1956.
10. M. Narand, J. Phys. et Rad. 9:81, 1948.
11. G. E. Blair and S. Urnes, Glastechn. Ber. 34:391, 1961.
12. A. I. Korelova, M. G. Degen, and O. S. Alekseeva, this collection, p. 65.
13. V. N. Filipovich, this collection, p. 9.
14. N. S. Andreev, D. A. Goganov, E. A. Porai-Koshits, and Yu. G. Sokolov, this collection, p. 47.
15. M. V. Vol'kenshtein and O. B. Ptitsyn, Zhur. Tekh. Fiz. 26(10), 1956.

# MICROSTRUCTURE OF TWO-COMPONENT LITHIUM SILICATE
# GLASSES AT VARIOUS CRYSTALLIZATION STAGES

## A. I. Korelova, M. G. Degen, and O. S. Alekseeva

The microstructures of original and crystallized lithium silicate glasses were studied with the aid of the optical and electron microscopes. To reveal the structure, the polished surface of the specimens was etched with 1-5% hydrofluoric acid for various times, depending on the chemical composition of the glass. The etched surface of the specimen was coated with a layer of silver and then examined and photographed by reflected light under the optical microscope at magnifications from 50 to 400, and at 700 by the immersion technique. For electron microscope studies the polished surface was etched and coated at an angle of 45° with a thin layer (50-100 A) of carbon and platinum with simultaneous evaporation [1]. The replicas were separated from the glass by immersion in 5% HF solution, washed in distilled water, and examined in the ÉM-3 electron microscope at magnifications of 4000-7000.

Parallel studies of polished and fractured surfaces of the same specimens showed a complete analogy of microstructure,* but in many cases the fractured surfaces exhibited irregularities which were caused by propagation of elastic waves during the fracture and which distorted the structure.

The lithium oxide content of the glasses ranged from 23 to 42 mole%. The glasses were crystallized by heat treatment at temperatures ranging from 480 to 1000°, and the treatment time was varied from 5 to 24 h, and even longer in some cases. The results of phase analysis of the crystals formed are given in [2].

Figures 1 and 2 show the microstructure of glass containing 23.4 mole% $Li_2O$ after various heat treatments. If the glass is cooled slowly when initially made, it is opaque and electron micrographs (Fig. 1a) show characteristic separation into two phases, with separated droplets about 0.1-0.2 $\mu$ in size. If the same glass is cooled rapidly, it remains transparent and, as Fig. 1b shows, it has a fine structure with microheterogeneities about 300 A in size. If this glass is kept at 480° for 96 h, it becomes opalescent; this is probably the result of phase separation, as is suggested by the rounded appearance of the heterogeneities in it. These heterogeneities are 0.05-0.15 $\mu$ in size (Fig. 1c). According to [2], $Li_2O \cdot 2SiO_2$ begins to crystallize in the glass under these conditions, but the crystals formed were so small and few that they could not be detected by means of the electron microscope.

Heat treatment of this glass at 620° for 5 h leads to complete crystallization of lithium disilicate with formation of large spherulites (150-250 $\mu$) having a characteristic orientation in the form of rays radiating from the center (Fig. 2a). Regions without structural orientation, probably corresponding to residual glass, may be seen between the spherulites and occasionally within some spherulites. Figure 2b shows details of the spherulite structure; the directed configuration and rounded outlines of the constituent crystalline formations may be seen. Comparison of Figs. 1c and 2b shows a resemblance in the nature of the structure, and this suggests that crystallization of the glass of this composition was preceded by separation into two phases; the regions richer in $Li_2O$ subsequently crystallized while retaining their initial droplet shape.

When glass of this composition is crystallized by consecutive treatments for 20 h each at 480 and 620°, the character of the microstructure remains the same as before (Fig. 2a), but the spherulites formed are considerably smaller (15-25 $\mu$).

---

*Volume crystallization took place in the lithium silicate glasses of the compositions studied.

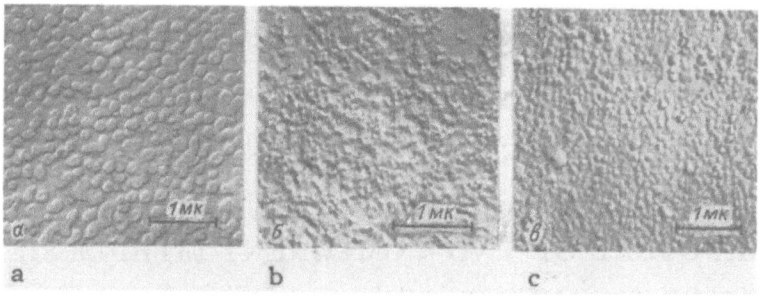

Fig. 1. Electron micrographs of the structure of lithium silicate glass (23.4 mole%
$Li_2O$). a) Slowly cooled, opaque; b) chilled, transparent; c) kept for 96 h at 480°.

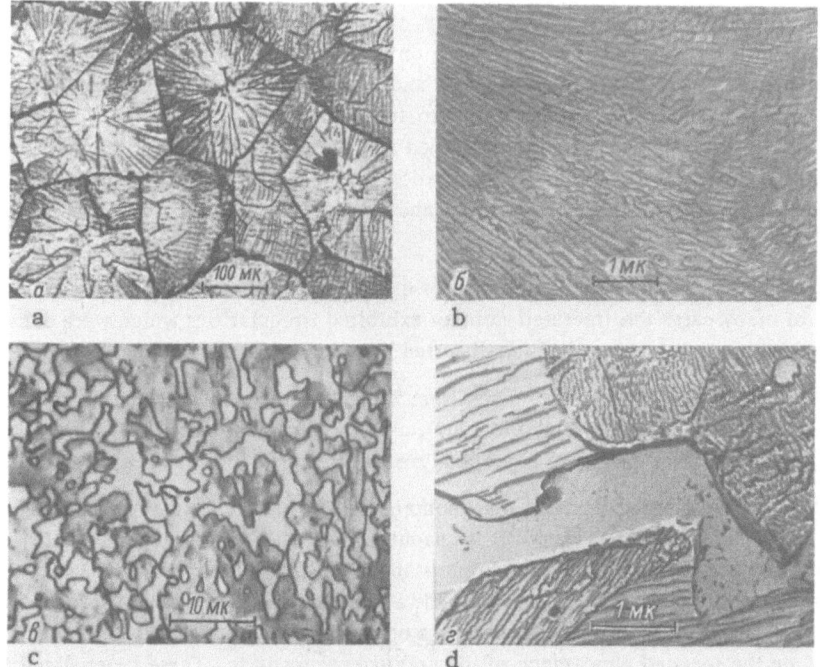

Fig. 2. Microstructure of crystallized lithium silicate glass (23.4 mole% $Li_2O$)
after heat treatment. a,b) 5 h at 620°; c,d) 24 h each at 480 and 630° and 18 h
at 1000°.

Heat treatment for 24 h each at 480 and 630° and for 18 h at 1000° leads to a considerable change of microstructure; this is due to crystallization of cristobalite from the residual vitreous silica under these conditions [2]. It is seen in Fig. 2c that the spherulite outlines disappear but numerous elongated cristobalite crystals (1.5-3) $\times$ (4-8) $\mu$ in size appear. The electron micrograph in Fig. 2d shows a portion of a spherulite with a directed structure containing embedded cristobalite crystals.

Microheterogeneities could not be detected by means of the electron microscope in the original glasses containing 37.1 and 41.6 mole% $Li_2O$. If these glasses are heated for 24 h at 480°, they crystallize with formation of $Li_2O \cdot 2SiO_2$ and $Li_2O \cdot SiO_2$. The microstructure of glass containing 37.1 mole% $Li_2O$ and crystallized at 480° is shown in Fig. 3a. In this case, dendritic spherulites 30-40 $\mu$ in size are formed, with interpenetration of lithium di- and metasilicate crystals (Fig. 3b). After double heat treatment of this glass (24 h each at 480 and 630°), the spherulites are 15-30 $\mu$ in size, and the structure of the dendritic branches becomes finer (Fig. 3c). If this glass or glass containing 41.6 mole% $Li_2O$ is additionally treated for 24 h at 960°, new crystalline formations

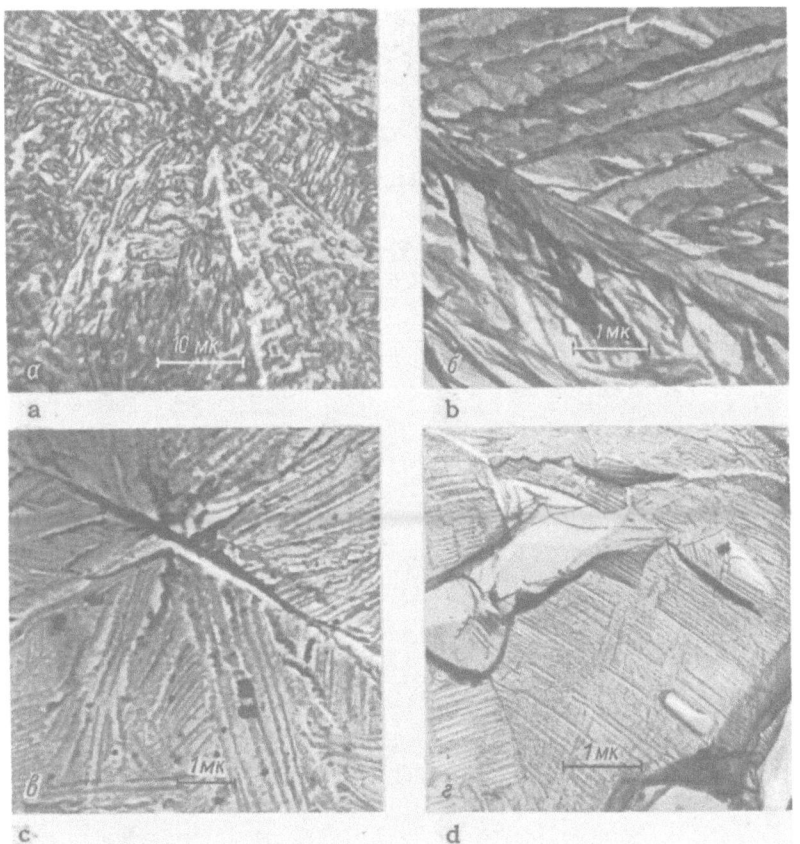

Fig. 3. Microstructure of crystallized lithium silicate glass (37.1 mole% $Li_2O$) after heat treatment. a,b) 24 h at 480°; c) 24 h each at 480 and 630°; d) 24 h each at 480, 630, and 960°.

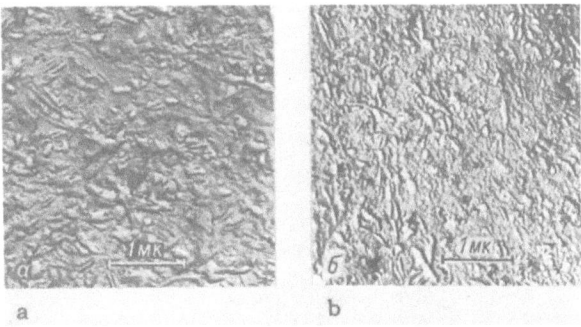

Fig. 4. Microstructure of crystallized lithium silicate glass (41.6 mole% $Li_2O$) after heat treatment. a) 24 h at 480°; b) 24 h each at 480 and 630°.

several microns in size appear in them (Fig. 3d), evidently as the result of recrystallization of lithium silicates.

Crystallization of glass containing 41.6 mole% $Li_2O$ (24 h at 480°) results in the formation of a microcrystalline structure composed of elongated interconnected crystals 0.15-0.3 $\mu$ long and about 0.15 $\mu$ across (Fig. 4a). If this glass is additionally heated for 24 h at 630°, the structure becomes even finer (Fig. 4b). In this case the formation of a microcrystalline structure is due to simultaneous crystallization of roughly equal amounts of lithium di- and metasilicates [2].

The microhardness of these specimens was found to be higher than that of any other lithium silicate glasses studied by us [3].

LITERATURE CITED

1. G. S. Gritsaenko and N. F. Samotin, Zap. Vsesoyuz. Min. Obshch. Ser. II, Part 91, No. 1, 1962.
2. A. M. Kalinina, V. N. Filipovich, V. A. Kolesova, and I. A. Bondar', this collection, p.53.
3. A. I. Korelova, O. S. Alekseeva, and M. G. Degen, Opt.-Mekh. Prom. 29:9, 1962.

# LITHIUM GLASSES AND SOME OF THEIR CRYSTALLIZATION CHARACTERISTICS

## N. A. Shmeleva and N. M. Ivanova

We investigated crystal formation in lithium glasses starting from the binary system $Li_2O-SiO_2$, with gradual complication of the composition by introduction of photosensitive additives: $(Ag + CeO_2)$, $K_2O$, and $Al_2O_3$. The glasses were crystallized in thin layers. The glass was powdered down to grains 0.1-0.5 mm in size and crystallized in 5-cm$^3$ porcelain crucibles for 5 min at 800°.

The first investigations were carried out with glasses I, II, and III, the compositions of which are given in Table 1.

The calculated content of possible $Li_2Si_2O_5$ is 100% for glass I, and 70 and 38% for glasses II and III.

In glasses I-III crystallized at 800°, the crystals have radial orientation at the fragment surfaces, but there is no definite orientation at a depth of 30-50 $\mu$.

In glasses I and II every fragment, regardless of shape and size, contains a central gas cavity occupying up to one fifth of the fragment volume (Fig. 1).

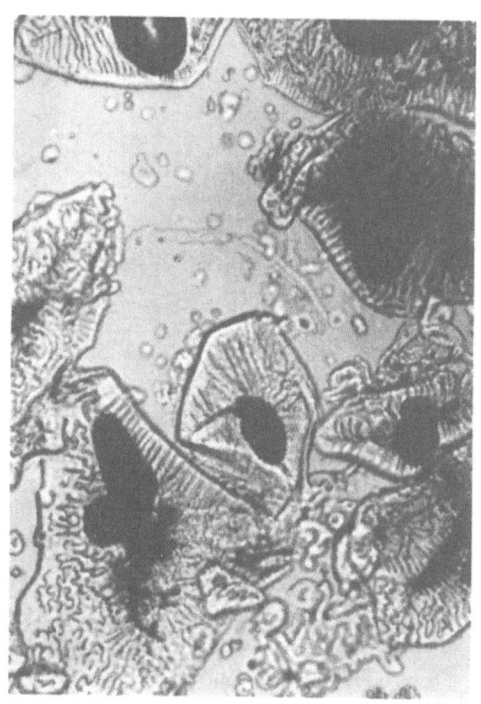

Fig. 1. Glass I, magnification 180.

Crystallization always begins at the fragment surface, forming a rigid skelton which determines the entire course of subsequent crystallization of the glass, with formation of extension zones at the centers of the fragments, where cavities in the form of bubbles are produced (because of the different specific volumes of the glass and the crystalline phase). These cavities have diverse shapes, from large rounded bubbles to dendritic and very fine threads with a negative relief. The bubbles are often surrounded by dark fields where the microcavities enter the crystalline mass.

The distribution of gas cavities in glasses I-III and If-IIIf is shown schematically in Fig. 2; it is seen that the volume of the gas cavities decreases with increase of the $SiO_2$ content in the glass, and in glass III they consist of thin dendrites instead of bubbles. In crystallization of glasses of the If-IIIf series (glasses I-III with photosensitive additions: 0.02% Ag and 0.05% $CeO_2$),

TABLE 1

| Glass | Molecular comp. | Composition, % by weight | | Weight % of possible $Li_2Si_2O_5$ | Density of $Li_2Si_2O_5$, g/cm$^3$ |
|---|---|---|---|---|---|
| | | $SiO_2$ | $Li_2O$ | | |
| I | $Li_2O \cdot 2SiO_2$ | 80 | 20 | 100 | Crystals, 2.45 |
| II | $Li_2O \cdot 3SiO_2$ | 86 | 14 | 70 | Glass, 2.30 |
| III | $Li_2O \cdot 6SiO_2$ | 92.4 | 7.6 | 38 | |

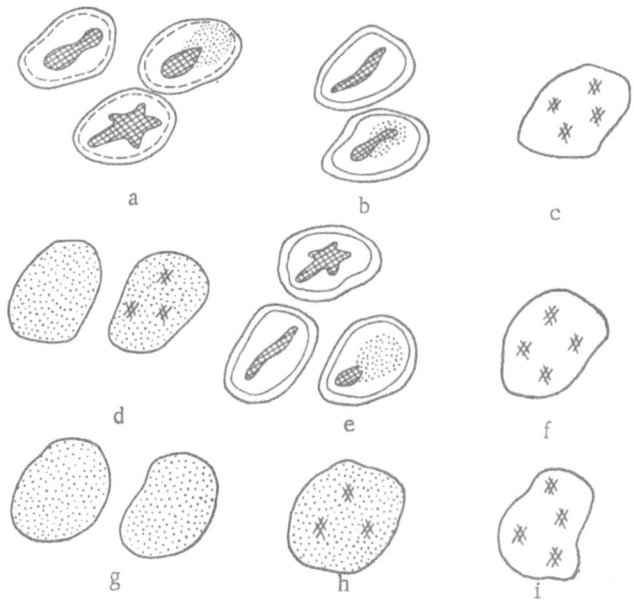

Fig. 2. Distribution of gas cavities in irradiated and non-irradiated glasses of series I-III, If-IIIf. a) Glass I, $N_{av} \simeq 1.550$; b) glass II, $N_{av} \simeq 1.530$; c) glass III, $N_{av} \simeq 1.505$; d) glass If (Ag + $CeO_2$), $N_{av} \simeq 1.543$; e) glass IIf, $N_{av} \simeq 1.528$; f) glass IIIf, $N_{av} \simeq 1.492$; g, h, i) glasses If, IIf, and IIIf after ultraviolet irradiation, with $N_{av}$ of <1.543, <1.528, and <1.531, respectively.

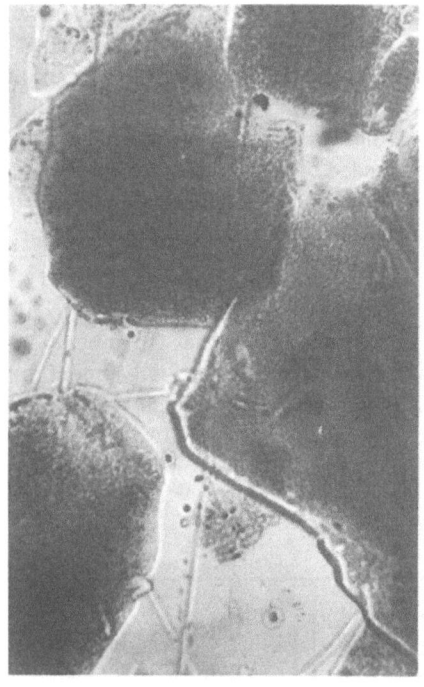

Fig. 3. Glass If, magnification 180.

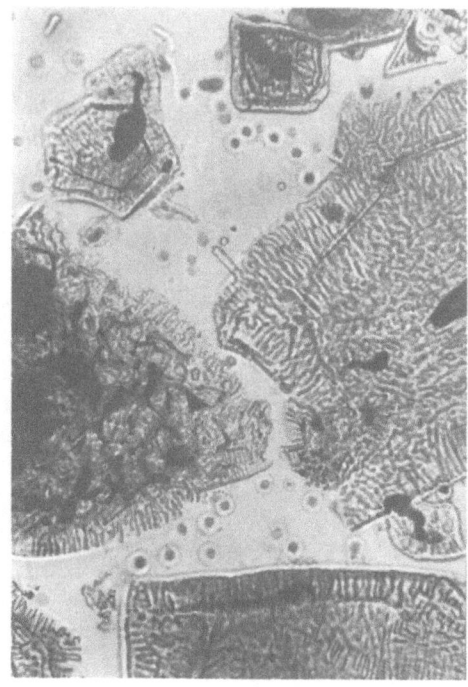

Fig. 4. Glass IIf, magnification 180.

TABLE 2

| Glass | Li$_2$O content, % | Crystallization conditions | N$_{av}$ of Li$_2$Si$_2$O$_5$ crystals |
|---|---|---|---|
| I | 20 | 600-800°, 10 h <br> 800°, 5 min | 1.553 <br> 1.543 - 1.552 |
| II | 14 | 800°, 5 min: <br> surface <br> volume | 1.528 - 1.531 <br> 1.522 - 1.525 |
| III | 7.6 | 800°, 5 min: <br> surface <br> volume | 1.505 - 1.507 <br> 1.495 - 1.501 |

Note. The crystallization tendency of Li$_2$Si$_2$O$_5$ is so great that despite the short crystallization time, well-defined Li$_2$Si$_2$O$_5$ crystals are formed. $N_g - N_p$ of the crystals varies from 0.005 to 0.011.

TABLE 3

| Crystallization conditions | If | | IIf | | IIIf | | Li$_2$Si$_2$O$_5$ standards | | | |
|---|---|---|---|---|---|---|---|---|---|---|
| | d, A | I, % | d, A | I, % | d, A | I, % | Roy-Osborn | | ASTM | |
| 800°, 5 min, not irradiated | 3.73 | 44 | 3.74 | 40 | 3.77 | 13 | – | – | – | – |
| | 3.67 | 66 | 3.64 | 80 | 3.64 | 13 | – | – | – | – |
| | 3.61 | 100 | 3.59 | 100 | 3.58 | 29 | – | – | – | – |
| | – | – | – | – | 3.35 | 100 | – | – | – | – |
| 800°, 5 min, irradiated | 3.74 | 43 | 3.74 | 47 | 3.71 | 14 | 3.73 | 80 | 3.75 | 33 |
| | 3.65 | 70 | 3.64 | 69 | 3.66 | 12 | 3.64 | 65 | 3.67 | 100 |
| | 3.59 | 100 | 3.60 | 100 | 3.57 | 22 | 3.58 | 100 | – | – |
| | – | – | – | – | 3.35 | 100 | – | – | – | – |

TABLE 4

| Principal lines, d in A | Glass 2 | | Glass 136 | | Glass 172 | |
|---|---|---|---|---|---|---|
| | nonirradiated, I, % | irradiated, I, % | nonirradiated, I, % | irradiated, I, % | nonirradiated, I, % | irradiated, I, % |
| 4.71 | 44 | 38 | 48 | 75 | 63 | 80 |
| 3.43 | 100 | 100 | 100 | 100 | 100 | 80 |
| 3.32 | 37 | 48 | 50 | 87 | 54 | 100 |

the structure changes sharply only in glass If (Fig. 2) and remains almost unchanged in IIf and IIIf.

In glass If, crystallization begins not at the surface but around the crystallization centers — colloidal silver — ensuring uniform distribution of the crystalline substance and gas cavities throughout the volume without formation of large cavities. In glass IIf, large gas cavities are retained, with a definite tendency to finer splitting (Figs. 3 and 4).

Previously irradiated glasses of the If-IIIf series exhibited crystallization throughout the volume at 800°, with very fine distribution of the gas cavities in glasses If and IIf (Fig. 2g, h) and retention of a coarse-grained structure in IIIf (Fig. 2i).

Figure 5 is a schematic representation of the most probable distribution of gas cavities under conditions of heterogeneous crystallization (glass If irradiated and nonirradiated, and glass IIf irradiated). In these cases,

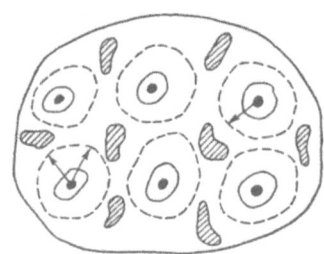

Fig. 5. Distribution of gas cavities and crystalline substance in glasses If (nonirradiated and irradiated) and IIf (irradiated).

zonal crystallization occurs around the silver centers with formation of a spherulite structure, leading to uniform distribution of extension microzones throughout the volume (hatched regions in Fig. 5).

The distribution of gas cavities after crystallization at 700 and 900° was the same as at 800°. With two-step crystallization (2 h at 500° and 5 min at 900°), dendritic gas cavities were formed.

The main crystalline phase in glasses of the I-III and If-IIIf series, irradiated and nonirradiated, is lithium disilicate and its solid solutions. In glasses III and IIIf, grains of secondary quartz with $N_{av}$ = 1.535 separate out together with the solid solutions.

It is known from the literature [1] that solid solutions of silica in lithium disilicate separate out in glasses of the composition $Li_2O \cdot (2-6)SiO_2$, and the average refractive index of the crystalline phase decreases with increase of the silica ratio of the glass.

Table 2 shows variations of $N_{av}$ of the $Li_2Si_2O_5$ crystals on the glass fragments in relation to the glass composition and the crystallization zone.

The observed differences between $N_{av}$ of $Li_2Si_2O_5$ crystals on the surface and within the volume indicate that the phase composition is heterogeneous even in fragments smaller than 0.3 mm. The lowest refractive indices are found at regions in immediate proximity to the gas bubbles or in regions of the finest penetration of the gas cavities into the crystalline mass.

This suggests that the gas cavities assist the incorporation of silica in the crystalline mass. Direct proof of enrichment of the bubble wall layer with silica is provided by the Becke line, which often diverges from the bubble contours in different directions.

The ionization curves of crystallized glasses of series I-III and If-IIIf, previously irradiated and nonirradiated, differed only in the intensities of the maxima in the low-angle region. The lines chosen for comparison corresponded to interplanar spacings d of 3.75, 3.67, and 3.59 A — the main lines in our x-ray diagrams and of the Roy-Osborn standards and the ASTM tables.

Interpretation of the x-ray diagrams of glasses with 20 and 14% $Li_2O$ showed the presence of $Li_2Si_2O_5$ (without $Li_2SiO_3$ or silica); the $Li_2Si_2O_5$ crystals, with N close to that of pure $Li_2Si_2O_5$ ($N_g$ = 1.558 and $N_p$ = 1.547), approximate to the ASTM standard with the main line at 3.6 A, while in the formation of solid solutions ($N_{av}$ = 1.550-1.496), the x-ray diagrams correspond to the Roy-Osborn standard with the main line at 3.58 A (Table 3).

The x-ray patterns of glasses III and IIIf (7.6% $Li_2O$) confirmed that the main phase is $\alpha$-quartz with the main line d = 3.35 A; $Li_2Si_2O_5$ was present as an impurity. The content of $\alpha$-quartz in crystallized glass IIIf was considerably higher after preliminary irradiation than in the nonirradiated specimen (as shown by the intensity ratio of the d = 3.35 A lines).

Thus, the principal phases in crystallization of glasses of the $Li_2O$-$SiO_2$ system are $Li_2Si_2O_5$ and quartz; free quartz is obtained only if there is considerable supersaturation with silica ($Li_2O \cdot 6SiO_2$), while in glasses of lower silica contents the quartz is present in the solid solution.

The degree of saturation of the solid solution with silica increases with increase of the excess $SiO_2$ (in re-relation to $Li_2Si_2O_5$) in the glass, in the presence of photosensitive additives, upon irradiation with ultraviolet light, with increase of the time of heat treatment in the extension zones, and also when the microcavities are distributed most finely in the crystalline substance.

Addition of 4% $K_2O$ in place of $Li_2O$ to glass If did not alter the diffraction pattern, which gave a principal line at 3.67 A for nonirradiated specimens and at 3.59 A for irradiated specimens. The composition was complicated further by introduction of 8% $Al_2O_3$. Three glasses, Nos. 2, 136, and 172, were studied; they were of the same composition but differed in the melting conditions. The ratio of the main components, $SiO_2$ : $Li_2O$ (without a correction for $Al_2O_3$ or for replacement of $Li_2O$ by 4% $K_2O$), was 85.2 : 14.8; i.e., it approximated to glass II with the composition $Li_2O \cdot 3SiO_2$. Introduction of 8% $Al_2O_3$ sharply altered the phase composition and texture of the crystallized glasses. The crystallization rate fell considerably in both irradiated and nonirradiated specimens. After 2 h of heat treatment at 800°, the irradiated specimens crystallized throughout the volume, forming a border 10 $\mu$ thick at the edge with $N_{av} = 1.540$. Within the fragments the mass was microcrystalline with $N_{av} = 1.525$, and without visible gas cavities. In the nonirradiated specimens the interior of each fragment had a dark brown color, indicating the formation of microcavities.

The diffraction curves of the last three specimens (Table 4) differ sharply from the curves for glasses without alumina: in addition to the principal line with d = 3.43 A, lines with 4.7 and 3.32 A are clearly seen. These lines were assigned to $Li_2SiO_3$ after an examination of the entire range of interplanar spacings. The considerable increase in the intensity of the 3.32 A line is apparently due to the presence of an additional phase with the same interplanar spacing.

In irradiated specimens the intensity of the principal line, 3.43 A, decreases, and that of the 3.32 A line increases and reaches the maximum (I = 100) in specimen 172 (glass made under reducing conditions), apparently corresponding to the principal quartz line, 3.35 A (100). Increase of the crystallization temperature to 850° shifts the principal line to 3.46 A (I = 100), which approaches the principal spodumene line, 3.48 A.

Thus, when glasses with 8% $Al_2O_3$ crystallize at temperatures below 850° the formation of another series of solid solutions of the type of spodumene (eucryptite) with quartz or other forms of $SiO_2$ is to be expected, rather than the formation of lithium disilicate. Differences in the saturation of the solid solutions with silica determine the diverse properties of the crystallized glasses.

It is likely that loose structures would be obtained, as with 8% $Al_2O_3$ in the glass the maximum spodumene content cannot exceed 30% by weight. Crystallization throughout the volume can be expected only if solid solutions with $SiO_2$ are formed, pure forms of silica separate out, and gas cavities are present.

We note in conclusion that study of the crystallization process in small glass fragments made it possible to follow the course of formation of microcavities by way of bubbles, large cavities, and intermediate dendritic forms. We demonstrated the presence of extension microzones where the material is ruptured with formation of microcavities. In the extension zones the nature of the crystallization alters somewhat toward formation of solid solutions with higher $SiO_2$ contents or separation of free $SiO_2$.

Introduction of photosensitive additives (Ag + $CeO_2$) into the glass favors the finest distribution of the gas cavities or extension microzones throughout the glass volume; this effect is attained in glass If, with 20% $Li_2O$, directly by crystallization, but in glass IIf only by crystallization after irradiation. Very fine distribution of gas cavities in the crystalline substance is the cause of the brown coloration of varying intensity in the fragments, which gives the impression of a "black" substance which moves with displacement of the extension zones.

Crystallization of $SiO_2$ in the form of cristobalite, tridymite, or quartz occurs at the gas cavities. Preliminary irradiation shifts the crystallization equilibrium toward a higher quartz content.

LITERATURE CITED

1.  M. A. Matveev and V. V. Velya, Steklo i Keramika No. 10, 14, 1959.

*The Lithium Aluminosilicate System*

The papers in this section are concerned with investigations of glasses in the $Li_2O$-$Al_2O_3$-$SiO_2$ system, close to spodumene in composition and containing various amounts of $TiO_2$ as the crystallization catalyst.

These glasses are of great practical importance, as they yield crystalline glass materials, including transparent ones, with low positive, zero, and negative coefficients of thermal expansion.

Fourteen of the communications in the section are concerned with investigations of a glass of the same composition (13) in which the $Li_2O$ content is somewhat lower and the $Al_2O_3$ content somewhat higher than in spodumene. The amount of catalyst $TiO_2$ was varied from 0 to 11% by weight and was 5% in most of the investigations.

The purpose of these investigations was to elucidate the complex mechanism of catalyzed crystallization with glass of one composition.

The conclusions, based on the use of numerous different methods of investigation, are sometimes contradictory. This is because of the complex nature of the processes occurring during the precrystallization period, and also because of our incomplete knowledge of the $Li_2O$-$Al_2O_3$-$SiO_2$ system.

# MECHANISM OF THE CATALYZED CRYSTALLIZATION OF GLASSES IN THE $Li_2O-Al_2O_3-SiO_2$ SYSTEM WITH TITANIUM DIOXIDE

## E. V. Podushko and A. B. Kozlova

**1.** We define catalyzed crystallization as volume crystallization occurring in the presence of a catalyst, which may be one of the main glass components or an additional component specially introduced as the catalyst.

The mechanism of catalyzed crystallization was studied with glasses of the $Li_2O-Al_2O_3-SiO_2$ system, with addition of titanium dioxide, from 2.0 to 11.0% by weight, as catalyst. The glass compositions were close to spodumene. Certain amounts of various additives were introduced to facilitate the melting and to prevent crystallization of the final glass during cooling.

**2.** Catalyzed crystallization comprises two main stages. The catalyst acts during the first stage, and conditions are created for subsequent formation of crystallization centers simultaneously in very large numbers throughout the volume. During the second stage crystallization centers arise simultaneously throughout the volume and crystals of the main phase grow. The investigation covered both stages.

In the examination of the catalyst action, we investigated the problem of the stage during the melting process at which this action occurs and how it depends on the catalyst content and the heat-treatment conditions. We did not investigate the manner in which the catalyst creates conditions for subsequent nucleation uniformly throughout the glass volume. This question can probably be answered on general lines, as follows.

The catalyst creates structural defects — phase separation, discontinuities, etc., i.e., places with weakened energy bonds, where crystallization centers arise under certain conditions. The number of these defects and their defectiveness depend on the composition of the glass and the catalyst content, and determine the number of nucleating centers and therefore the size of the crystals of the main crystalline phase growing on them.

The question of the composition of the crystallization centers is still not clear. Some investigators believe that the crystallization centers are catalyst particles, i.e., in glasses catalyzed by titanium dioxide, rutile or some titanate in the crystalline state. However, no experiments confirming this viewpoint are known at present. Moreover, in the opinion of Bobovich, who investigated these glasses by means of Raman spectroscopy, the original glasses catalyzed by titanium dioxide contain neither rutile nor any titanate in the crystalline state, regardless of the thermal history (heat treatment). We consider that the crystallization centers arising on structural defects correspond to the composition of the main crystalline phase.

In a study of the second stage, we investigated the conditions of nucleation and crystal growth in relation to the heat treatment and catalyst content, and the influence of heat treatment on the character and certain properties of the crystallized glass.

We believe that there are no grounds for regarding volume crystallization in the presence of a catalyst (nucleation and crystal growth) as proceeding in any manner other than by the Tammann mechanism, and that the well-known Tammann graph should be applied in this case.

**3.** The following five series of experiments were carried out to determine when the catalyst exerts its action. Glasses of the same composition but with different $TiO_2$ contents were melted in 3-liter quartz pots. The pots, containing the glass at 1480-1500°, were then put into an electric furnace heated to 900°. The furnace was regulated so as to ensure the required cooling conditions, excluding any rise of the glass temperature in the pots to 500°. The temperature of the glass was measured during the cooling by means of a thermocouple

TABLE 1. Characteristics of the Glasses after Cooling

| TiO$_2$ content, % by weight | Cooling conditions | | | | |
|---|---|---|---|---|---|
| | No. 1, held at 820° | No. 2, held at 740° | No. 3, held at 715° | No. 4, held at 700° | No. 5, no holding time |
| 3.0 | A | A | A | A | A |
| 5.0 | A | C | A | A | A |
| 7.0 | B | B | C | C | C |
| 9.0 | B | B | - | - | C |
| 11.0 | B | B | - | - | C |

A - Transparent glass, not crystallized

B - Opaque, stonelike

C - Opaque, waxlike

immersed in it. The cooling conditions were as follows: the temperature was held at 820° for the first series, at 740° for the second, at 715° for the third, and at 700° for the fourth. Glasses of the fifth series were cooled without a holding time. These holding temperatures were chosen on the following grounds: crystals of the main phase grow at 700-740°, and at 800-850° transparent crystallized glass becomes opaque.

Three of the series (Nos. 1, 2, and 5) each consisted of six glasses with the following TiO$_2$ contents: first, 3.0%; second and third, 5.0%; fourth, fifth, and sixth, 7.0, 9.0, and 11.0%. The other two series (Nos. 3 and 4) each consisted of four glasses with the following TiO$_2$ contents: first, 3.0%; second and third, 5.0%; fourth, 7.0%. The thermocouple was always immersed in glass with 5.0% TiO$_2$. The main results of the five series of experiments are given in Table 1.

From Table 1 it follows that: 1) with 3.0% TiO$_2$ catalyzed crystallization does not occur, regardless of the cooling conditions; 2) with 5.0% TiO$_2$ catalyzed crystallization occurs only if the glass is held at 740° during the cooling; 3) with 7.0, 9.0, and 11.0% TiO$_2$ catalyzed crystallization occurs during cooling by any of the procedures used.

The principal crystalline phase in all the crystallized glasses was β-spodumene, i.e., the phase obtained as the result of catalyzed crystallization of glasses in this system under the usual conditions. The thermal expansion of glasses crystallized under conditions of falling temperature is also the same as that of glasses crystallized under the usual conditions. The elongation curve of one of the opaque waxlike glasses is shown in Fig. 1.

It was thus shown experimentally that catalyzed crystallization can be effected only during cooling of liquid glass; therefore, the catalyst (titanium dioxide in this instance) acts as a source of crystallization centers at a fairly high temperature, in any event higher than the maximum holding temperature in these experiments. This is apparently the temperature of formation of the glass structure in the liquid state. This conclusion is confirmed by the results of the following experiment. Glass containing 5.0% TiO$_2$ was poured into water at 1500°. No differences were noticed between the catalyzed crystallization of this glass and a normally annealed glass.

The following should also be noted. Glasses of the first series contained spherulites 6-10 mm in diameter, distributed throughout the glass 10-20 mm apart. These spherulites were observed in both uncrystallized and crystallized glasses. They were found to consist of β-spodumene. Despite the fact that the spherulites and the rest of the mass in crystallized glasses of this series had the same composition, the process of their formation was different. Spherulites are crystals which usually arise in glasses which tend to crystallize under unsuitable cooling conditions. Their formation is not connected in any way with the presence of catalysts. When a glass is cooled, the spherulites form first and do not influence the development of catalyzed crystallization. During catalyzed crystallization of a glass containing spherulites, the latter are unchanged and remain embedded in the crystallized mass, differing from the latter in appearance.

The possibility of catalyzed crystallization during cooling leads to the conclusion that the cooling glass passes through a temperature range where both the nucleation rate and the rate of growth of the crystals of the main phase on the resultant nuclei are different from zero; the results of this series of experiments can therefore be explained by the Tammann mechanism. The suggested schemes for different TiO$_2$ contents are shown in Fig. 2.

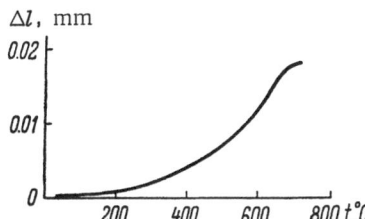

Fig. 1. Effect of temperature on elongation of a specimen 50 mm long, crystallized during cooling.

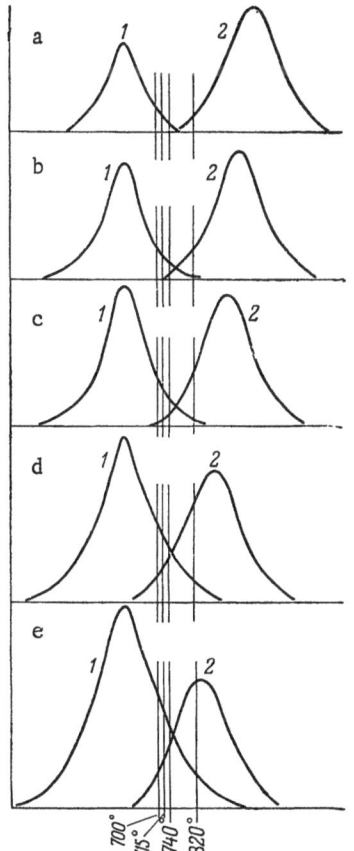

Fig. 2. Effect of temperature on the nucleation rate (1) and rate of crystal growth (2) in glasses with different $TiO_2$ contents. a) 3.0%; b) 5.0%; c) 7.0%; d) 9.0%; e) 11.0%.

We never succeeded in obtaining a transparent crystallized glass by catalyzed crystallization during cooling; this is readily explained on the grounds that the crystals of the main phase grow simultaneously with formation of crystallization centers. In this case, the crystals are larger than the permissible size for a transparent crystallized glass.

**4.** The following technique was adopted for studying the conditions for nucleation and growth of crystals of the main phase in relation to the heat-treatment conditions and catalyst content. The glass, annealed and cooled to room temperature, was transferred in ceramic boats into a gradient furnace and kept there for 24 h. After the heat treatment in the boat, the following principal zones were determined: zone of the original glass, precrystallization zone, zone of transparent crystallized glass, and zone of opaque crystallized glass.

Glass in the precrystallization zone is somewhat darker than the original and the transparent crystallized glass. The transition from transparent to opaque crystallized glass is fairly sharp. The transition temperature from transparent to opaque glass is defined as the temperature of transparency loss. The temperature of transparency loss was taken as the main criterion of the process. This is because the temperature of transparency loss is determined both by the size and by the number of the crystals and, therefore, by the number of nuclei. It also depends on the exposure time. This influence was not taken into account, because the same exposure time was used in all the experiments with the gradient furnace.

Preliminary heat treatment has a significant effect on the temperature of transparency loss. The experiments showed that for each composition there is an optimum heat-treatment temperature (or temperature range) $T_{opt}$ at which the temperature of transparency loss has the highest possible value. The maximum possible temperature of transparency loss is defined as the temperature above which loss of transparency is associated with the formation of a new crystalline phase (or phases) by recrystallization of the phase of the transparent state, or with the growth of a secondary crystalline phase. The titanate detected by Kind in a crystal-optical study of glasses crystallized above the maximum possible temperature of transparency loss [2] is such a secondary phase. Changes in the composition of the main crystalline phase alter the properties of the material. Figure 3 shows elongation curves of a transparent crystallized glass and of an opaque glass of the same composition, which lost its transparency above the maximum temperature of transparency loss. Alekseev and Florinskaya [3] also found a change in the composition of the crystalline phase in the transition to opaque glass.

Preliminary heat treatment below $T_{opt}$ lowers the temperature of transparency loss, while treatment above $T_{opt}$ not only lowers the temperature of transparency loss but, in a number of cases, leads to volume turbidity in the glass. Determination of the coefficient of expansion of such turbid glass showed that volume crystallization occurs in this case, with the formation of $\beta$-spodumene as the main crystalline phase.

Since the temperature of transparency loss is determined ultimately by the number of nuclei, it follows that during preliminary thermal treatment at a temperature close to $T_{opt}$ crystallization centers arise, and on further rise of temperature, crystals of the main phase grow on them (or they grow themselves).

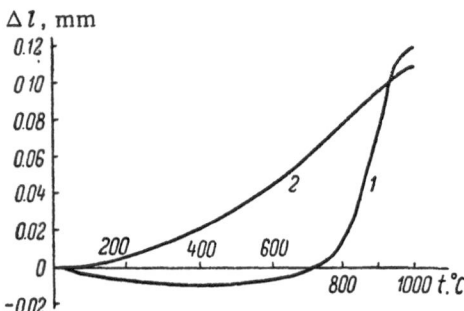
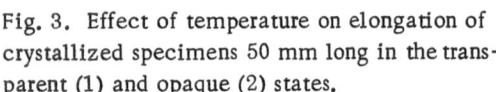

Fig. 3. Effect of temperature on elongation of crystallized specimens 50 mm long in the transparent (1) and opaque (2) states.

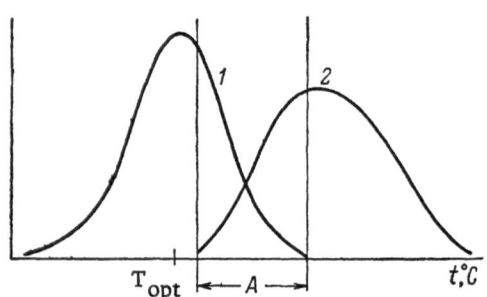

Fig. 4. Effect of temperature on the nucleation rate (1) and rate of crystal growth (2).

TABLE 2. Density and Refractive Index of Glasses in the $Li_2O$-$Al_2O_3$-$SiO_2$ System

| Glass | Density, g/cm³ | | Refractive index | |
|---|---|---|---|---|
| | before | after | before | after |
| | crystallization | | crystallization | |
| Original glass | 2.441 | 2.489 | 1.5402 | 1.5476 |
| Treated for 292 h at 600° | 2.456 | 2.489 | 1.5415 | 1.5474 |
| Treated for 292 h at 635° | 2.461 | 2.488 | 1.5425 | 1.5472 |

The temperature region close to $T_{opt}$, where the crystallization centers are formed, i.e., where the process determining the further development of crystallization occurs, is termed by us the precrystallization region, and thermal treatment in this temperature region is termed precrystallizational thermal treatment. The precrystallization region has no sharp temperature boundaries and apparently coincides with the annealing range of a given glass.

At $T_{opt}$, the nucleation rate is maximal; it decreases both with fall and with rise of temperature (curve 1, Fig. 4). Volume crystallization of glass during preliminary thermal treatment at temperatures above $T_{opt}$ occurs because, in this case, the glass is in a temperature region where the nucleation rate and the rate of crystal growth both differ from zero. This is the region A in Fig. 4. The character of crystallization in region A is the same whether the glass enters it directly from the liquid state or already cooled.

**5.** The amount of catalyst present in the glass, titanium dioxide in this instance, has a considerable effect on the temperature of transparency loss. This temperature rises with increasing amount of catalyst. This is because the number of centers increases and the crystals therefore decrease in size. By increase of the catalyst content in the glass it is possible to raise the nucleation rate to a value at which it is no longer necessary to hold the glass in the precrystallization zone. The necessary amount of nuclei ensuring the maximum temperature of transparency loss is formed during cooling of the glass and during repeated heating, when the glass passes through the precrystallization region.

The amount of catalyst with which the temperature of transparency loss becomes virtually independent of preliminary thermal treatment can be easily determined by the same method as $T_{opt}$.

**6.** The significance of precrystallizational thermal treatment is not confined to nucleation. Certain physical properties of the glass, such as density, refraction, etc., change during precrystallizational thermal treatment. This question was investigated in detail by Buzhinskii [4]. We must point out, however, that we did not detect any changes in the coefficient of thermal expansion simultaneously with changes in a number of physical properties when the glass was held for 580 h at 635°, i.e., close to $T_{opt}$. Tudorovskaya [5] did not find any change of the thermal effect in the course of differential thermal analysis (DTA) of this glass. The fact that the coefficient of expansion and the thermal effect remain unchanged shows that there is no crystal growth during precrystallizational thermal treatment. At the same time, changes of certain physical properties, and especially the increased

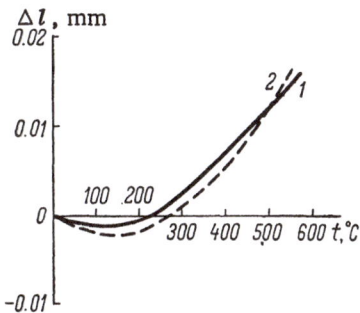

Fig. 5. Effect of temperature on elongation of crystallized specimens. 1) Without precrystallizational treatment; 2) after treatment for 580 h at 635°.

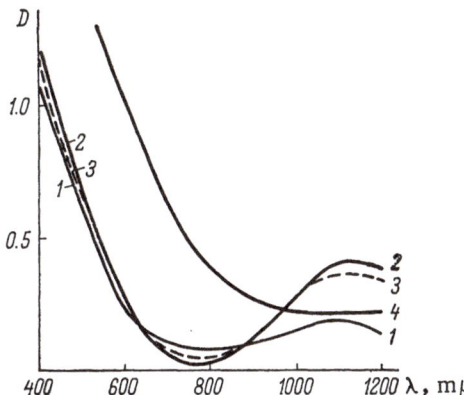

Fig. 6. Dependence of absorption on wavelength. 1) Original glass; 2) crystallized glass previously treated for 20 h at 640°; 2) glass crystallized without preliminary treatment; 4) glass treated for 20 h at 640°.

color of the glass corresponding to iron titanates, indicate that structural changes involving an increase in the amount of these compounds occur during this period.

The iron titanate content of the glass should increase if the concentration of $TiO_2$ or iron oxides increases in certain regions owing to release of $TiO_2$ from the glass network. We consider that liberation of $TiO_2$ from the glass network occurs as the result of ordering of the structure and formation of spodumene groupings during precrystallizational thermal treatment. We consider that such spodumene groupings (not yet crystals) can form around the crystallization centers as the latter arise.

Since the $TiO_2$ content can increase only to a certain limit, determined by the composition of the glass, the coloration increase must also tend to a certain limit. Kondrat'ev and Podushko [6] showed experimentally the existence of a limiting value of the optical density during precrystallizational thermal treatment. It may be noted that Vertsner [7] also detected, with the aid of the electron microscope, an increase of structural order during precrystallizational thermal treatment.

**7.** The influence of precrystallizational treatment on certain properties of crystallized glass was investigated for glasses of the $Li_2O-Al_2O_3-SiO_2$ system with addition of 0.5% titanium dioxide by weight as catalyst. The original glass was subjected to prolonged precrystallizational treatment at a temperature close to $T_{opt}$ and then crystallized simultaneously with the original untreated glass in the same electric furnace. The experimental values of certain physical properties of these glasses are given in Table 2 and in Fig. 5 and Fig. 6.

The data in Table 2 and Figs. 5 and 6 show that the properties studied do not depend on precrystallizational thermal treatment of crystallized glass. As has already been shown, precrystallizational thermal treatment determines the number of nuclei and the crystal size, but not the amount or composition of the crystalline phase, which depend on the crystallization time and temperature. Since the properties in question are determined by the amount and composition of the crystalline phase, it is evident that they are not affected by precrystallizational thermal treatment. The conclusion that certain properties are independent of precrystallizational thermal treatment cannot be extended to all properties or to glasses in other systems.

Properties determined by the size and amount of crystals, such as mechanical properties, must of course depend on the precrystallizational thermal treatment. In glasses of the system studied only one crystalline phase, in the main, separates out during crystallization. If more than one phase separates out simultaneously during crystallization, as in the system $MgO-Al_2O_3-SiO_2$, precrystallizational thermal treatment makes it possible to separate these phases and to vary their proportions within certain limits. In such cases properties determined by the composition and amounts of the crystalline phases must also depend on precrystallizational thermal treatment.

**8.** The mechanism proposed for catalyzed crystallization of glasses in the $Li_2O-Al_2O_3-SiO_2$ system with titanium dioxide is not, of course, completely rigorous. However, in the first approximation it gives a correct picture of the process and provides answers to a number of questions arising in studies of various aspects of the catalyzed crystallization of inorganic glasses.

## LITERATURE CITED

1. Ya. S. Bobovich, this collection, p. 93.
2. N. E. Kind, this collection, p. 111.
3. A. G. Alekseev and L. A. Fedorova, this collection, p. 90.
4. I. M. Buzhinskii, E. I. Sabaeva, and A. N. Khomyakov, this collection, p. 133.
5. N. A. Tudorovskaya and A. I. Sherstyuk, this collection, p. 126.
6. Yu. N. Kondrat'ev and E. V. Podushko, this collection, p. 107.
7. V. N. Vertsner and L. V. Degteva, this collection, p. 86.

# USE OF THE ÉM-7 ELECTRON MICROSCOPE
# FOR INVESTIGATION OF CRYSTAL LATTICES
# AND OBSERVATION OF DISLOCATIONS IN THEM

## V. N. Vertsner, Yu. M. Vorona, and G. S. Zhdanov

The resolving power of the best electron microscopes now in production reaches 8 A, and in some of the most favorable cases it is possible to obtain photomicrographs with resolution down to 4.5 A. The electron microscope is therefore capable of producing visual images of structural microdetails close to the atomic in size.

In the course of a study of the relationship between the propagation of cracks and the microstructure of glass, Menter investigated very thin glass films under the electron microscope. A crack about 10 A in width was found in one of the photomicrographs. This led to the idea of observing under the electron microscope the crystal lattices of certain substances with relatively large interplanar spacings [1]. The images which he obtained of the crystal planes of various phthalocyanines allowed direct measurement of the interplanar spacings and made it possible to observe dislocations in the crystals. Several workers [2-4] subsequently used the method of direct observation of crystal lattices for investigating defects in them. As a result of these investigations, the requirements to which the crystals prepared for lattice observation must conform were formulated.

The main requirements are the following: 1) the planes of the lattice to be resolved must be almost parallel to the incident electron beam and must have a high enough structure factor to give satisfactory contrast; 2) the crystals must be thin in the direction of the electron beam in order to weaken the inelastic scattering effect; 3) the crystals must not be destroyed in the beam.

We used the ÉM-5 electron microscope with the optical system of the ÉM-7 instrument for observing crystal planes in copper phthalocyanine crystals. We succeeded in raising the resolving power of the microscope to 10 A by using an improved high-voltage source and by careful adjustment of the instrument. In addition, increased excitation in the objective lens to 4000 ampere-turns led to considerable decrease of astigmatism and of spherical and chromatic aberration. The accelerating voltage was 60 kV. A diaphragm 30 $\mu$ in diameter was used in the condenser lens at a beam current of 20 $\mu$A. The diameter of the diaphragm in the objective lens was 50 $\mu$. The electron magnification was 53,000 or 67,000, and the exposure was 8-10 sec.

Copper phthalocyanine crystals convenient for investigation under the electron microscope were prepared by vacuum sublimation of industrial phthalocyanine powder onto carbon films with numerous holes, previously heated to 150° [5]. The appearance of these crystals is shown in Fig. 1. The crystal edges lie on the carbon film while the middle hangs over the hole. Additional scattering of electrons in the support was thus avoided. The crystals are thin plates with the (201) plane most developed, and therefore the (001) planes with a spacing of 12.6 A lie almost parallel to the electron beam. Crystal planes with a spacing of the order of 10 A are not visible on the screen of the electron microscope, and therefore focusing of the object was performed by a sharp image of the crystal edges or by the disappearance of the diffraction fringe. It was not possible to tell until after development whether the planes were resolved. Statistical analysis shows that small regions of the lattice with 12.6 A spacing were resolved in 50% of the photographs. In the other

Fig. 1. Copper phthalocyanine crystals. Magnification 53,000.

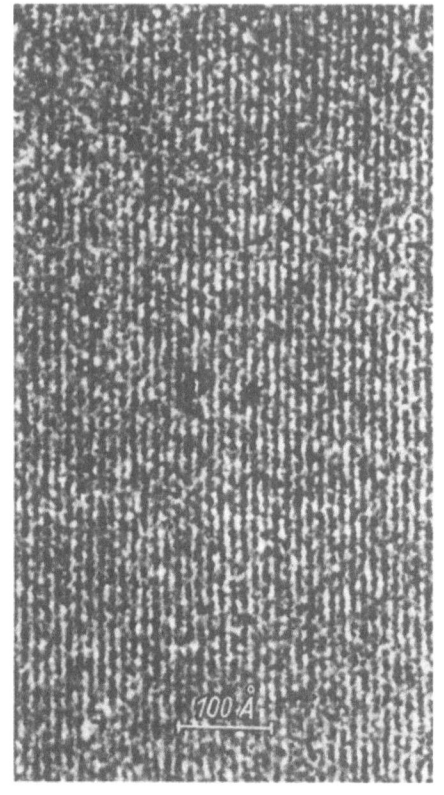

Fig. 2. Copper phthalocyanine crystals. Magnification 1,200,000. The crystal planes are parallel to the crystal edge.

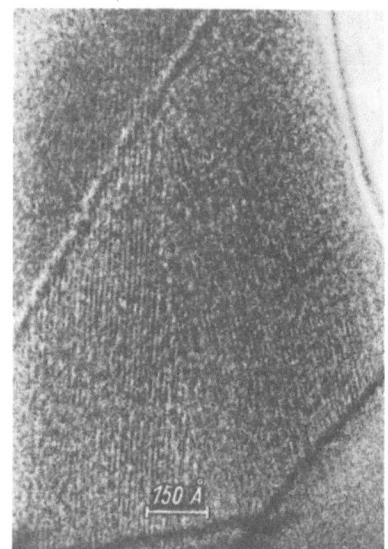

Fig. 3. Copper phthalocyanine crystals. Magnification 670,000. The crystal planes converge at an angle of 15°.

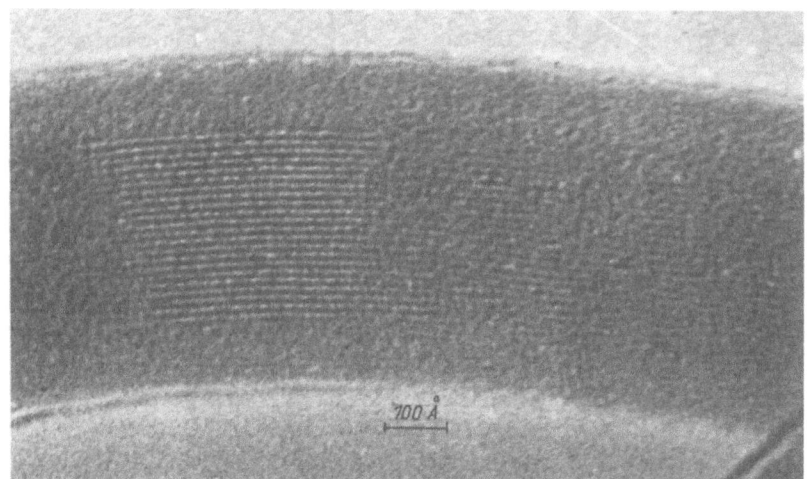

Fig. 4. Bent copper phthalocyanine crystal. Magnification 1,100,000.

photographs the planes were not resolved owing to displacement of the crystals during exposure or to destruction and contamination of the crystals under electron bombardment.

Crystals or regions of crystals with resolved planes are shown in Figs. 2-4. Each line corresponds to a photographic projection of the (001) plane formed by copper phthalocyanine molecules. The interplanar spacings were measured from electron diffraction patterns of the crystals or from the known magnification. The crystal planes are usually parallel to the crystal edge and have perfect structure (Fig. 2). However, dislocations are sometimes seen. For example, in the crystal shown in Fig. 3, the planes converge at an angle of 15°. A photograph of a slightly bent crystal is shown in Fig. 4. It is seen that the crystal planes reproduce the external form of the crystal.

Use of the ÉM-7 electron microscope is certainly promising for investigation of the structure of glassceramics because of its high resolving power.

## LITERATURE CITED

1.   J. W. Menter, Proc. Roy. Soc. A236:119, 1956.
2.   G. A. Bassett, J. W. Menter, and D. W. Pashloy, Proc. Roy. Soc. A246:345, 1958.
3.   S. Ogawa, D. Watanabe, and T. Komoda, Acta Cryst. 11(12):872, 1958.
4.   T. Komoda, E. Suito, and N. Uyeda, Nature 181(4605):332, 1958.
5.   H. Espagne, Compt. rend. Acad. Sci. 247(14):992, 1958.

# ELECTRON MICROSCOPE INVESTIGATION
# OF THE CATALYZED CRYSTALLIZATION OF GLASSES

## V. N. Vertsner and L. V. Degteva

Glass 13 with an admixture of titanium (not over 10%) as catalyst was used for the electron microscope investigation. The investigations were carried out on the original glass and on the transparent and opaque crystallized glasses obtained from it as a result of heat treatment. The technique used was that of replicas obtained from freshly fractured glass surfaces. We used both carbon replicas shadowed by chromium or platinum–palladium alloy and carbon replicas with preliminary shadowing, i.e., replicas taken from fractured surfaces previously coated obliquely with a thin layer of platinum and palladium. The replicas were separated with the aid of a thick gelatin film. The latter was subsequently removed by thorough washing of the replica in hot water.

Much attention was devoted to development of the method. At first replicas from the same glasses were inspected repeatedly in order to eliminate errors introduced in the preparation technique. Although replicas from the same glass made by the two above-mentioned methods revealed a similar structure under the electron microscope, the replica quality was better with preliminary shadowing. This method was therefore preferred.

Electron microscope investigations, usually carried out at electron-optical magnifications of 6500, showed that each of the glasses had a characteristic structure, distinct from the others.

A fracture surface of the original glass has a homogeneous structure; Fig. 1 shows that it is covered with spherulites 200-250 A in diameter.

Crystallized transparent glass obtained by heat treatment at about 750° consists of indistinctly faceted crystals 0.05-0.2 $\mu$ in size (Fig. 2). Spherulites of the kind observed in the original glass may be seen on their surfaces. Most of these spherulites are joined in short chains of three or four.

In the opaque crystallized glasses obtained when the original glass is heated to 800° and higher, structural elements similar to those found in the crystallized transparent glass can be seen on the fracture surface (Fig. 3). However, the crystal size increases to 0.3-1 $\mu$ and the outlines become sharper. It appears that the material undergoes recrystallization which leads to formation of larger particles. The spherulite chains in the opaque glass are usually longer, often consisting of four or more units.

At the same time we studied the structure of glasses subjected to prolonged heat treatment at temperatures about 100° below the crystallization temperature. It was found that the structure was changed only in the case of glass heated for 300 h at 635°. Electron micrographs of replicas taken from different regions of the fractured glass surface are shown in Figs. 4 and 5; it is seen that in some regions there is a peculiar phase separation (Fig. 5a), whereas in others the process has gone further and the structure resembles that of transparent crystallized glass (Fig. 5b). However, in the latter case the crystallization is still incomplete and the relief is considerably smoothed out. These micrographs give an indication of changes in the structure of the glass during the period before crystallization.

We also studied the effect of varying $TiO_2$ content on the structure of glass 13. It was found that the glass structure changes only when the $TiO_2$ content is raised to 11%. The micrographs in Figs. 4 and 6 correspond to glasses without $TiO_2$ and with 11% $TiO_2$. It is seen that glasses without titanium and with less than 11% $TiO_2$ are structureless, whereas crystallization occurs in glass with a high $TiO_2$ content.

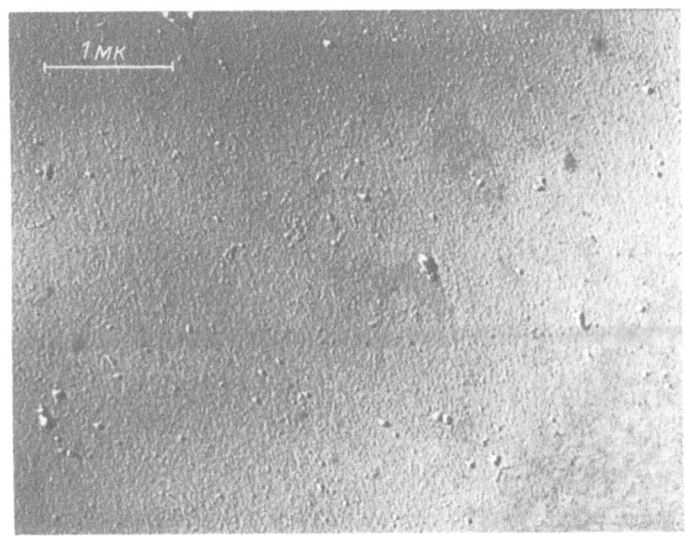

Fig. 1. The original glass.

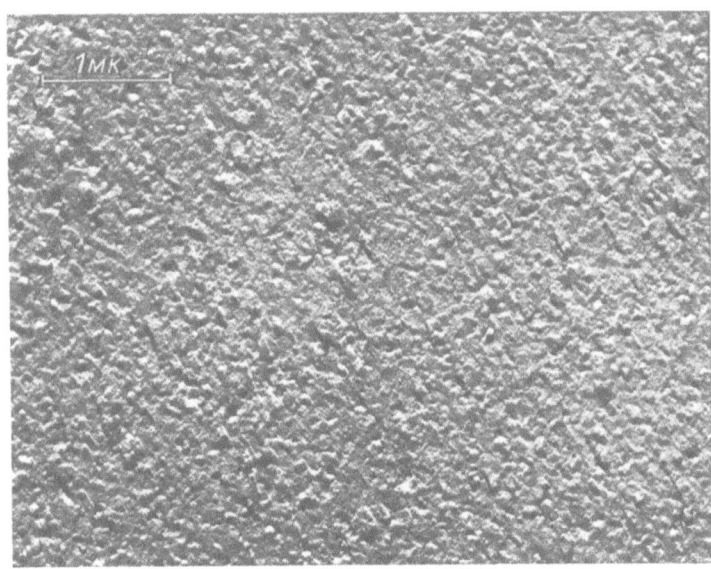

Fig. 2. The original glass heated to 750°.

Fig. 3. The original glass heated to 800°.

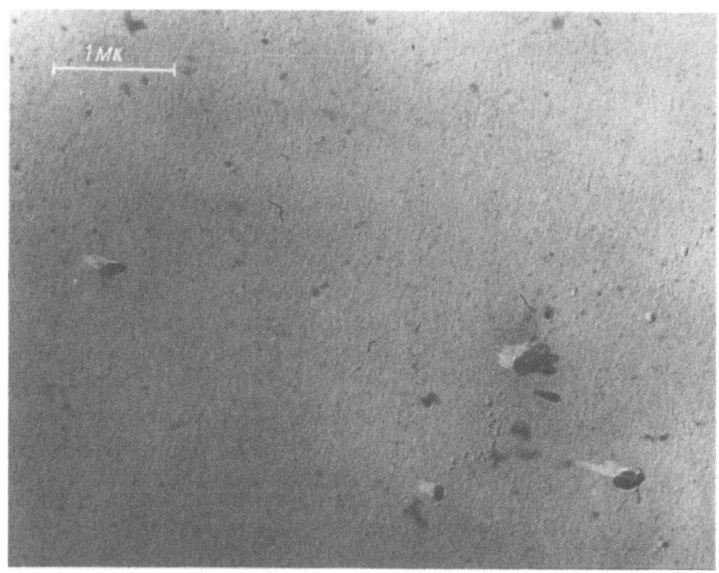

Fig. 4. Glass without TiO$_2$.

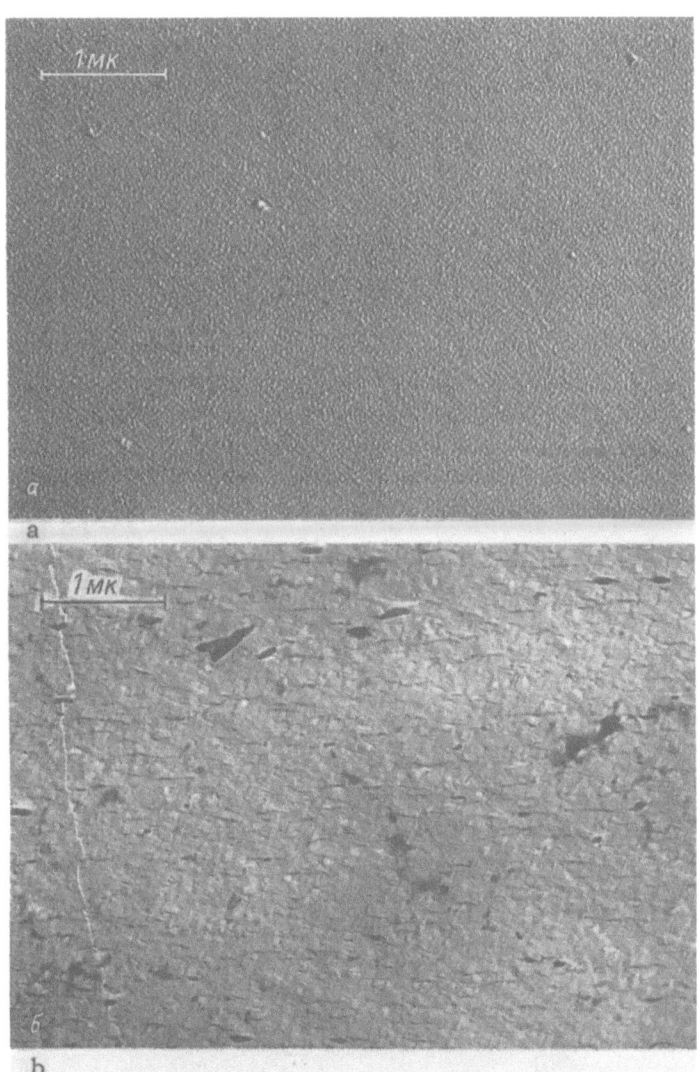

Fig. 5. Glass (a and b) heated for 300°h at 635°

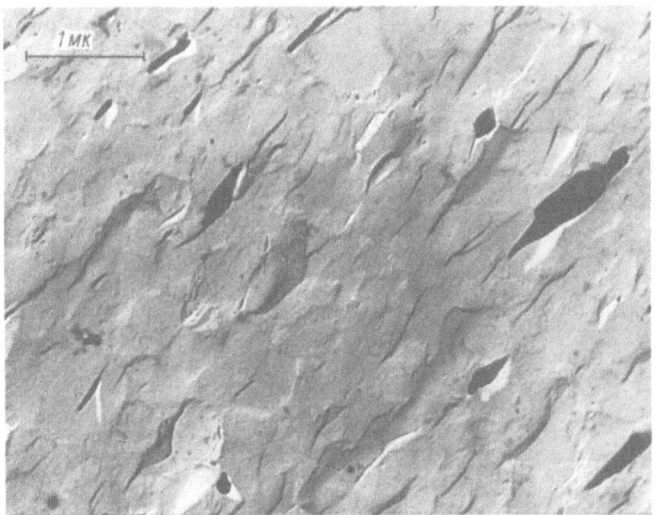

Fig. 6. Glass with 11% TiO$_2$.

# X-RAY DIFFRACTION INVESTIGATION
# OF THE CATALYZED CRYSTALLIZATION OF GLASSES

## A. G. Alekseev and L. A. Fedorova

The x-ray study of catalyzed crystallization was carried out with glass 13 containing, as catalyst, from 0 to 11% $TiO_2$ by weight and small amounts of oxides of elements from groups I, II, and III of the periodic system.

Investigation of crystallized glasses by the x-ray structural method showed that the presence of small amounts of additionally introduced oxides has virtually no effect on the phase composition. The function of $TiO_2$ in these glasses is apparently to initiate volume crystallization. The suggestion made by certain investigators in the literature, that such glasses contain rutile crystallites as crystallization centers, was not confirmed by x-ray diffraction studies of specimens with high $TiO_2$ contents.

The effect of heat treatment on crystallization processes was studied with glass 13 containing 5% $TiO_2$. Glass plates $170 \times 20 \times 5$ mm in size were put into a gradient furnace with a temperature range of 530-950° and kept there for 24 h. The plate surface and deeper layers were then investigated by reflection of x rays from narrow zones situated consecutively along the temperature-variation axis.

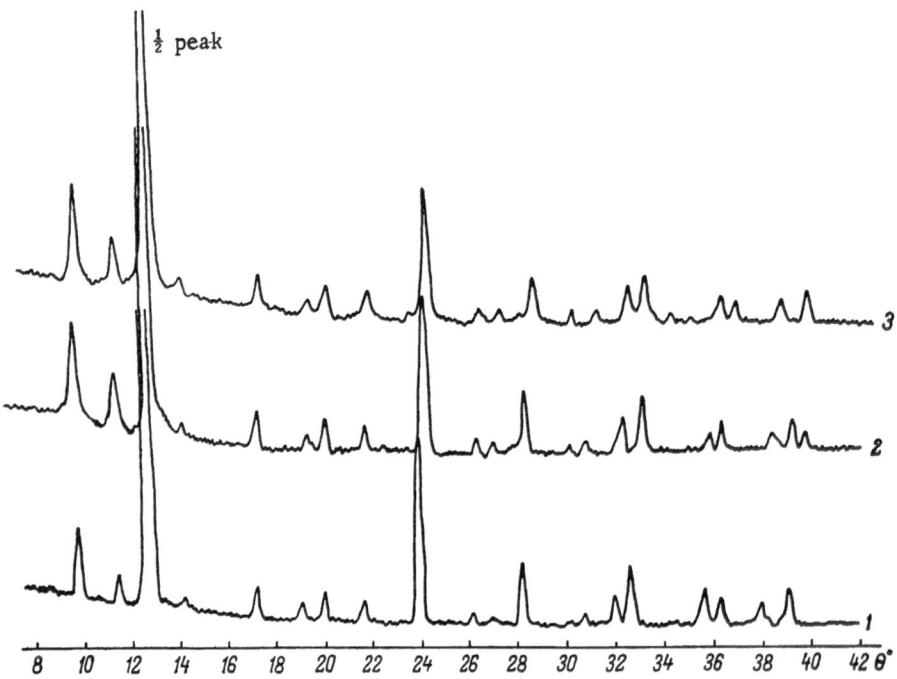

Fig. 1. X-ray diffraction curves obtained from crystallization products of glasses with the composition of eucryptite (1), spodumene, heated at relatively low temperatures (2), and petalite with addition of 5% $TiO_2$ (3).

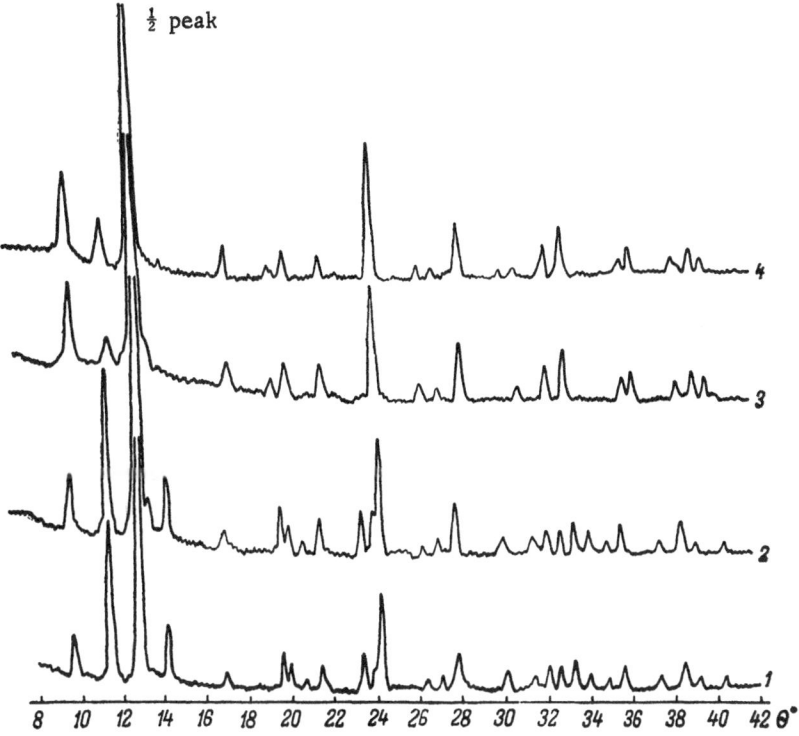

½ peak

8 10 12 14 16 18 20 22 24 26 28 30 32 34 36 38 40 42 θ°

Fig. 2. X-ray diffraction curves obtained from crystallized glasses. 1) Glass with the composition of spodumene, crystallized at high temperatures; 2) opaque crystallized glass 13 with 5% $TiO_2$; 3) transparent crystallized glass 13 with 5% $TiO_2$; 4) glass with the composition of spodumene, crystallized at low temperatures.

The results of the x-ray investigation show that the heating temperature influences the course of crystallization of the glass, as follows.

Zones of the specimen heated at temperatures below 700° give diffraction patterns characteristic of glass with diffuse maxima. Crystalline maxima appear on the diffraction diagrams obtained from zones heated at about 700°. The amount of crystalline phase increases sharply over a fairly narrow temperature range, and the subsequent increase with rise of temperature is slight. Assessment of the amount of crystalline phase by the x-ray technique gave a value of 60-70%.

The crystallized glass remains transparent when heated to about 800°, while at higher temperatures it becomes opaque.

The x-ray diffraction curves of transparent (3) and opaque (2) crystallized glass 13 with 5% $TiO_2$ are given in Fig. 2. Although the differences between the curves are not large, they are confirmed by numerous experiments and provide adequate evidence that different main crystalline phases are present in opaque and transparent crystallized glasses.

Preliminary heat treatment of the original glasses at temperatures below the crystallization temperature increases somewhat the temperature range in which the crystallized glasses remain transparent, but has no effect on the phase composition.

The crystallization of glass with the composition of spodumene was studied at various temperatures by means of x-ray diffraction in order to identify the crystalline phases separating out in glass 13 with 5% $TiO_2$. It was found that at relatively low temperatures (about 800°), glass of the spodumene composition deposits crystals identical with those found in transparent crystallized glass 13 with 5% $TiO_2$ (Fig. 2, 4), while at higher temperatures the crystals deposited are the same as those present in the opaque glass (Fig. 2, 1).

The latter are satisfactorily identified with the known high-temperature form of spodumene, whereas the main crystalline phase in transparent crystalline glass 13 with 5% $TiO_2$ and glass with the composition of spodumene heated at low temperatures is not identified with any of the known forms of spodumene, or of eucryptite or petalite. It is possible that an unknown form of spodumene, stable at temperatures below the stability range of $\beta$-spodumene, may be present. Some support for this view is provided by the fact that similar crystals were recorded in an x-ray diffraction study of specimens of low-temperature spodumene heated at 950° for 24 h.

It must be noted, on the other hand, that the x-ray diagrams given by these glasses and by the crystallization products of glasses with the composition of petalite and eucryptite exhibit the characteristics of solid solutions (Fig. 1).

Particular interest attaches to the possibility of using x-ray diffraction for following the microtransition from the glassy to the crystalline state, i.e., the conversion of a nucleus into a crystal. We investigated for this purpose a considerable number of specimens of glass 13 with 5% $TiO_2$ at the earliest crystallization stages.

With increase of the time or temperature of heat treatment, the x-ray diffraction diagrams first remain unchanged and then at a certain point in time they undergo a sharp qualitative change: sharp peaks characteristic of crystals appear on the diffuse x-ray scattering curve. No appreciable broadening of the peaks on the x-ray diagrams was observed in any of the cases studied.

These results can be explained on the hypothesis that the formation of a crystal from a nucleus occurs stepwise, i.e., at a considerably higher rate than crystal growth. The crystals formed are not smaller than 100 A.

# INVESTIGATION OF CATALYZED CRYSTALLIZATION
# BY MEANS OF RAMAN SPECTROSCOPY

## Ya. S. Bobovich

This investigation was carried out on specimens of a complex glassy system with the composition of spodumene, with a small addition of titanium dioxide. This oxide is usually believed to play an important part in the formation of a new material with specific properties as the result of heat treatment of specimens of this system [1]. The object of the present work was to elucidate the role of titanium dioxide. Interpretation of the spectra was based on the results of earlier investigations in which it was shown that the characteristic bands for structures consisting of joined titanium-oxygen octahedrons lie in the ~600-630 cm$^{-1}$ region, and those corresponding to titanium-oxygen tetrahedrons in the ~875-960 cm$^{-1}$ region [2-6]. The main experimental data are presented in the figure in the form of spectra of specimens of the original glass after various heat treatments.

Spectrum 1 of the original glass greatly resembles the spectra obtained earlier from normal three-component silicate glasses with small contents of $TiO_2$ [2]. As previously, the presence of $TiO_2$ is manifested in the ~900 cm$^{-1}$ band. On the high-frequency side of this there is a neighboring band of similar intensity, undoubtedly associated with vibrations of the aluminosilicate framework. This is shown by the analogy with the spectra of two-component alkali silicate glasses.

Prolonged exposure of the specimen at 600° led to some relative weakening of these bands (spectrum 2). Further heat treatment, at progressively higher temperatures, weakened these bands still further, and eventually they disappeared almost entirely in spectrum 4. At the same time, other interesting details may be observed. First, a fairly broad band becomes progressively more distinct in the ~600 cm$^{-1}$ region; second, the ~500 cm$^{-1}$ band exhibits fine structure; third, a weak band appears in the ~220 cm$^{-1}$ region; fourth, a strong band arises in the low-frequency region (120 cm$^{-1}$).

The last spectrum in Fig. 1 relates to a specimen after the full heat treatment, when it acquired a series of new macroscopic properties. It is seen that its spectrum consists of a series of more or less narrow lines and bands. In the internal vibration region (approximately from 200 cm$^{-1}$ upward), it is very similar to the spectra of crystals with octahedral titanium coordination [2-6]. It is especially noteworthy that the line in the low-frequency region becomes very narrow and strong. It is also interesting that the weak band at ~220 cm$^{-1}$ does not appear in the spectrum of this specimen. Thus, the structure of the specimen does not alter steadily with intensification of the heat treatment.

The following tentative interpretation is offered for the behavior of the Raman spectra of the specimens studied. As the specimen is heated, the ~900 cm$^{-1}$ band, ascribed to vibrations of linked $TiO_4$ tetrahedrons, becomes steadily weaker and a band at ~600 cm$^{-1}$, characterizing vibrations of $TiO_6$ octahedrons, appears at the same time; it may be concluded from this that titanium gradually passes from tetrahedral to octahedral coordination. This structural rearrangement is completed after prolonged heat treatment at ~660°, as is indicated by the almost complete disappearance of the high-frequency band. At the same time, the form of these (and the other) bands remains unchanged; they remain broad as before, which is highly characteristic of the glassy state. Hence, it may be asserted that crystallites, defined as g e o m e t r i c a l l y  o r d e r e d  regions of definite composition, are not present in the volume of the specimen at this stage of heat treatment. The results of electron microscope studies led Vertsner and Degtev [7] to the same conclusion [7].

The foregoing considerations raise the question whether glassy substances composed of interlinked octa-

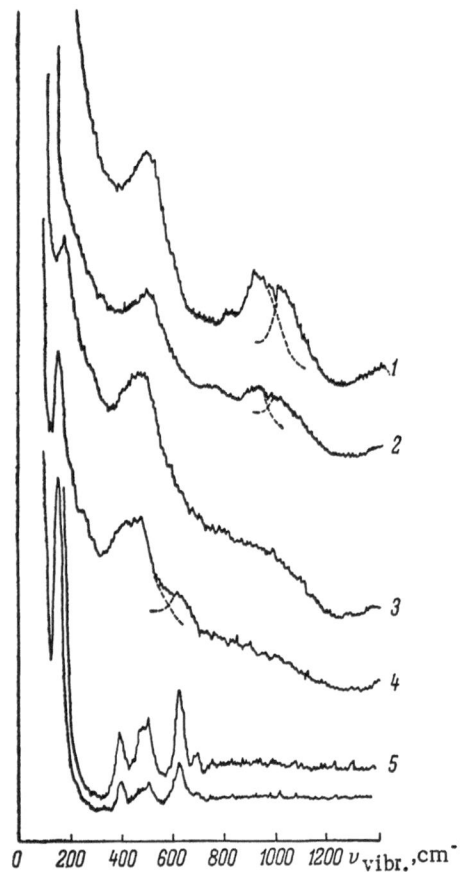

Raman spectra of the glass specimens. 1) Original specimen; 2-4) specimens heated to 600, 630, and 660°, respectively; 5) specimen after the full heat treatment. In the latter case, the spectrum was recorded with two different amplifications.

hedrons exist at all. Brady showed by x-ray analysis that a two-component tellurite glass has this structure [8]. However, in this glass, in contrast to crystalline $TeO_2$, the $TeO_6$ octahedrons are joined mainly by the corners; this ensures flexibility of the structural network — an essential condition for vitrification. The formation of tellurite glass by the mechanism put forward by Brady is not accompanied by significant shifts of the main bands in the Raman spectrum [9]; it is therefore likely that the titanium-containing phase in the present specimen has an analogous structure.

If so, conversion of titanium into octahedral coordination requires a source of oxygen ions. This may be lithium oxide, which passes into the titanium-containing structural network, or a small part of the aluminum oxide, in which aluminum was initially present in octahedral oxygen environment and changed its coordination to tetrahedral under the influence of heat treatment. Both hypotheses are consistent with the sharp weakening of the highest-frequency band which relates, as already noted, to vibrations of the aluminosilicate framework. Of course, simultaneous transfer of both oxides is not excluded.

Thus, the first (precrystallizational) stage of the process culminates in separation of a glassy component consisting of linked $TiO_6$ octahedrons, and in a certain structural alteration in the main phase. With regard to the subsequent course of the process, only more or less likely hypotheses may be put forward at this stage on the basis of Raman spectra.

It is known that titanium dioxide is incapable of forming a stable glass [10]. This means that further heating of the glass must inevitably influence this structural component first. The effect may be removal of aluminum—oxygen and lithium—oxygen radicals from the titanium-containing network, and as a result the $TiO_6$ octahedrons become joined partly by the edges. This is nothing but o r d e r i n g  o f their geometrical arrangement, i.e., formation of crystallites fully analogous to rutile, or of more complex composition. The crystallites or the liberated aluminum—oxygen and lithium—oxygen radicals (most probably joined into crystals of submicroscopic dimensions) may become crystallization centers which initiate intensive crystallization in the bulk of the substance, with approximately the composition of spodumene. It is also possible that heat treatment is accompanied by formation of crystallites of the composition $m\,Al_2O_3 \cdot n\,TiO_2$ and that these become the crystallization centers. This is suggested by the results of Kind's investigations of the crystallization products of similar specimens [11]. The ultimate outcome of the second stage of the process, whatever its mechanism, is reflected in sharp narrowing of the Raman lines, which is an essential characteristic of Raman spectra of crystalline bodies.

The absence of the weak band at ~220 $cm^{-1}$ in the spectrum of the crystalline glass specimen is noteworthy. It is most reliably assigned to deformational vibrations of the lattice. Ordering of the lattice should influence such vibrations first. Moreover, it is quite obvious that when the octahedrons are joined predominantly by the corners there are more degrees of freedom for various deformational vibrations. Therefore, the disappearance of the ~220 $cm^{-1}$ band is an additional argument in favor of the point of view put forward above.

In conclusion, we must note that the appearance of a very strong and narrow line in the low-frequency region of the spectrum of a specimen after full heat treatment cannot be interpreted as yet. No such line is present in the spectrum of rutile [5] or of powdered $TiO_2$. It is unlikely to appear in the spectra of more complex titanium-containing crystals in which titanium may, in principle, separate out. It is also doubtful whether this line can be assigned to vibrations of crystallizing spodumene, although this possibility cannot be totally rejected, so that the question of the origin of this line remains open. It may be that the small size of the crystals is responsible for its appearance. The narrowness of the line may indicate that the crystal size has not reached the critical value at which the lines begin to broaden.

## LITERATURE CITED

1.  A. I. Berezhnoi, Photosensitive Glasses and Crystalline Glass Materials of the Pyroceram Type, Moscow, 1960.
2.  Ya. S. Bobovich and T. P. Tulub, Opt.-Mekh. Prom. No. 9, 40, 1961.
3.  Ya. S. Bobovich, Optika i Spektroskopiya, 10:418, 1961.
4.  Ya. S. Bobovich and É. V. Bursiyan, Optika i Spektroskopiya 11:131, 1961.
5.  P. S. Narayanan, Proc. Indian Acad. Sci. A37:411, 1953.
6.  P. S. Narayanan and K. Vedam, Zs. Phys. 163:158, 1961.
7.  V. N. Vertsner and L. V. Degteva, this collection, p.86.
8.  G. W. Brady, J. Chem. Phys. 27:300, 1957.
9.  V. P. Cheremisinov and V. P. Zlomanov, Optika i Spektroskopiya 12:208, 1962.
10. P. L. Baynton, H. Rawson, and J. E. Stanworth, Travaux du IV Congrès Internat. du Verre, Paris, 1956.
11. N. E. Kind, this collection, p.111.

# INFRARED SPECTRA OF GLASSY AND CRYSTALLIZED SILICATES OF THE $Li_2O-Al_2O_3-SiO_2+TiO_2$ SYSTEM AND THEIR RELATIONSHIP TO STRUCTURE

## V. A. Florinskaya, E. V. Podushko, I. N. Gonek, and É. F. Cherneva

We investigated infrared reflection spectra of glass 13 with additions of $TiO_2$ in the course of crystallization, in the 7-14 $\mu$ range. The aim of the investigation was to determine what crystals are deposited in glasses of this system as the result of heat treatment, how crystals deposited in transparent crystallized glass differ from crystals in opaque glass of the same melting, how small amounts of various oxides added to glass of the same basic composition influence the composition or structure of the crystalline phase, and to find whether there is a latent period in which the glass is prepared for crystallization.

To identify the crystals deposited in glasses as the result of heat treatment, it is necessary to have standard spectra of all the principal silicates (both three- and two-component) in the $Li_2O-SiO_2-Al_2O_3$ system. Such spectra are not available at this time. Therefore, infrared spectroscopy can at present give only limited information on the problem.

The spectra of a number of native minerals were taken to give an approximate guide to the system. These spectra are shown in Fig. 1. The abscissa gives the wavelength in microns, and the ordinate represents the intensity ratio of the light reflected from the mineral surface and the incident light. Curve 6 represents the spectrum of the surface crust of glass with the composition of spodumene, crystallized when cast. Crystal-optical analysis indicates that the crystallization product is "high-temperature spodumene." This spectrum coincides with the spectrum of the inner layers of glass with the composition of spodumene, melted and crystallized in the art-glass plant. According to the plant data, the crystallization product is β-spodumene. Curve 4 relates to the crystallization product of glass obtained from the art-glass plant and identified there as β-eucryptite. Comparison of all the curves in Fig. 1 indicates that the spectra of all the silicates exhibit considerable differences from each other, and therefore these crystals can be easily identified spectroscopically in glass crystallization products.

Curves 1-4 in Fig. 2 are spectra of glass 13 with 5% $TiO_2$, crystallized in different ways. Comparison of the curves shows that none of the crystallized glasses contains low-temperature α-spodumene, and that different crystalline phases are deposited in transparent and opaque glasses, the same crystalline phase being largely deposited in the transparent crystallized glass as in pure crystallized spodumene glass (the so-called β-spodumene). The fine structure of the band at 10 $\mu$ in the spectrum of the original glass exhibits maxima at the same wavelengths as the maxima of transparent crystallized glass. The second bands, between 13 and 14 $\mu$, similarly coincide in position. This suggests that the original glass of this melting already contained nuclei (crystallites) of the crystals which were deposited in the transparent crystallized glass.

Figure 3 shows spectra of transparent crystallized glasses of the same chemical composition, differing only by small additions of various oxides (with only one additive in each glass). Comparison shows that the selective reflection maxima in the spectra of all the glasses are at exactly the same wavelengths as in the case of the so-called β-spodumene, but the intensity varies. In some spectra the short-wave maximum is the stronger; in others the long-wave maximum, and some spectra have only a single maximum.

96

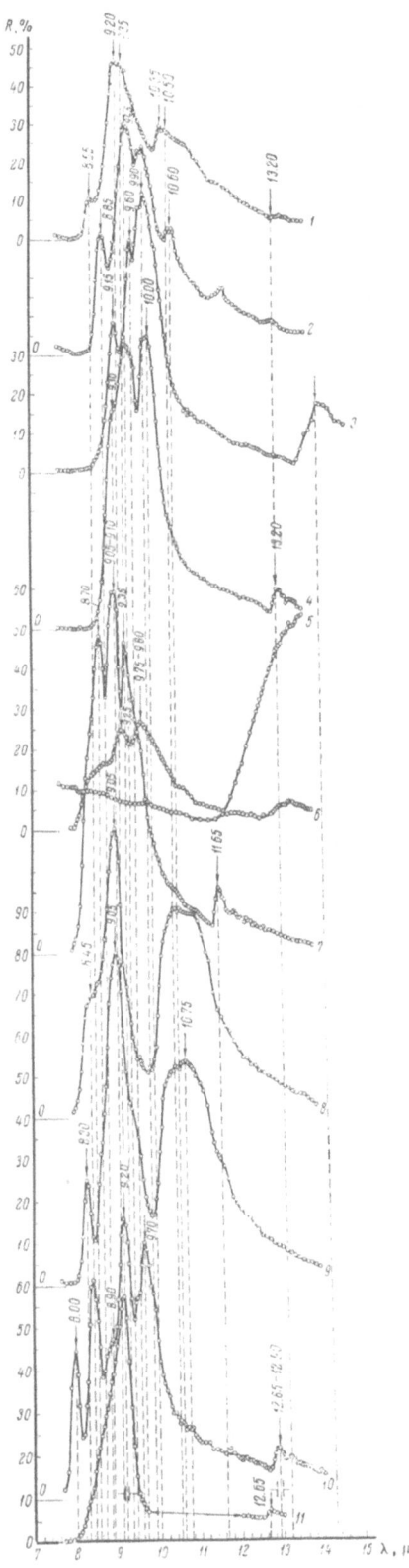

Fig. 1. Infrared reflection spectra of minerals. 1) Jadeite, $NaAlSi_2O_6$; 2) aegirite, $NaFeSiO_4$; 3) nepheline, $NaAlSiO_4$; 4) $\beta$-eucryptite, $LiAlSiO_4$; 5) rutile, $TiO_2$; 6) high-temperature spodumene ($\beta$-spodumene); 7) low-temperature spodumene ($\alpha$-spodumene); 8) low-temperature spodumene ($\alpha$-spodumene) of different orientation ($LiAlSi_2O_6$); 9) nonoriented $\alpha$-spodumene; 10) petalite, $LiAlSi_2O_6$; 11) cristobalite, $SiO_2$.

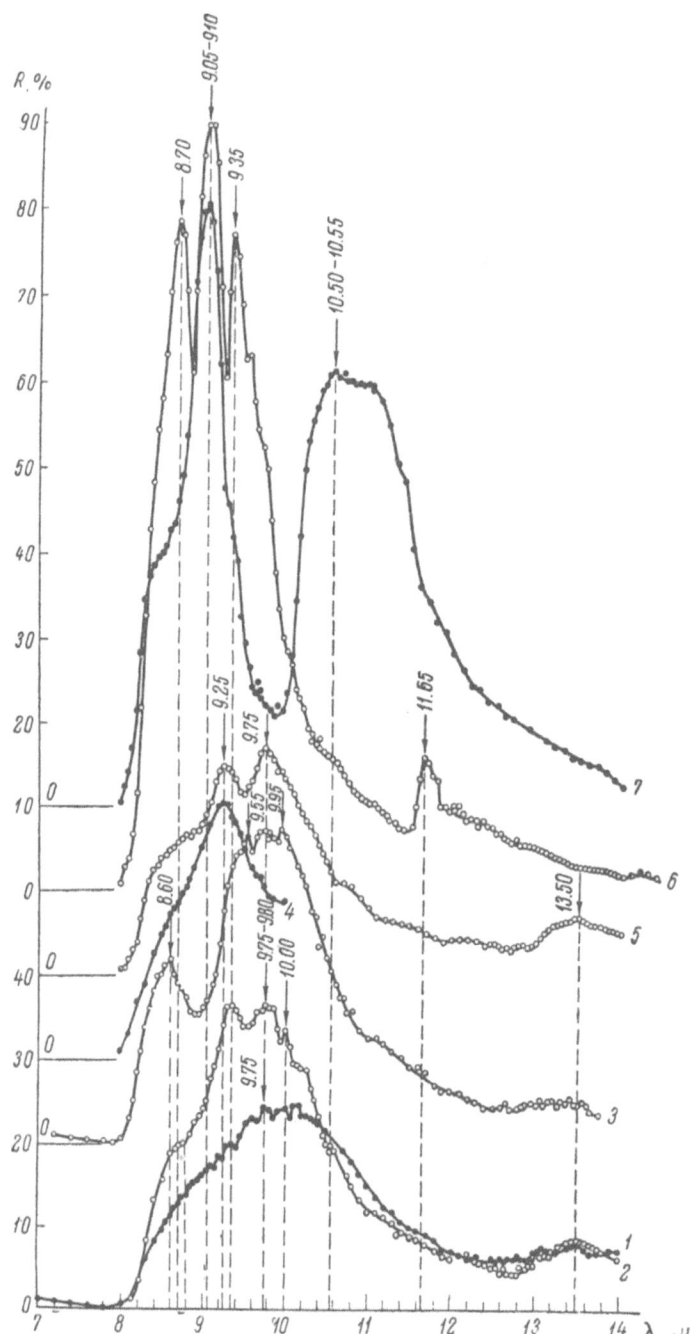

Fig. 2. Infrared reflection spectra of spodumene and of crystallized glasses close to spodumene in composition (melting No. 1). 1) Original glass 13 with 5% $TiO_2$; 2) the same, transparent crystallized glass; 3,4) the same, opaque glasses crystallized at different temperatures (curve 4 relates to the surface crust); 5) crystallized glass with the composition of spodumene; 6,7) spectra of low-temperature spodumene cut parallel to and perpendicular to one of the optical axes of the crystal.

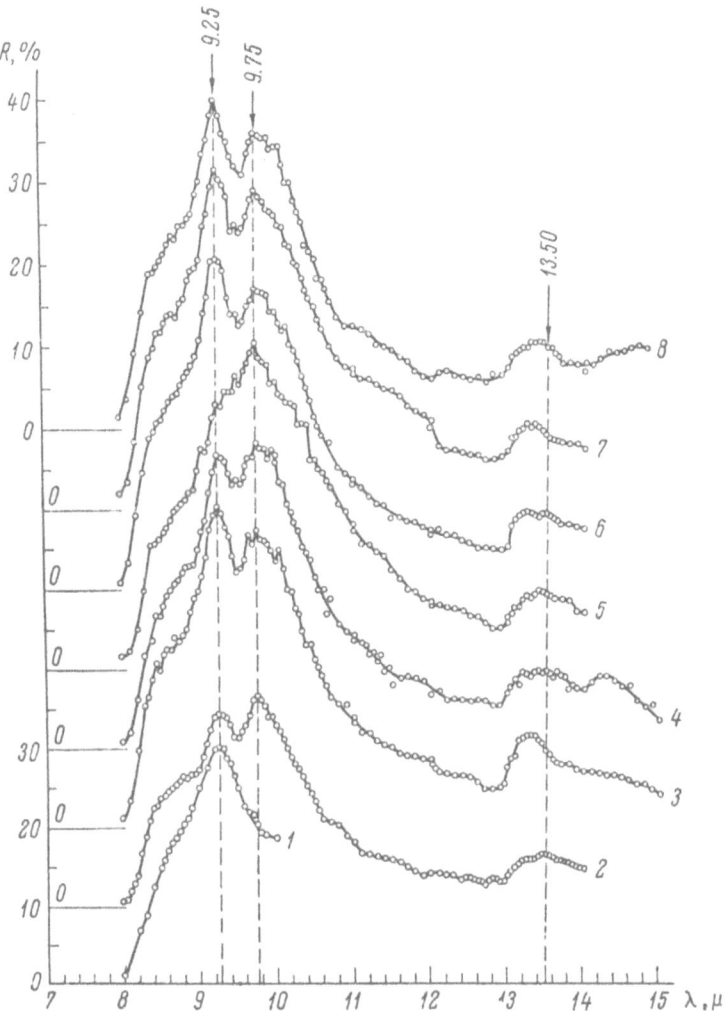

Fig. 3. Infrared spectra of glass 13 with 5% TiO$_2$, containing different added oxides. 1) Curve 4 of Fig. 2; 2) pure crystallized spodumene glass; 3-8) glass 13 with 5% TiO$_2$ and different added oxides.

This fact indicates that either the crystallization products have nothing in common with β-spodumene and that the coincidence of the maxima is accidental, or the substance generally described as β-spodumene, giving a spectrum with two maxima in the region of 9-10 μ (at 9.3 and 9.8 μ), is really a mixture of two different types of crystals. The crystal-optical technique either cannot reveal the difference between these crystals, or cannot detect one of these types because of their small size.

We consider that the same crystalline phases are deposited both in glasses with additions and in pure spodumene glass, and that we are concerned with a mixture of crystals of two different types. It is not yet clear whether the deposited crystals are different forms of the same compound, namely spodumene, or silicates differing in chemical composition.

Let us now examine the difference between transparent and opaque crystalline glasses of the same or similar composition. Comparison of spectra 1 and 3 (Fig. 4), relating to glasses from the same melting, shows that the opaque glass does not contain the same crystalline phase as is present in the transparent glass. At the same time, it follows from a comparison of curves 1 and 2 that the crystalline phase in glass represented by curve 2 is the same as in the transparent glass. It follows that the loss of transparency of crystallized glass may be due to entirely different causes in different cases.

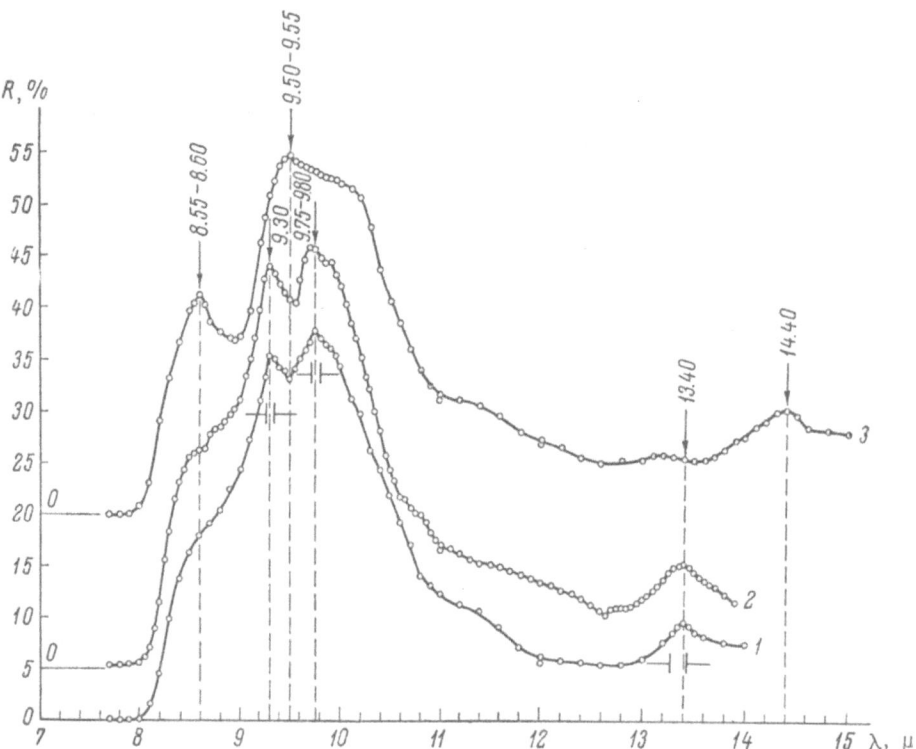

Fig. 4. Infrared reflection spectra of crystallized glass 13 with 5 and 2% $TiO_2$ (melting No. 2). 1) Transparent crystallized glass 13 with 5% $TiO_2$ (t = 719°, $\tau$ = 24 h); 2) opaque crystallized glass 13 with 2% $TiO_2$ (t = 770-775°, $\tau$ = 90 h); 3) opaque crystallized glass 13 with 5% $TiO_2$ (t = 828°, $\tau$ = 24 h).

These crystallized glasses have very low reflection coefficients. The reflection coefficient of minerals and well-crystallized glasses is of the order of 80-90% at the band maxima (Fig. 1). It is about 26% for the original glass 13 with 5% $TiO_2$, and only 34-36% for transparent and opaque crystallized glasses. In the opinion of some of the present authors, this may be an indication of a low content of crystalline phase in the transparent crystallized material.

In view of the fact that crystallized glasses differing in structure are obtained as the result of heat treatment of the same glass at different temperatures, it was of interest to investigate the structure of glasses over a wide temperature range, from temperatures close to the melting point to temperatures considerably below the annealing range. Long polished glass plates were used for this purpose; they were subjected to heat treatment in gradient furnaces. Spectra were recorded at different regions of each plate, corresponding to different temperatures. In each case the spectrum was recorded twice: at the outer layers, in contact with the air during crystallization, and at inner layers at a depth of 0.1 mm from the surface.

Reflection results for the inner layers are shown in Fig. 5. After treatment at 719-800°, when the crystallized glasses remain transparent, the main crystallization products are the same two silicates (maxima at 9.25-9.30 and 9.75-9.80 $\mu$) as were deposited during crystallization of glass with the composition of spodumene. In addition to these two silicates, there are always admixtures of some other silicates.

At higher temperatures, when the glass becomes turbid, a new crystalline phase appears with a characteristic band at 8.55-8.60 $\mu$. The intensity of the spectral band increases with increasing turbidity, but in the 805-822° range the main crystalline phases are the same as those detected in the transparent crystallized glass (maxima at 9.25-9.30 and 9.75-9.80 $\mu$). At yet higher temperatures, the old phases are partially retained but the phase giving a maximum at 9.50-9.55 $\mu$ becomes predominant.

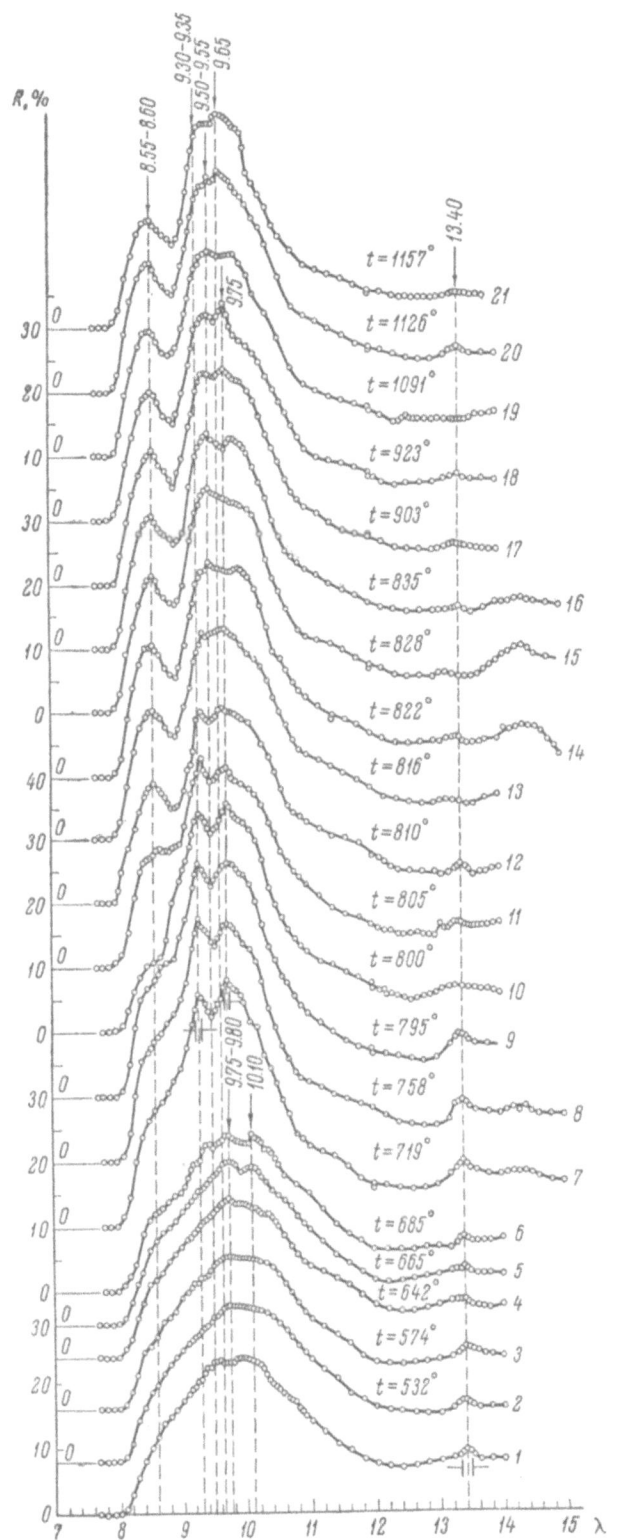

Fig. 5. Infrared reflection spectra of inner layers of glass 13 with 5% TiO$_2$ after different heat treatments. 1) Original glass (melting No. 2); 2-6) heated uncrystallized glasses (only diffuse rings seen in the x-ray diffraction patterns); 7-10) transparent crystallized glasses (first traces of turbidity in specimen 10); 11-13) turbid crystallized glasses; 14-21) opaque crystallized glasses. Heat treatment times: 24 h for specimens 2-18, 9-10 h for specimens 19-21.

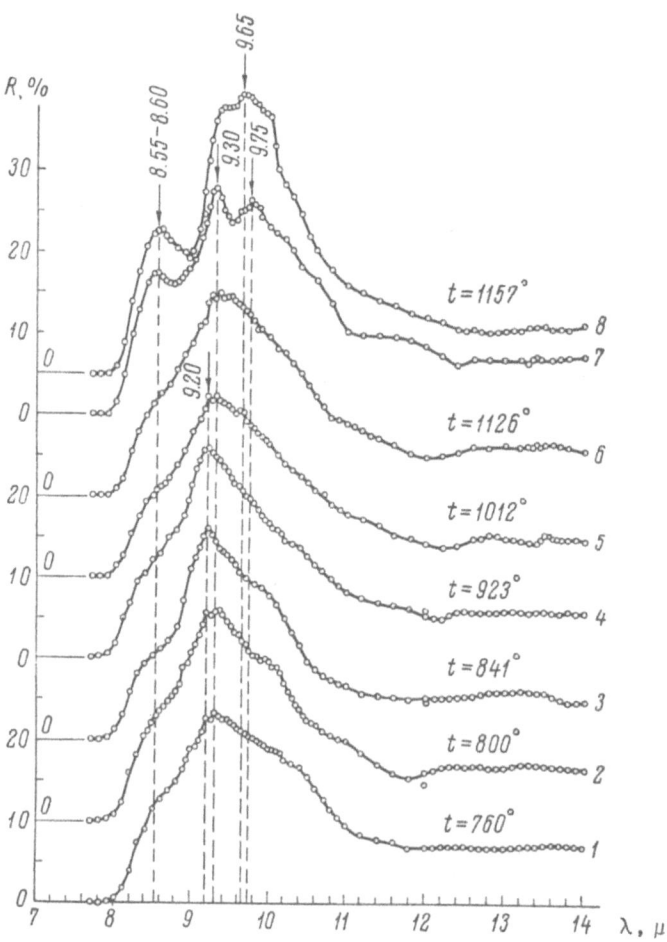

Fig. 6. Infrared reflection of the same glasses as in Fig. 5.
1-7) Crust regions of the glasses; 8) region under the crust at a
depth of 0.1 mm.

It is not clear whether the spectral bands at 8.55-8.60 and 9.50-9.55 $\mu$ relate to the same or to different crystalline phases. It may be supposed that both bands belong to the spectrum of the same silicate, which causes opacification of specimens 10-13, but the amount present in these specimens is relatively small, so that the 9.50 $\mu$ band does not appear. The 8.6 $\mu$ band appears earlier because it is at a considerable distance from the other bands. At higher temperatures the new crystalline phase becomes one of the main crystalline components, and therefore the 9.5 $\mu$ band also appears. An alternative interpretation of the results is that the bands at 8.6 and 9.5 $\mu$ belong to different silicates. It is not yet clear which interpretation is correct. In the 1126-1157° range there is yet another crystalline phase giving a selective reflection maximum at 9.65 $\mu$.

It is still uncertain whether all the crystalline phases found are solid solutions or various chemical compounds.

Figure 6 shows the spectra of crust regions of the same glasses as in Fig. 5. One phase, with a reflection maximum at 9.2 $\mu$, was deposited along almost the entire specimen. Other silicates were deposited only at the highest temperature. The silicates which exist in transparent crystallized glass predominate among these. Comparison of curves 8 and 7 shows that the silicates crystallizing on the outside and inside the specimen are different.

Let us now examine curves 1-6 (Fig. 5). It is seen that the main reflection maxima for the original glass, heated and uncrystallized, and transparent crystallized glasses of composition 13 with 5% $TiO_2$ coincide in

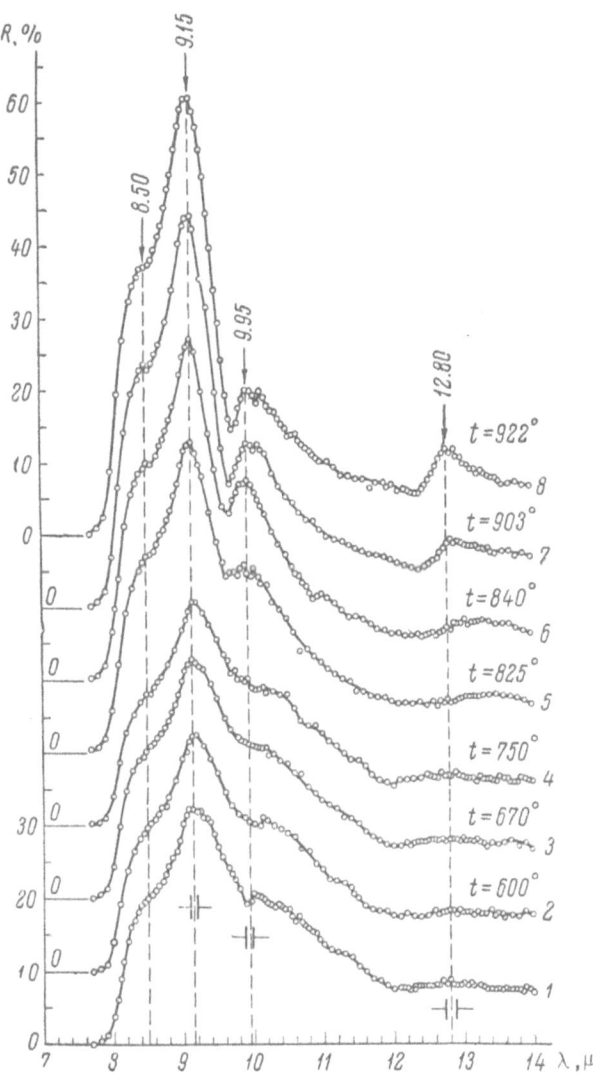

Fig. 7. Infrared reflection spectra of crust regions of glass with
the composition of petalite containing 5% $TiO_2$. 1) Original
glass; 2-4) heated uncrystallized glasses (the x-ray diffraction
pattern of specimen 4 gives a diffuse maximum); 5-7) transparent crystallized glasses; 8) crystallized glass with first traces
of turbidity ($\tau$ = 24 h).

position. This indicates that nuclei of the crystals deposited in transparent crystallized glass are already present
in heated glass and even in the original glass. During heat treatment at the temperatures in question, the glass
undergoes preparation for crystallization. The higher the temperature the more pronounced is the process of
preparation for crystallization.

Figure 7 shows the reflection spectra (surface layers) of glass with the composition of petalite with addition of 5% $TiO_2$. All the spectra in Fig. 7 are of similar character. The principal maximum is at 9.15 $\mu$ in
each case. This shows that both the original glass and the heated uncrystallized glasses already contained considerable amounts of nuclei of the crystals formed in the transparent crystallized glass. The crystallization product in transparent glasses is either silica or a high-silica silicate, to which the bands at 9.15 and 12.80 $\mu$ may
be assigned, and some other silicates giving a band in the 9.9-10.2 $\mu$ region. Figure 7 shows that crystallization
of these glasses is preceded by a period of latent preparation of the structure.

A somewhat different picture is obtained at higher heat-treatment temperatures (Fig. 8). In this case,
some metastable forms of silica are deposited, giving an intense band at 9.0 $\mu$. This silica has time to pass

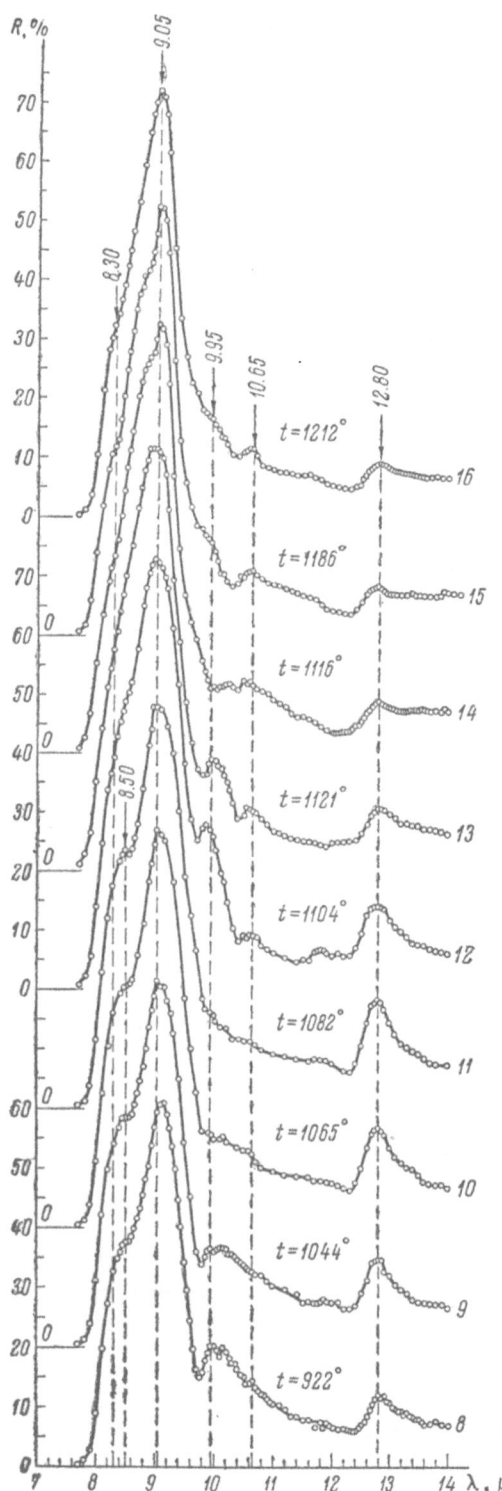

Fig. 8. Infrared reflection spectra of crust regions of glass with the composition of petalite containing 5% TiO$_2$. 8) Crystallized glass with the first signs of turbidity ($\tau$ = 24 h); 9) turbid crystallized glass; 10-16) opaque crystallized glasses ($\tau$ = 9-10 h).

partially into other forms, so that the strongest band is either shifted toward shorter wavelengths, as in spectrum 13, or splits as in spectra 14 and 15. In addition to silica, several silicates of different compositions are deposited. It is likely that the bands at 10.5-10.6 $\mu$ in spectra 12-16 relate either to lithium aluminosilicates or to silicates with a high lithium content. It is not clear why the band at 12.8 $\mu$, which should most likely be regarded as the second principal silica band in spectra 9-12, has especially high intensity.

The following conclusions may be drawn from the foregoing results.

1. Infrared spectroscopy should be widely used for investigation of processes occurring in the heat treatment of glass.

2. The crystallization products in titanium-containing transparent crystallized glasses close to spodumene in composition are the same crystals as in pure crystallized spodumene glass. The region of their formation is below 800°.

3. The loss of transparency of crystallized glasses of the same or similar composition may be due to several causes, in particular: a) appearance of crystalline phases differing from the phases deposited in transparent glass; b) larger size of the same crystals as are deposited in transparent glass.

4. The thermal conditions in crystallization of glass must be strictly regulated.

5. Small amounts of added oxides may influence the composition of crystalline phases in crystallized glasses.

6. Crystallization of glass is preceded by a period of latent structure preparation.

# INVESTIGATION OF THE CATALYZED CRYSTALLIZATION
# OF CERTAIN GLASSES BY MEANS OF INFRARED SPECTROSCOPY

## O. P. Girin

This communication contains the results of a preliminary study of the infrared absorption spectra of various original and crystallized glasses based on glass 13. The spectra were recorded in the range from 1500 to 400 cm$^{-1}$ with the aid of the Unicam SP-100 automatic double-beam spectrophotometer. The specimens were prepared by application of fine powders (average particle size about 1 $\mu$) onto a polished window made from a KBr crystal. The precision in the determination of the band positions was about 10 cm$^{-1}$.

TiO$_2$ was added to the glasses as a crystallization catalyst. It was of interest to study variations in the spectra of the glasses in relation to the amount of TiO$_2$ and to attempt to detect various modifications of spodumene crystals in the crystallized glasses. Accordingly, the following spectra were obtained: that of glass with the composition of spodumene, crystallized glass of the same composition, presumably consisting of $\beta$-spodumene, and the mineral $\alpha$-spodumene (Fig. 1). According to [1], the band with a maximum at 750 cm$^{-1}$ is assigned to vibrations of AlO$_4$ tetrahedrons and its appearance characterizes the change of the coordination number of Al from 6 to 4. In the spectrum of the glass (curve 1), the 750 cm$^{-1}$ band is weak and diffuse, and the structure of the 1000 cm$^{-1}$ band corresponding to the silicate framework has disappeared. It may be concluded that the spectral indications of glass crystallization and separation of $\beta$-spodumene in the glass are the appearance of a sharp band at 750 cm$^{-1}$ and of structure in the band at 1000 cm$^{-1}$.

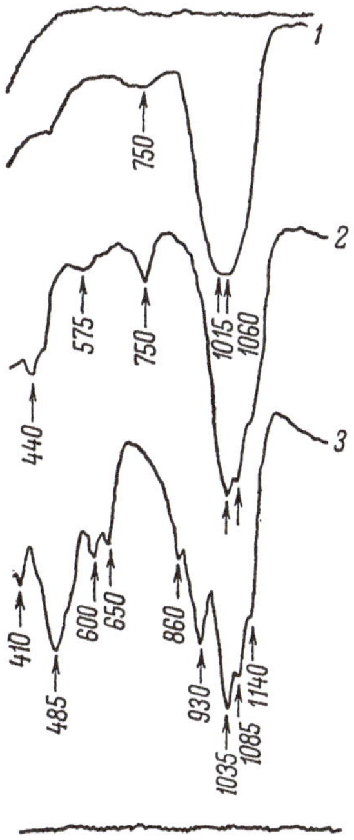

Fig. 1. Infrared absorption spectra. 1) Glass of spodumene composition; 2) crystallized glass, presumably $\beta$-spodumene; 3) $\alpha$-spodumene mineral.

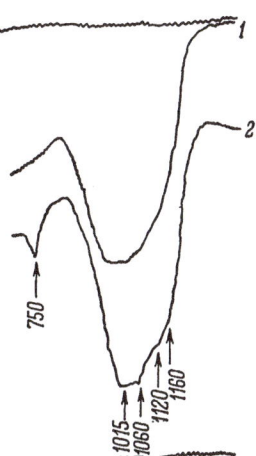

Fig. 2. Infrared spectrum of glass 13 with 7% TiO$_2$. 1) Original; 2) completely crystallized.

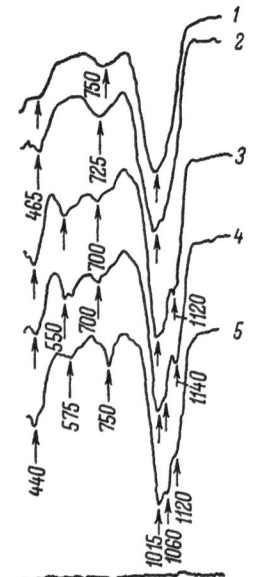

Fig. 3. Spectra of glass 13 specimens with 5% $TiO_2$ subjected to different heat treatments. 1) 600°; 2) 660°; 3) 850°; 4) 900°; 5) absorption spectrum of β-spodumene.

The infrared spectrum of rutile ($TiO_2$) was obtained; however, it is meager and the weak band at 420 cm$^{-1}$ and the very broad band with a maximum at 650 cm$^{-1}$ are hardly suitable for spectroscopic determination of the presence of rutile in glasses.

Variations of the spectra of glass 13 in relation to the $TiO_2$ content and heat treatment were studied with 24 specimens, in which the amount of $TiO_2$ was varied from 0 to 11%. The spectra of all the specimens are similar to each other and to those of spodumene glasses. Figure 2 shows typical spectra of the original and completely crystallized glasses containing 7% $TiO_2$. Comparison of spectra of crystallized glasses without $TiO_2$ and with 11% $TiO_2$ showed that the amount of $TiO_2$ has little effect on the spectrum: in both cases β-spodumene separates out. Spectra of the original glasses showed that with 11% $TiO_2$ partial crystallization begins even without heat treatment; here, again, β-spodumene is deposited. This indicates that $TiO_2$ causes partial crystallization of the glass even during the melting process; i.e., it confirms the catalytic function of titanium.

Figure 3 gives the spectra of four specimens of glass 13 with 5% $TiO_2$, subjected to different heat treatments at temperatures from 600° (curve 1) to 900° (curve 4). It follows from Fig. 3 that the 750 cm$^{-1}$ band, which is attributed to vibrations of $AlO_4$ tetrahedrons and to the presence of β-spodumene, is shifted into the long-wave region even at relatively low heat-treatment temperatures (curve 2). This may mean that some sort of structural rearrangement accompanied by a change in the coordination number of aluminum occurs in these glasses. In that case, it is difficult to speak of the deposition of α- or β-spodumene crystals in the glasses.

LITERATURE CITED

1. V. A. Kolesova, Optika i Spektroskopiya 6:38, 1959.

# INVESTIGATION OF CATALYZED CRYSTALLIZATION
# BY ABSORPTION CHANGES

## Yu. N. Kondrat'ev and E. V. Podushko

The purpose of this work was to elucidate the nature of the processes taking place in glass during crystallization. The optical density of some glasses in the $SiO_2$-$Al_2O_3$-$Li_2O$ and $SiO_2$-$MgO$-$Li_2O$-$Al_2O_3$ systems containing added $TiO_2$ varies considerably during the crystallization periods and also before crystallization. This density variation is due, on the one hand, to changes of color intensity and, on the other, to appearance or disappearance of opalescence. The density may increase or decrease depending on the process occurring in the glass.

An apparatus for recording changes of optical density was assembled. A beam of white or monochromatic light passes from a light source through the specimen in the furnace and is received by a photocell, the photocurrent of which is recorded by means of an ÉPP-09 potentiometer.

The investigation was carried out with glasses of the system $Li_2O$-$Al_2O_3$-$SiO_2$ with additions of $TiO_2$ and with variable amounts of $Li_2O$, in the region close to spodumene. Two temperature regions were investigated: a region where the glass does not crystallize, and a region where crystallization and further structural changes occur.

In the 20-500° range, the density increases slightly with temperature, and this increase is reversible; i.e., the density falls to the initial value when the temperature is lowered. This may be attributed to broadening of the absorption band at high temperatures owing to increase of polarization in the colorants. At temperatures above 565°, the optical density of the glass begins to increase irreversibly with time (Fig. 1). The figure shows that the curves converge at long exposure times, and the density subsequently remains unchanged or the changes are comparable with experimental error.

Simple mathematical treatment shows that these curves may be represented by the general equation

$$D_t = D_p + (D_0 - D_p) e^{-K_1 t}, \tag{1}$$

where $D_t$ is the density at a given instant; $D_p$ is the equilibrium density, taken from the asymptotic value; $D_0$ is the initial density; $K_1$ is a constant which depends on the temperature; t is the time.

The constant $K_1$ has a definite physical meaning — it is the rate constant for variation of the optical density at a given temperature. To find $K_1$, we can plot the function (Fig. 2):

$$\ln \frac{D_p - D_0}{D_p - D_t} = K_1 t. \tag{2}$$

The equation of a reaction of the first order is of the form

$$\ln \frac{a}{a - x} = K_2 t, \tag{3}$$

where a and x are concentrations, $K_2$ is the reaction rate constant, and t is the time.

It is easy to show that Eq. (2) is analogous to Eq. (3), and it follows that the process in question involves a monomolecular reaction.

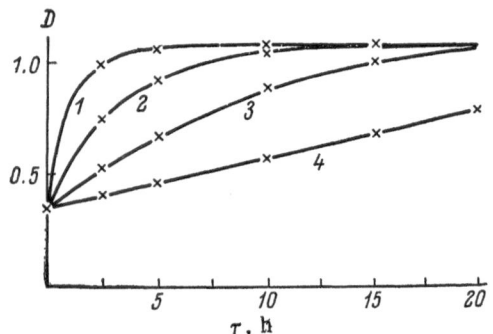

Fig. 1. Variations of optical density with time and temperature. 1) 640°; 2) 620°; 3) 595°; 4) 565°.

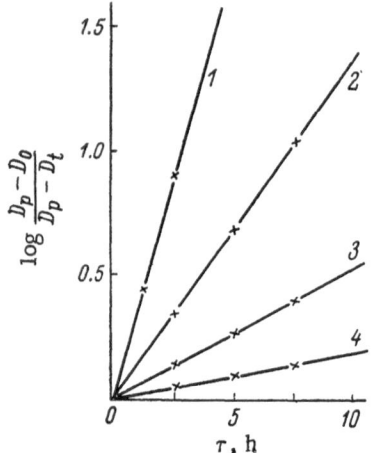

Fig. 2. Plots of the function $\log \dfrac{Dp - D_0}{Dp - Dt} = Kt$ at different temperatures. 1) 640°; 2) 620°; 3) 595°; 4) 565°.

Hence K should obey the Arrhenius equation

$$\ln K = \ln Pz - \frac{E_a}{RT},\qquad(4)$$

where P is a steric factor; z is the total number of collisions; $E_a$ is the activation energy; R is the gas constant; T is the absolute temperature. Figure 3 shows that this simple dependence holds for our glasses.

The calculated activation energies have the following values:

| $Li_2O$ content, wt.%..... | 6.0 | 7.5 | 8.0 |
|---|---|---|---|
| $E_a$, kcal/mole....... | 43.1 | 61.5 | 66.5 |

It is interesting to note that the activation energy of $Li^{1+}$ displacement, calculated for these glasses from conductivity data, is 16-19 kcal/mole, which is less by a factor of 2.5 than the activation energies found. The latter values might be likened to the activation energy of viscous flow [1]. Crystallization consists of two processes: diffusion of light, network-modifying ions and diffusion of network elements; the final stage of these processes is the formation of regions of greater order than in the original glass. It is evident that the precrystallization period includes a process of structural preparation, governed by the stage making the highest energy demands − the stage of covalent diffusion.

How is this change of optical density to be explained? It is known that the color of glasses containing $TiO_2$ and $Fe_2O_3$ simultaneously is much more intense than that of glasses containing the separate components, even in the same amounts. Many investigators attribute this effect to the strong antipolarizing action of quadrivalent titanium on the oxygen atoms of iron oxide, or to the formation of iron titanates [2-4].

It is possible to correlate the values of the activation energy with the coordination of titanium in the glass. Some workers consider that titanium is present in the $TiO_6$ state in glass [5,6]. Others assume that titanium exists in the form of $TiO_4$ groups [7,8]. In view of the intensity of the disorienting thermal motion and of the fact that the two oxygens in rutile are nonequivalent to the other four, it may be concluded that,at the high temperature of glass melting, titanium is present in tetrahedral coordination in the form of $TiO_4$ groups. With fairly rapid chilling the structure is frozen and the $TiO_4$ groups enter a state of metastable equilibrium. On further heating, i.e., when energy is transferred to the glass, this metastable equilibrium becomes a state of true equilibrium and $TiO_6$ groups are formed. The spectral absorption curves (Fig. 4) of the original and treated glasses show that absorption in the violet region of the spectrum increases with the time of treatment. This band corresponds to the coloring iron − titanium complex.

The reaction described has the following mechanism: the $TiO_4$ groups are converted into $TiO_6$ groups, transferring iron from the $FeO_6$ position into the $FeO_4$ position. Because of the low concentration of iron, the bimolecular reaction is pseudo-monomolecular in character. Consequently, three facts may be adduced in favor of this mechanism: first, the pseudo-monomolecular character of the reaction; second, the high value of the activation energy, indicating the switching of stable Al-O-Ti and Si-O-Ti bonds; third, the spectral absorption curves.

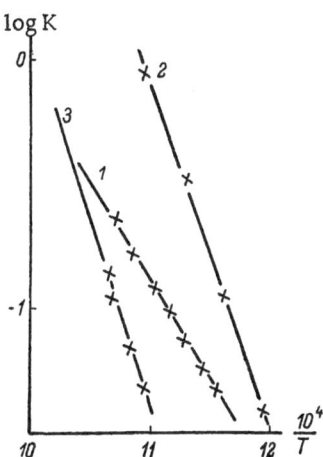

Fig. 3. Dependence of the reaction rate constant on the reciprocal absolute temperature with different contents of $Li_2O$ by weight. 1) 6.0%; 2) 7.5%; 3) 8.0%.

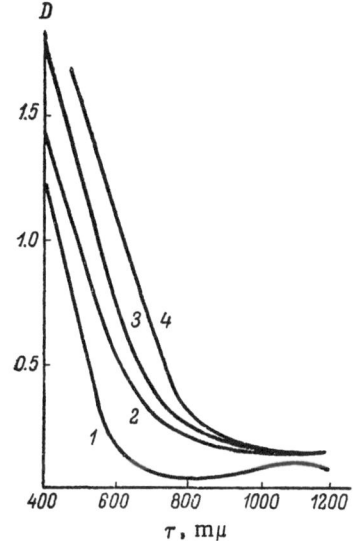

Fig. 4. Spectral absorption curves of glass with 7.5% $Li_2O$ for different times of heating. 1) Original glass; 2) 1.0 h; 3) 2.5 h; 4) 20 h.

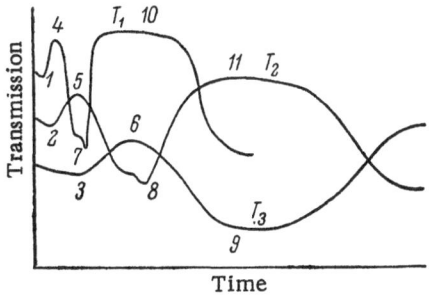

Fig. 5. Variations of the transmission of glass with time at high temperatures. $T_1 > T_2 > T_3$.

In the high-temperature region the glass begins to crystallize, and its transmission increases. Figure 5 represents the character of the transmission rather than the actual experimental curves. The considerable complexity of the curves indicates that the processes occurring in this temperature region are diverse. Interpretation of the curves is difficult because we do not have any one constant parameter up to a given time instant. The temperature alters as the specimen is heated; this process takes 1.5-2.0 min, but subsequently the temperature of the specimen rises (sometimes by 50-60°) at the instant of crystallization. Despite this, the main stages of the changes taking place in the glass on heating may be indicated.

The first minimum (points 1-3) may be attributed to broadening of the absorption band when the glass is heated. This conclusion follows from the reversibility of the effect. Further, it is seen that the glass becomes considerably clearer, and the nature of the clearing is strongly dependent on the temperature (points 4-6). This effect is irreversible, and it is suggested that it is associated with nonequilibrium heating processes. No such clearing occurs in experiments in which the specimens are heated slowly to a high temperature. Figure 5 shows that the clearing is followed by abrupt darkening of the specimen, and the rate of darkening depends on the temperature. Analysis of the spectral absorption curves indicates that this effect is associated with the precrystallization process described above [Fig. 1, Eq. (1)]. The inflection, which is quite distinct in experiments at high temperatures, is interesting; this point marks a sharp rise in the temperature of the specimen, indicating the start of crystallization. It is significant that the density at points 7-9 is the same, within the limits of experimental error. Beyond the inflection, the transmission of the glass increases, which evidently indicates a decrease of the concentration of the coloring complexes which were formed during the precrystallization period. This may possibly be due to transition of titanium from glassy $TiO_6$ to crystalline $TiO_6$ and formation of rutile or colorless titanates.

Kind's investigations [9] show that a phase with a very high refractive index, 1.750, is indeed formed in the high-temperature region. It is quite possible that some titanium-containing phase has this refractive index.

The subsequent fall of transmission (regions beginning at points 10 and 11) is due to increasing opalescence and conversion of the transparent material into an opaque one. This opalescence is apparently associated with separation of a new crystalline phase or recrystallization of an existing one.

The method proposed here for investigation of the kinetics of precrystallization and crystallization processes is convenient because the absorption of glass is a property which is sensitive to structural changes; in view of the fact that glass-making materials contain a colored indicator (iron in the present instance), it is always possible to investigate such processes by this method. In particular, the processes described above may be associated with nucleation, with formation of a structure intermediate between glass and crystal, and with transition of the glass into a crystalline-glass state.

## LITERATURE CITED

1.  V. T. Slavyanskii, "Calculation of the activation energy of viscous flow of alkali silicate glasses from known chemical composition," in The Structure of Glass, Vol. 2, Consultants Bureau, New York, 1960, p. 289.
2.  W. Weil, Colored Glass, New York, 1956.
3.  A. Dietzel, Naturwiss. 29:537, 1941.
4.  V. V. Vargin, Doklady Akad. Nauk SSSR 103(1):105, 1955.
5.  R. C. Turnbull and W. C. Lawrence, J. Am. Ceram. Soc. 35(2):48, 1952.
6.  I. Minoru, Yogyo Kyokai Shi 67(717), 1959.
7.  A. A. Appen, Certain General Relationships between the Properties and Composition of Silicate Glasses, Doctoral Dissertation, Leningrad, 1952.
8.  Ya. S. Bobovich and T. P. Tulub, Opt.-Mekh. Prom. No. 9, 40, 1961.
9.  N. E. Kind, this collection, p. 111.

# INVESTIGATION OF THE PRODUCTS OF CATALYZED CRYSTALLIZATION BY OPTICAL CRYSTALLOGRAPHY

## N. E. Kind

Glass 13 was investigated. Titanium dioxide, in amounts from 1 to 11% $TiO_2$ by weight, was added as the catalyst. All the glasses were made in a semiproduction furnace. A series of the glasses was then subjected to heat treatment under laboratory conditions in special thermostatic furnaces.

The heat-treated and untreated glasses were placed in boats into a gradient furnace for crystallization and kept there for 24 h in the 600-1400° range. The specimens were examined visually for a general assessment of the visible crystallization range and for determination of the liquidus boundary.

The polished-section and immersion methods were used for determining the variations of glass structure with temperature and for establishing the crystalline phases. A glass with the composition of spodumene, specially made for the purpose, was also investigated similarly.

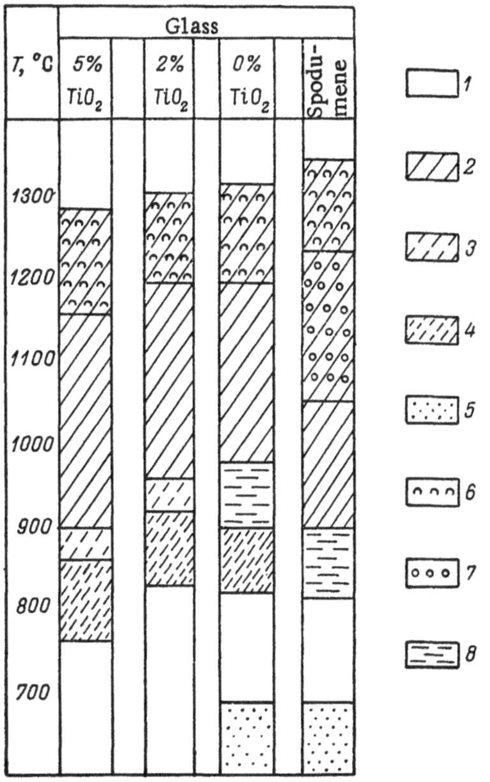

Experiments on crystallization of the glasses and glassy spodumene made it possible to establish their liquidus temperatures; these were 1380° for the spodumene glass, and 1320, 1310, and 1290° for glass 13 containing 0, 2, and 5% $TiO_2$. In spodumene glass visible crystallization after 24 h in the gradient furnace began at 860-880°. In glass 13 with additions of $TiO_2$, distinct crystallization began at about 900° and higher. At lower temperatures the latter glasses had different degrees of opalescence or were transparent (Fig. 1).

Examination of glass specimens held under conditions close to their upper crystallization temperatures showed that in spodumene glass the primary crystalline phase is high-temperature spodumene. The primary crystalline phase in glass 13 is also high-temperature spodumene.

Immersion examination of glass 13 specimens containing 0, 2, and 5% $TiO_2$, previously heat-treated and untreated and then subjected to crystallization for 24 h in a gradient furnace at temperatures up to 1350°, showed that low-temperature spodumene with a fibrous structure is formed in glasses with 0 and 2% $TiO_2$ at low crystallization temperatures (900-1200°). Above 1200°, the main crystalline phase (spodumene) becomes interpenetrated by numerous acicular crystals with a somewhat higher refractive index.

Spodumene crystals were not found in glass with 5%

Fig. 1. Glass crystallizability diagram. 1) Transparent; 2) crystallized; 3) strongly opalescent; 4) weakly opalescent; 5) smoky; 6) surface fusion; 7) swelling; 8) cracking.

Fig. 2. Electron micrographs of crystals isolated from glass 13 with 11% $TiO_2$.
1) Prismatic crystals, magnification 8000; 2) laminar rhomboidal crystals, magnification 13,000.

$TiO_2$ at low crystallization temperatures (900-1250°). This may have been caused either by the very fine crystallization of spodumene under these conditions or by the fact that the fragments observed by immersion microscopy were penetrated by numerous minute prismatic crystals with direct extinction, a high refractive index, and high birefringence. The largest crystals were up to 0.01-0.02 mm long. The crystals became somewhat larger at higher crystallization temperatures. It may be assumed that this crystalline phase is also present in the glass at lower temperatures, but in such cases the crystals are too small to be distinguishable under the microscope. Above 1250°, this microcrystalline phase disappeared and was replaced by crystals of pure spodumene and spodumene crystals penetrated by acicular crystals, with similar refractive indices.

Because of the small size of the prismatic crystals found at temperatures up to 1250°, they could not be identified by optical crystallography. As the unknown phase resembled rutile in crystal form and in its high refractive index, and as deposition of rutile in these glasses was possible, an attempt was made to isolate it chemically from the glass.

For determination of the composition of the unknown crystalline phase, we used specimens of glass 13 containing 11% $TiO_2$, in which similar fine crystals with high $n_g'$ were observed under the microscope. The crushed glass was treated twice with 20% HF and heated to decomposition. After decantation and additional treatment with 15% HCl, the crystals were washed with water until neutral and examined under the microscope. They were then fused with sodium carbonate, dissolved in acid, and analyzed.* They were found to consist of $TiO_2$ and $Al_2O_3$.

The analytical results are given in Table 1.

The observed deviations between the results of separate experiments may be attributed to difficulties in the analysis of such a microcrystalline phase, but the results are of the same order.

Treatment of similar glasses containing 9, 7, and 3% $TiO_2$ gave entirely different results. In this case the isolated crystals again consisted of $TiO_2$ and $Al_2O_3$, but the ratio of the oxides was different, with a higher $Al_2O_3$ content (Table 2).

* The analyses were performed by K. A. Yakovleva.

TABLE 1. Chemical Composition of Crystals Deposited during Heat Treatment of Glass 13 with 11% TiO by Weight

| Sample No. | Wt. % | | Mole % | |
|---|---|---|---|---|
| | TiO$_2$ | Al$_2$O$_3$ | TiO$_2$ | Al$_2$O$_3$ |
| 1 | 25.4 | 75.6 | 30.2 | 69.8 |
| 2 | 33.0 | 67.0 | 38.8 | 61.2 |
| 3 | 37.0 | 63.0 | 42.6 | 57.4 |

TABLE 2. Chemical Composition of Crystals Deposited during Heat Treatment of Glass 13 with 9, 7, and 3% TiO$_2$ by Weight

| Ti$_2$O content in glass, wt.% | Sample No. | Wt. % | | Mole % | |
|---|---|---|---|---|---|
| | | TiO$_2$ | Al$_2$O$_3$ | TiO$_2$ | Al$_2$O$_3$ |
| 9 | 1 | 10.7 | 83.3 | 13.3 | 86.7 |
| 7 | 1 | 10.1 | 90.9 | 12.5 | 87.5 |
| 7 | 2 | 12.6 | 87.4 | 15.5 | 84.5 |
| 3 | 1 | 15.7 | 84.3 | 18.2 | 81.8 |
| 3 | 2 | 14.5 | 85.5 | 17.8 | 82.2 |

Table 2 shows that the results agree in order of magnitude and differ considerably from the data in Table 1.

These results are in agreement with V. N. Vertsner's electron microscope data, according to which the structure of glass 13 varies sharply with the TiO$_2$ content. Compositions with up to 9% TiO$_2$ have the same structure. The structure of the composition with 11% TiO$_2$ is quite distinct [1].

Thus, isolation of the crystals and chemical determination of their composition showed that glass 13 containing from 3 to 11% TiO$_2$ deposits crystals in a definite temperature range; the composition of these crystals can be represented by the ratio xTiO$_2$ : yAl$_2$O$_3$. The oxide ratio is variable and depends on the TiO$_2$ content in the glass.

According to the phase diagram of the Al$_2$O$_3$-TiO$_2$ system [2], it contains one binary compound, Al$_2$O$_3$ · TiO$_2$. Hence the oxide ratios found may be attributed either to simultaneous deposition of several crystalline phases (TiO$_2$, Al$_2$O$_3$, TiO$_2$ · Al$_2$O$_3$) in various proportions in accordance with the TiO$_2$ content in the glass, or to formation of new aluminum titanates.

Electron micrographs of crystals (isolated from glass 13 with 11% TiO$_2$) distinctly reveal two types of crystals — prismatic and laminar rhomboidal (Fig. 2).

High-temperature spodumene is therefore the primary crystalline phase deposited over the entire visible crystallization range of glass 13. Additions of more than 3% TiO$_2$ alter the nature of the crystallization entirely and lead to deposition of aluminum titanates of variable composition.

LITERATURE CITED

1. V. N. Vertsner and L. V. Degteva, this collection, p.86.
2. D. S. Belyankin, V. V. Lapin, and N. A. Toropov, Physicochemical Systems of Silicate Technology, Promstroiizdat, 1954.

# INVESTIGATION OF THE CRYSTALLIZATION PROCESS
# BY THE COLORED INDICATOR METHOD
# AND BY THE LEACHING METHOD

## V. V. Vargin

The absorption spectra of colored glasses and leachability determinations do not provide direct data on the processes taking place during crystallization of glass. These are indirect methods, the results of which help in the interpretation of data obtained in studies of the crystallization process by other techniques. We used these methods for investigating the crystallization of glass 13 with 5% $TiO_2$.

For all our determinations, the glasses were cooled after they had been made and then subjected to additional heat treatment at temperatures from 600 to 1000° for 24 h.

For investigating the crystallization of glasses by the colored indicator method, we used $Co^{2+}$ and $Ni^{2+}$ ions, which are often used as indicators of the structure of glass and also of crystals and solutions. It is known that the spectral absorption of cobalt and nickel ions alters sharply with change of their coordination number (from 4 to 6), while the latter itself depends on the structure of the main substance [1-3].

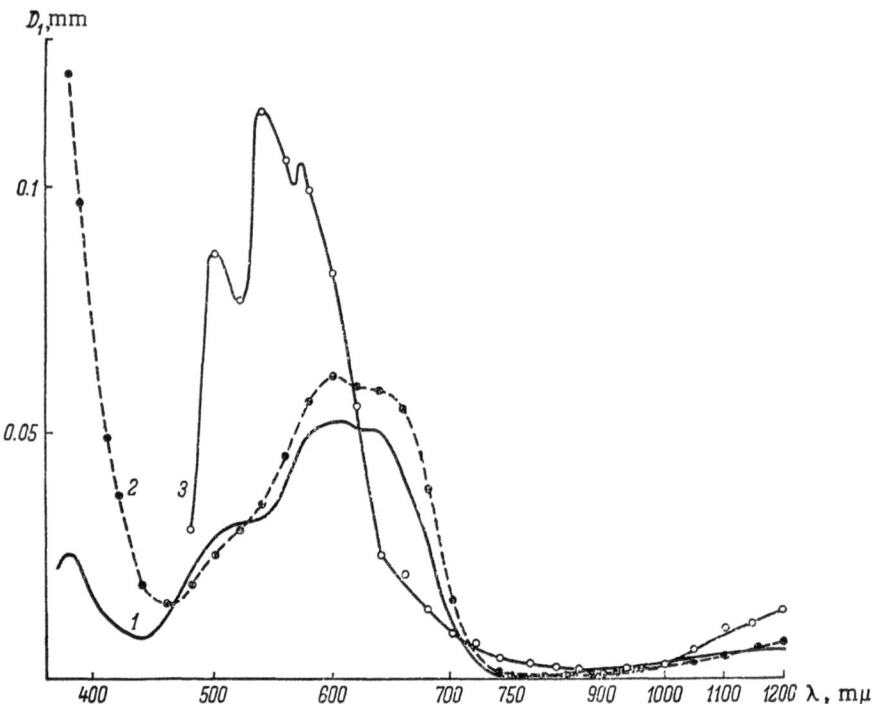

Fig. 1. Effect of the temperature of heat treatment on the spectral absorption of glass 13 colored by $Co^{2+}$. 1) Not treated; 2) treated at 640°; 3) treated at 725°.

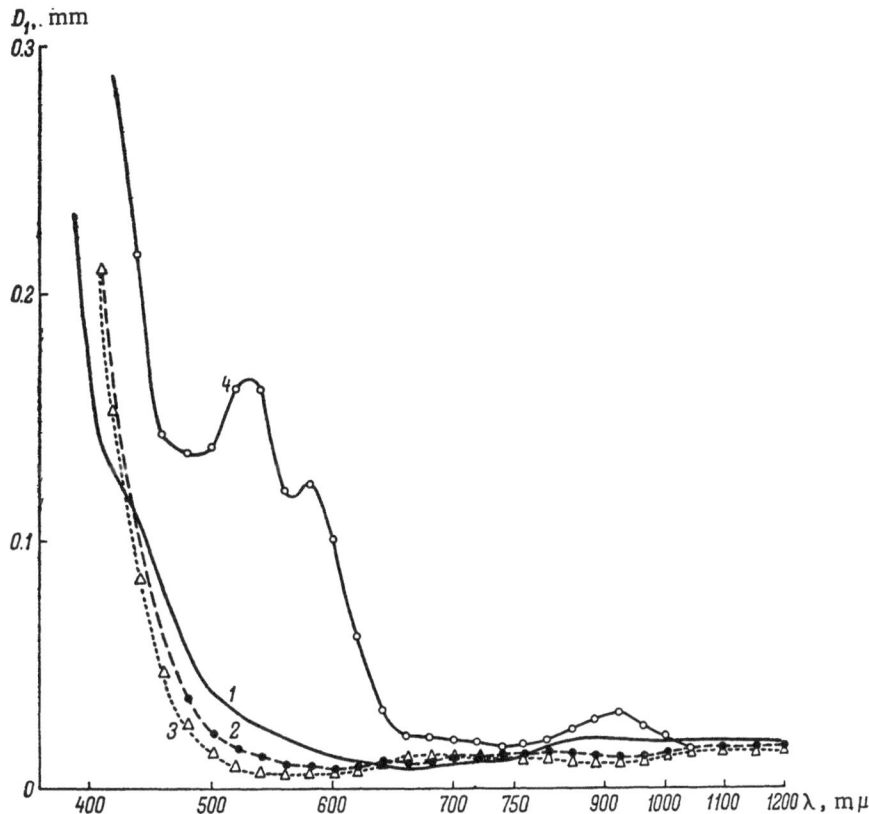

Fig. 2. Effect of the temperature of heat treatment on the spectral absorption of glass 13 colored by $Ni^{2+}$. 1) Not treated; 2) treated at 640°; 3) at 655°; 4) at 725°.

The amounts of the indicators added to the glass were 0.03% CoO and 0.1% NiO. Polished specimens were prepared from the glasses, and the spectral absorption was determined with a photoelectric spectrophotometer.[*]

Glass colored by $Co^{2+}$ and not subjected to heat treatment exhibits spectral absorption characteristics of cobalt in sixfold coordination. After heat treatment at 640°, there is little change in the spectral absorption; the maxima on the absorption curve become less distinct (Fig. 1).

If the temperature of heat treatment is raised to 725°, when the glass crystallizes but remains transparent (as was shown by other methods), the spectral absorption changes sharply (Fig. 1). The course of the absorption curve becomes characteristic of $Co^{2+}$ in fourfold coordination. The absorption curves in Fig. 1 are very similar to those of minerals containing $Co^{2+}$ in sixfold and fourfold coordination [4].

Glass colored by nickel and not subjected to heat treatment is yellow, and its spectral absorption is characteristic of $Ni^{2+}$ in sixfold coordination (Fig. 2) [5]. As in the case of glass colored by cobalt, the spectral absorption changes little as the result of heat treatment at 640° and 655°. However, the absorption curve becomes steeper. When the temperature of heat treatment is raised to 725°, the color of the glass changes from yellow to purple and the spectral absorption alters sharply (Fig. 2). This course of the spectral absorption curve is characteristic of $Ni^{2+}$ in fourfold coordination [5].

The transition of cobalt and nickel ions from one coordination state to another in alkali silicate glasses containing boric anhydride, aluminum oxide, or gallium oxide corresponds to the change in the coordination of boron, aluminum, and gallium ions in these glasses. This transition is especially pronounced in glasses containing aluminum.

---

[*] The spectral absorption determinations were carried out by V. I. Skorospelova.

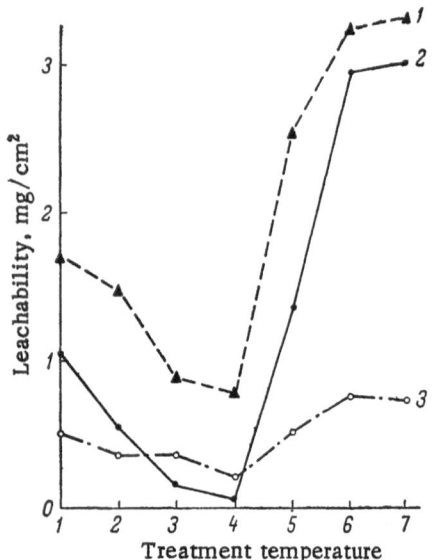

Fig. 3. Effect of the temperature of heat treatment on the leachability of glass 13 by various solutions. 1) HF; 2) HCl; 3) NaOH.

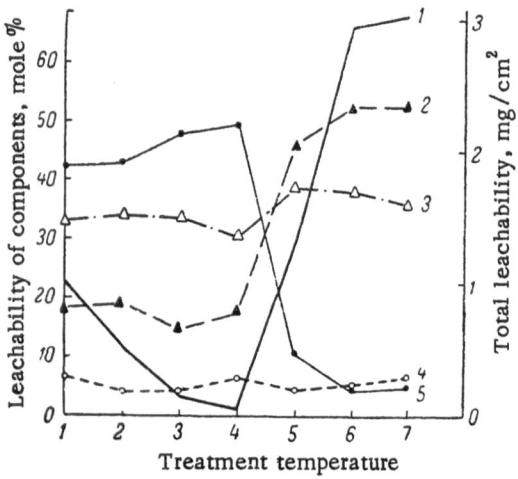

Fig. 4. Effect of the temperature of heat treatment on the leachability of glass 13 by 20% HCl solution and on the chemical composition of the extracts. 1) Total leachability; 2) $Li_2O$; 3) $Al_2O_3$; 4) $TiO_2$; 5) $SiO_2$.

We investigated glasses of the $Na_2O-Al_2O_3-SiO_2$ system colored by nickel. It was shown that the coordination of $Ni^{2+}$ in the glasses depends on the coordination of aluminum (on $Na_2O:Al_2O_3$). At ratios higher than unity the color of the glass is purple and $Ni^{2+}$ is in fourfold coordination. At $Na_2O:Al_2O_3$ ratios less than unity the glass is yellow, corresponding to nickel ions in sixfold coordination [4]. If even small amounts of aluminum are present in the glass in sixfold coordination, $Ni^{2+}$ is also present in sixfold coordination only [6].

In the lithium aluminosilicate glass investigated the $Li_2O$ content is lower than the $Al_2O_3$ content, and therefore a part of the aluminum is in sixfold coordination. Cobalt and nickel ions in this glass are also in sixfold coordination. During the period preceding crystallization of the glass (treatment temperature 655°), certain definite changes take place in the spectral absorption (Figs. 1 and 2) indicating structural changes in the glass, but the coordination of cobalt and nickel remains unchanged.

When the glass crystallizes (treatment temperature 725° and higher), its color changes, as already stated, from yellow to purple and the coordination number of the cobalt and nickel ions changes to four.

It must be pointed out that glasses of the $Li_2O-Al_2O_3-SiO_2$ system cannot have $Co^{2+}$ or $Ni^{2+}$ in fourfold coordination, whatever the ratio of the components. The change of the coordination number to four and the corresponding color change can be due only to the fact that these ions pass into the crystalline phase, in which aluminum is in fourfold coordination. Apparently, in this glass the crystalline phase consists of β-eucryptite and β-spodumene, in which aluminum is in fourfold coordination [7].

For the leachability investigations, the glasses were used in the form of plates polished on all sides, 20 × 20 × 3 mm in size.† Leaching was effected by 1% cold HF solution for 1 h, 20% HCl solution at the boil for 5 h, 4% NaOH solution at the boil for 1 h, and water at the boil for 200 h. The weight loss of the plates during the experiments was determined. Treatment with water did not give reliable results, as the weight loss during 200 h of boiling did not exceed 1.5 mg; therefore, data on the leachability of the glasses by water are omitted.

The results for all the other reagents are given in Fig. 3. The graph shows that the leachability greatly depends on the temperature of heat treatment. Leachability by all the reagents decreases with rise of the treatment temperature up to the temperature at which the glass crystallizes without loss of transparency (Fig. 3, point 4). This is especially prominent in treatment with HCl, when the leachability diminishes by a factor of about 20. On further increase of the heat-treatment temperature, the leachability rises sharply and at 900° it is 50 times the minimum leachability (Fig. 3, point 6).

†K. A. Yakovleva performed the chemical analyses.

116

Chemical analysis of the hydrochloric acid extracts showed that their composition varied little and did not depend on which glass, original or crystallized opaque, was leached (treatments 1-4, Fig. 4). The ratio of the oxides in these extracts is fairly close, apart from silica, to that in the original glass. The silica content in the hydrochloric acid extracts is lower because silica is only slightly soluble in HCl and remains on the surface of the leached specimens in the form of a protective film.

Therefore, the processes taking place during the precrystallization period of heat treatment (Fig. 4, points 2 and 3) and during the formation of transparent crystallized glass (point 4) result only in a general decrease of leachability without any significant changes in the chemical composition of the extracts.

When the heat-treatment temperature is raised and the glass becomes first opal (Fig. 4, point 5) and then opaque (points 6 and 7), the total leachability and the ratio of the oxides in the hydrochloric acid extracts alter sharply.

The silica content of the extracts falls to a very low value (about 5 mole%) while the lithium oxide content rises sharply. The aluminum oxide content also increases a little. The high total leachability and the predominance of lithium and aluminum in the extracts show that at high heat-treatment temperatures a different crystalline phase, chemically less stable, is formed. It is possible that at heat-treatment temperatures of 850-1000° a compound of the lithium aluminate type, readily soluble in hydrochloric acid, is formed.

Thus, it was shown with the aid of colored indicators and by investigations of the leachability of lithium aluminosilicate glass containing titanium dioxide as catalyst that during the precrystallization period (treatment temperature 630-655°) a structural change takes place, the color of glass containing $Co^{2+}$ or $Ni^{2+}$ changes somewhat, and the chemical stability of the glass increases. The crystalline phase which forms in the glass without loss of transparency (treatment temperature 725-760°) is apparently $\beta$-eucryptite or $\beta$-spodumene (transition of $Co^{2+}$ and $Ni^{2+}$ into fourfold coordination). If the temperature of heat treatment is raised further, when the glass becomes opal (850°) and opaque milk-white (900-1000°), compounds easily leached out by acids are formed, and lithium and aluminum are dissolved out preferentially.

## LITERATURE CITED

1. T. I. Veinberg, Zhur. Fiz. Khim. 36(1):81, 1962.
2. A. A. Kefeli, E. I. Galant, and N. I. Vlasova, Zhur. Neorg. Khim. 5(8):1768, 1960.
3. V. V. Vargin and T. I. Veinberg, "Coloration of glasses in relation to their structure," in The Structure of Glass, Vol. 2, Consultants Bureau, New York, 1960, p. 331.
4. E. G. Valyashko and S. V. Grum-Grzhimailo, Trudy Instituta Kristallografii AN SSSR 3(8), 1953.
5. V. V. Vargin, Production of Colored Glass, Gizlegprom, 1940.
6. V. V. Vargin, Zhur. Priklad. Khim. 35(7):1618, 1962.
7. H. Saalfeld, Ber. Dtsch. keram. Ges. 38(7):281, 286, 1961.

# VISCOSIMETRIC INVESTIGATION OF THE KINETICS
# OF CATALYZED SUBMICROCRYSTALLIZATION

## G. T. Petrovskii and S. V. Nemilov

The numerous investigations in the field of glass crystallization are concerned mainly with elucidation of the nature of the separating phases and the rate of their separation. Undeservedly little attention is devoted to energetic activation factors of crystallization. The crystallization rate of silicates is directly connected with the viscosity, as is indicated by Leont'eva's empirical formula [1]:

$$v = \frac{k_1}{\eta} + k_2 \log \eta.$$

Analysis of the formula shows that if the viscosity, expressed, after Myuller [2], in terms of the free activation energy is substituted into it, an exponential relationship is found to exist between the crystallization rate and the activation energy. The logarithmic constant contains quantities, little dependent on the temperature, which determine the entropy and preexponential factors of the crystallization rate:

$$\log v = \log\left[k_1 \cdot k_2\left(\frac{E_\eta}{RT} + \frac{S_\eta}{R}\right)\right] - \frac{S_\eta}{R} - \frac{E_\eta}{RT}.$$

This relationship is quite understandable, since the rate of growth of the crystalline phase is directly determined by the energy barrier of activated switching of covalent $Si-O-Si$ bonds, the magnitude of which is the same in viscous flow as in the crystallization process.

In our opinion, two activation quantities must be taken into consideration in investigations of the crystallization process: the activation energy of covalent bond switching (the activation energy of covalent diffusion of the basic structural unit) and the activation energy of diffusion of the modifying unit (the ion). The latter is always less than the former; therefore, with the corresponding preexponential factors taken into account, we can say that the composition of the crystallizing phase is determined not only by the minimum free energy but also by the ratio of the rates of covalent and ionic diffusion. Nuclei of both stable and metastable crystalline phases may form and separate out in glasses, depending on the cooling rate and temperature conditions; this was noted by Myuller [3] and Florinskaya [4].

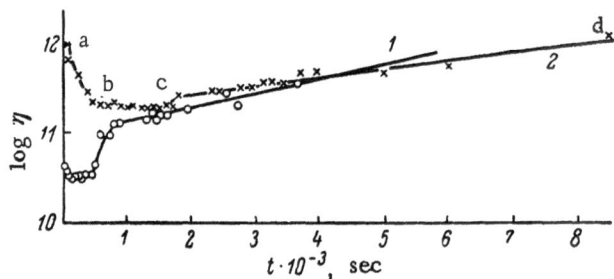

Fig. 1. Variations of viscosity with time during heat treatment of submicrocrystallizing glass. 1) 754°; 2) 784°.

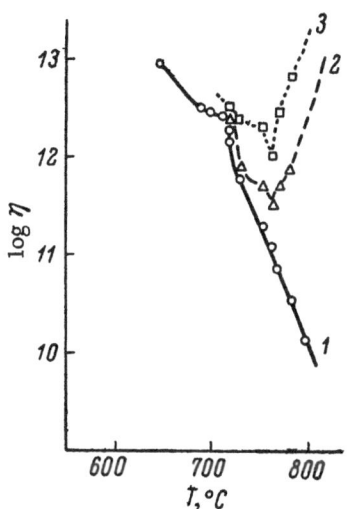

Fig. 2. Effect of temperature on the viscosity of the original glass melt and of the crystallization products. 1) Viscosity of the melt during the induction period; 2) viscosity 3000 sec after the start of crystallization; 3) viscosity 7000 sec after the start of crystallization.

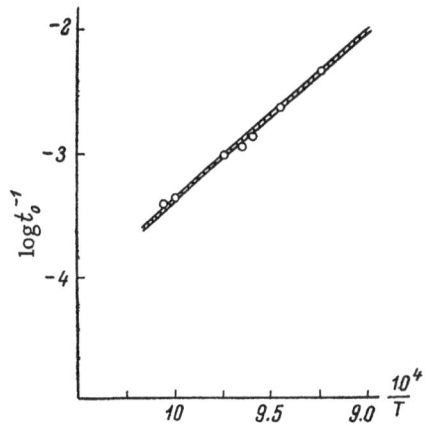

Fig. 3. Effect of temperature on the induction period. E = 62 ± 3 kcal/mole.

If the metastable phase which separates out is dispersed uniformly in the glass, and the glass is heated at temperatures which according to Myuller [5] ensure excitation of valence vibrations of the covalent bonds in the ordered framework, these nuclei begin to dissipate from the surface and the structure becomes stable. The size of the crystals formed is determined by the temperature and energy conditions. The rate of the process is always limited by covalent bond switching, which has a high energy demand.

We investigated catalyzed submicrocrystallization by a viscosimetric method, since the viscosity is a property which is intimately connected with the nature of the covalent glass framework.

The viscosity of glasses can be measured rapidly and with high precision by determination of the rate of penetration of a rod into the softened glass [6], even if the viscosity varies with time.

In our experiments specimens of glass 13, which belongs to the lithium aluminosilicate system with additions of titanium dioxide, were placed in a furnace heated to a definite temperature and the viscosity determinations were started immediately. Figure 1 illustrates the variations of viscosity during an experiment. The regions ab of the curves correspond to the fall of viscosity as the specimen is heated through; the magnitude of the resultant deformation corresponds to the elastic deformation of the initial measurements, and therefore this effect was excluded during the subsequent course of the experiment. After the specimen has reached the required temperature, the viscosity remains constant for some time (the induction period, region bc) and then increases, rapidly at first and then linearly with time (region cd). The duration of the induction period and the rate of the subsequent viscosity increase both depend considerably on the temperature.

The viscosity during the induction period characterizes the viscosity of the original melt, which is subsequently transformed into a microcrystalline product. These constant viscosity values are plotted against the temperature in Fig. 2, which also shows the viscosity increase after the induction period. The glass softens very little in the 640-715° range, but above 715° the viscosity drops sharply.

The method allowed us to determine with sufficient accuracy the length of the induction period and the rate of the subsequent viscosity increase only in the 720-800° range. Above 800° the microcrystallization of glass is accompanied by a thermal effect which distorts the results.

The abrupt drop of viscosity at 715°, which is considerably higher than the softening temperature of the glass (640°), is a sign of sudden excitation of the rigid, previously frozen covalent bonds. The structure of these bonds is stronger and more ordered than that of the principal bonds of the covalent glass framework, and therefore their excitation over a narrow temperature range is very reminiscent of fusion. Apparently at 715° dissipation of the metastable nuclei begins, leading to intensification of submicrocrystallization.

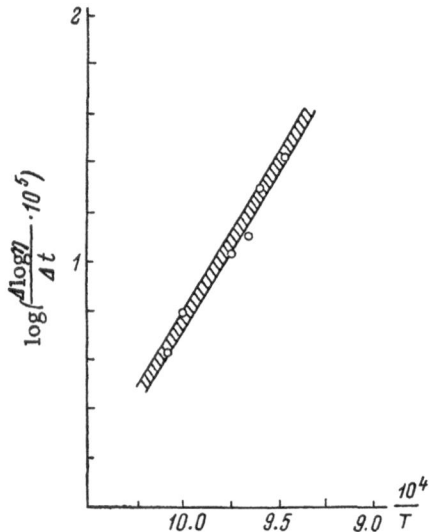

Fig. 4. Effect of temperature on the rate of volume crystallization. E = 60 ± 5 kcal/mole.

Figures 3 and 4 show that the induction period and the rate of catalyzed crystallization are exponential functions of the temperature. The activation energies were determined from the temperature coefficients of the linear plots. The processes taking place during the induction period have an activation energy of 60 ± 3 kcal/mole, and the subsequent crystallization processes, 60 ± 5 kcal/mole. The free energy of activation of viscous flow, calculated for the working viscosity range of $10^{10.5}$-$10^{11.5}$ poises, is 61 ± 3 kcal/mole (experimental and theoretical error). The similarity of the activation energies of viscous flow, of the processes taking place during the induction period, and of the crystallization rate indicates that the processes are all of the same type. It is quite evident that the step requiring the most energy in all three processes is switching of the covalent bonds or, if we consider the phenomenological aspect, covalent transformational diffusion, the nature of which must be defined separately in the case of catalyzed crystallization.

According to the results obtained by Bobovich in a study of the Raman spectra of the original glass 13 and its crystallization products, the coordination number of titanium changes from 4 to 6 during heat treatment. This change of the coordination number in the solid phase is due to a radical change in the bond hybridization and demands the formation of a new symmetrical oxygen coordination. Interaction of the polar-covalent lithium oxide and glass-forming aluminum oxide during the melting of the glass leads to appearance in the glass network of structural aluminum units with covalent coordination of four or six rather than three oxygen bridge bonds. The temperature conditions here appear to favor the coexistence of titanium in the lower coordination and of aluminum in excess coordination above the normal valence (we are considering the case of covalent and not geometrical coordination).

Cooling of the glass is accompanied by formation of nuclei containing titanium mainly in fourfold coordination. Above 715° excitation of titanium−oxygen covalent bonds takes place, and the titanium structural units tend to restore the sixfold coordination which is the equilibrium coordination at that temperature, capturing excess oxygen atoms from aluminum. This should be accompanied by diffusion of oxygen bridge bonds from aluminum to titanium, since titanium binds oxygen by principal hybrid valence forces and the weaker oxygen−aluminum donor − acceptor interaction cannot prevent this energetically.

It is most probable that transfer of oxygen bridge bonds from aluminum to titanium is effected not by displacement of the aluminum structural units as a whole but by transfer of these bridges from one atom to another by a covalent transformational diffusion mechanism; the composition of the glassy matrix phase alters and it becomes capable of crystallization. A necessary condition for this mechanism is the formation of a sufficient concentration of aluminum groups capable of transferring oxygen around the titanium centers. The principal feature of the induction period is the preparation of this environment.

Thus, viscosimetric investigations at high temperatures of the kinetics of catalyzed submicrocrystallization have shown that the process consists of two stages: an induction period and the crystallization process itself.

The energy characteristics indicate that the processes taking place during the induction period, like the processes of crystallization itself, are limited by switching of covalent bonds in the glass framework and not by ionic diffusion processes.

A mechanism is proposed for catalyzed crystallization, in which an important part is played by covalent transformational diffusion.

### LITERATURE CITED

1. A. A. Leontjewa, Acta physicochim. URSS 14(2):245, 1941.
2. R. L. Myuller, Zhur. Priklad. Khim. 28:363,1077, 1955.

3. R. L. Myuller, Zhur. Fiz. Khim. 30:1146, 1956.

4. V. A. Florinskaya and R. S. Pechenkina, "Investigation of crystallization products of glasses in the system $Na_2O$-$SiO_2$ by infrared spectroscopy," in The Structure of Glass, Vol. 2, Consultants Bureau, New York, 1960, p. 135; V. A. Florinskaya, "Infrared reflection spectra of sodium silicate glasses and their relationship to structure," ibid.,p. 154; Zhur. Strukt. Khim. 2(2):183, 1961.

5. R. L. Myuller, Zhur. Fiz. Khim. 28:1954, 2170, 1954.

6. S. V. Nemilov and G. T. Petrovskii, Zhur. Priklad. Khim. 36(1), 1963.

# THEORETICAL INVESTIGATION OF THE POSSIBILITY
# OF USING DIFFERENTIAL THERMAL ANALYSIS
# FOR QUANTITATIVE STUDIES OF CRYSTALLIZATION

## A. G. Vlasov and A. I. Sherstyuk

Differential thermal analysis (DTA) in the form used up to now does not give sufficiently accurate quantitative characteristics of the process studied; in particular, it is not possible to determine the amount of reacted phase, which is especially important in studies of the nature and dynamics of the crystallization process. In the case under consideration, crystallization proceeds over a fairly wide temperature range, which complicates the problem considerably.

The equation giving the temperature change at any point in the specimen can be written in the following general form:

$$\frac{\partial T}{\partial t} = b_1 \Delta T + \frac{1}{\rho_1 c_1} F(t, \ T),$$

(1)

where $b = \lambda_1 / \rho_1 c_1$; $\lambda_1$ is the thermal conductivity of the substance; $\rho_1$ is the density; $c_1$ is the specific heat; $\Delta = \frac{\partial^2}{\partial x^2} + \frac{\partial^2}{\partial y^2} + \frac{\partial^2}{\partial z^2}$; $F(t,T)$ is the quantity of heat produced per second in unit volume, which characterizes the crystallization rate.

For an inert substance we have

$$\frac{\partial T_{in}}{\partial t} = b_2 \Delta T_{in}$$

(2)

and the material constants of the substances are chosen to be equal ($b_1 = b_2 = b$). For convenient analysis, spherical specimens and reference standards are postulated, and T is assumed to be a function of the specimen radius only.

The experimental unit provides for linear heating of the surface of a metal block containing the test specimen; therefore the boundary conditions are $T|_{r=R} = \varepsilon t$, where R is the radius of the sphere and $\varepsilon$ is the rate of heating, maintained constant during the experiment.

The active and inert substances are heated under identical conditions; therefore we can immediately write the equation for the temperature difference observed experimentally:

$$\frac{\partial \theta}{\partial t} = b \left( \frac{\partial^2 \theta}{\partial r^2} + \frac{2}{r} \frac{\partial \theta}{\partial r} \right) + \frac{1}{\rho c} F(t, \ T),$$

(3)

$$\theta = T - T_{in}$$

with the boundary condition $\theta = 0$ with r = R at any time instant.

We are really faced with solving the reverse problem of the heat conductivity theory; i.e., we have to find the function $F(T, t)$ from the temperature change, which is much more difficult than the solution of the direct problem. Moreover, the function $F(T, t)$ depends not only on the temperature at a given instant but also on the relative amount of the phase which has not yet crystallized at that instant.

We start with the solution of the problem of the temperature change within the control substance or, analogously, within the test specimen before the start of the heat effect. We take $F(T, t) = 0$ in Eq. (1).

From the time instant $t_{st} = R^2/\pi^2 b$, a quasi-stationary state is established in the specimen, when the temperature at any point in the specimen varies linearly with the time at the rate $\varepsilon$, in accordance with the equation

$$T(r, t) = \varepsilon \left( t - \frac{R^2 - r^2}{6b} \right). \tag{4}$$

Under our conditions (R = 5 mm, b = $5.9 \cdot 10^{-2}$ cm$^2$/sec), $t_{st}$ = 4.4 sec. In practice, the heat effect usually begins considerably later than $t_{st}$, and therefore by that time Eq. (4) can be assumed to be precisely valid. Analogous equations may be obtained if several layers of different substances are present.

The propagation velocity of the temperature front in the specimen before the start of the reaction is

$$\frac{dr}{dt} = -\frac{3b}{r} \bigg|_{T=\text{const}}. \tag{5}$$

The time of propagation of the front to the point r = 0 is

$$t_0 = \frac{R^2}{6b}. \tag{6}$$

For a spherical glass specimen, for which curves were obtained at R = 5 mm, $t_0$ = 7 sec, and a heating rate of 30°/min, the temperature difference $\Delta t$ between the surface and the center is approximately 3.5°, which is a little greater than the error of temperature measurement. Therefore our assumption that the temperature of the reacting substance is the same throughout is valid in the first approximation. Therefore the temperature of the start of the process can be determined by means of a thermocouple in the center of the inert substance.

As already stated, solution of the heat conductivity equation if nonstationary heat sources are present within the specimen presents considerable mathematical difficulties. However, we can investigate an approximate model for explaining the characteristic features in the course of the curves.

Let us consider the heating of the specimen as a single entity, with the same temperature established at every point within it; this corresponds to a sufficiently small specimen and high temperature conductivity. As was shown above, this assumption is fully justified for our apparatus. The heat flow received by the specimen with this assumption is proportional to the difference between the heater temperature and the average temperature of the specimen. In this case,

$$\frac{dT}{dt} = K(T_{\text{heater}} - T) + \frac{1}{m_0 c} \frac{dq}{dt}, \tag{7}$$

where $dq/dt$ is the total quantity of heat liberated per unit time during crystallization; $K = \lambda'/m_0 c$, where $\lambda'$ is a certain quantity, determined experimentally, proportional to the thermal conductivity; $m_0$ is the mass of the substance; c is the specific heat.

If all points in the specimen reach the temperature $T_r$ of the start of the reaction simultaneously, the expression for the quantity of heat liberated is

$$q = \sigma m_r = \sigma m_0 (1 - e^{-\alpha t}), \tag{8}$$

where $\alpha$ is the reaction rate (quantity of substance converted in unit time) and $\sigma$ is the heat of reaction, characterizing the relative amount of crystallized phase.

To compare curves for the same substance heated at different rates, we assume $\alpha$ in Eq. (8) to be equal to the average reaction rate. The latter depends on the heating rate only, as the temperature at the start of the reaction alters when the heating rate is changed: the temperature of the start (and hence $\alpha$) increases with increase of the heating rate.

With the above assumptions, the characteristic courses of the curves derived from Eq. (7) coincide with

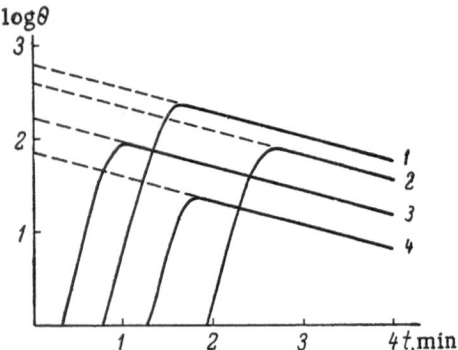

Typical plots of the logarithm of the temperature difference against time, obtained from thermograms. 1) Heating rate 19° per min, $\sigma = 100\%$; 2) heating rate 28°/min, $\sigma = 96\%$; 3) heating rate 21°/min, $\sigma = 70\%$; 4) heating rate 29°/min, $\sigma = 75\%$.

those of experimental differential thermograms. This suggests possible ways for approximate analysis of such thermograms. For the temperature difference we have

$$\frac{d\theta}{dt} + K\theta = \frac{1}{m_0 c}\frac{dq}{dt} \qquad (9)$$

when $\theta = T - T_{in}$.

Substitution gives

$$\theta = \frac{\delta}{\alpha - K}(e^{-Kt} - e^{-\alpha t}), \qquad (10)$$

where $\delta = \delta \alpha / c$ under the condition that $\theta = 0$ at the initial instant.

We find from (10) that maximum $\theta$ is reached at time

$$t_{max} = \frac{\ln\frac{\alpha}{K}}{\alpha - K} \qquad (11)$$

Since the reaction rate is usually higher than the rate of heat transfer, we can determine approximately the time of the end of the reaction. We write (10) in the form

$$\theta = \frac{\delta}{\alpha - K}e^{-Kt}(1 - e^{-(\alpha - K)t}). \qquad (12)$$

At time $t_f$, which we shall term the time of the end of the reaction, the second term in parentheses in Eq. (12) becomes negligibly small.

Pure loss of heat begins at the instant $t_f$. This instant can be determined experimentally with fair accuracy by plotting $\ln\theta$ against t and finding the point at which the plot becomes linear. When $t > t_f$,

$$\ln\theta(t) = \ln\theta_0 - Kt. \qquad (13)$$

In this region plots of $\ln\theta$ against t are linear, and the slope is equal to 2.3 K (see figure). Extrapolation of these lines to $t = 0$ gives

$$\ln\theta_0 = \ln\frac{\sigma\alpha}{c(\alpha - K)}, \qquad (14)$$

from which, if K is known, it is easy to find the relation between $\sigma$ and $\alpha$. The specific heat of the test substance is usually known.

The time $t_{max}$ of the maximum deviation of the curves can also be easily found from the graphs. Hence we can find the relation between $\alpha$ and K by graphical solution of the transcendental equation

$$\alpha = Ke^{t_{max}(\alpha - K)}. \qquad (15)$$

Thus, nearly all the thermal characteristics of the process in question can be determined from the plot of $\theta$ against t.

An especially important quantity is $\sigma$, the specific heat of reaction, which is proportional to the amount of uncrystallized phase. Comparison of the values of $\sigma$ found for different substances gives the degree of crystallization in relation to the preliminary treatment.

It should be noted that the dependence of $t_{max}$ on $\alpha$ weakens with increase of the rate of heating, and therefore the precision in the determination of $\alpha$ from the time of maximum deviation decreases. In practice the precision in determinations of $\alpha$ and $\sigma$ is $\pm 10\%$.

It is evident that the calculation can be improved considerably. The problem involves more precise solution of the heat conductivity equation with given boundary conditions and nonstationary heat sources distributed continuously throughout the volume of the specimen; the function $F(t, T)$ of the sources will have certain free parameters which must be determined experimentally.

# STUDY OF CATALYZED CRYSTALLIZATION
# BY DIFFERENTIAL THERMAL ANALYSIS

## N. A. Tudorovskaya and A. I. Sherstyuk

Crystallization of glasses is accompanied by liberation of heat, which is recorded in differential thermal analysis in the form of exothermic peaks on the curve representing the readings of a differential thermocouple.

In our investigation, the differential curve was recorded with the ÉPP-09 instrument with a $-1$ to $+3$ mV scale. The band speed in the instrument was 0.4 mm/sec, and the time for the carriage to traverse the whole scale was 2 sec. Under these conditions, the details in the differential curves can be easily distinguished even with the most rapid heat effects (lasting up to 1-2 min).

We studied the heat effects in crystallization of glasses in the $Li_2O$-$Al_2O_3$-$SiO_2$ system with the compositions of petalite, spodumene, and compositions close to the latter (mainly glass 13) with additions of not more than $10\%$ by weight of $TiO_2$ and certain other oxides. The glasses were studied both in the initial state and after various heat treatments. The glasses were made and heat-treated by E. V. Podushko. The glasses studied gave thermograms of the form shown in Fig. 1.

At the same rate of furnace heating, the thermograms of different glasses differ in $T_i$, the temperature of the start of the heat effect (at which the differential curve begins to diverge from the zero line), in the maximum temperature deviation $\Delta T_M$, and in the time $t_M$ during which the temperature deviation reaches a maximum; $t_M$ is read off from the start of the heat effect. The values of $T_i$, $\Delta T_M$, and $t_M$ were taken as the characteristics of the thermograms [1].

Glasses with the composition of spodumene and petalite each have one exothermic effect in the temperature range up to 1100°. The heat effect of the first glass begins at 830° and the greatest temperature deviation reaches 95° in about 2 min. The heat effect of the second glass begins at a higher temperature (910°) and the greatest temperature deviation reaches 70° in about 1.5 min. A number of glasses close to spodumene in composition exhibited two exothermic effects: one begins at 790-830°, depending on the composition, and the maximum temperature deviation reaches 60-70° in 1-2 min; the second effect begins at considerably higher temperatures, 1010-1040°, and the temperature deviation reaches only 5-10° in 1-2 min.

X-ray diffraction studies by A. G. Alekseev and mineralogical analysis by N. E. Kind showed that glass corresponding to spodumene in composition crystallizes completely and only high-temperature spodumene separates out in it. In glasses close to spodumene in composition high-temperature spodumene is again the principal phase to separate out.

It may be concluded that in glasses close to spodumene in composition the first heat effect is due to crystallization of high-temperature spodumene. The existence of a second effect at a high temperature indicates that a second, more refractory phase also crystallizes out in these glasses.

Thermograms were obtained at heating rates of 28 and 20 degrees/min for two glasses, one with $5\%$ $TiO_2$, one with the composition of spodumene. The percentage of the crystallized phase was calculated from these thermograms by the method described by Vlasov and Sherstyuk [1]. In the spodumene glass the amount of crystals which separated out was 96 and $100\%$, respectively; the values for the glass with $5\%$ $TiO_2$ were 73 and $75\%$. The composition of glass 13 with $5\%$ $TiO_2$ indicates that about $65\%$ spodumene should be obtained, which agrees to within $10\%$ with the calculated values. A precision of $10\%$ is obtained in calculations from thermo-

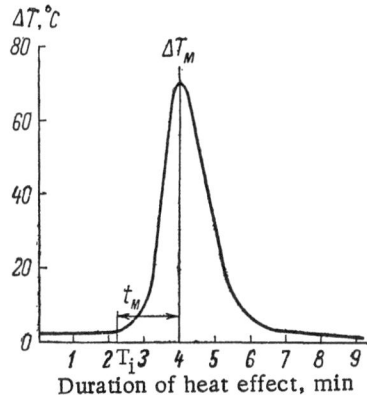

Fig. 1. Form of thermographic record.

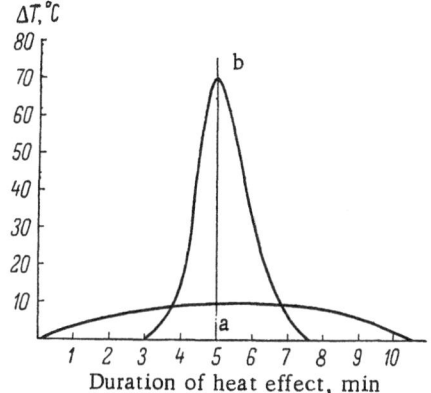

Fig. 2. Thermograms of glasses. a) Surface crystallization; b) volume crystallization.

TABLE 1

| Heating rate, degrees/min | $T_i$, °C | $\Delta T_M$, °C | $t_M$, sec | Heating rate, degrees/min | $T_i$, °C | $\Delta T_M$, °C | $t_M$, sec |
|---|---|---|---|---|---|---|---|
| 29 | 834 | 70 | 67 | 8 | 775 | 44 | 166 |
| 25 | 840 | 74 | 80 | 6 | 787 | 36 | 190 |
| 24 | 802 | 71 | 76 | 5.5 | 780 | 34 | 177 |
| 21 | 805 | 72 | 65 | 4.8 | 777 | 17 | 192 |
| 15 | 765 | 71 | 150 | 2 | 759 | 4 | 670 |

TABLE 2

| $TiO_2$ content, % by wt. | $T_i$, °C | $\Delta T_M$, °C | $t_M$, sec | $TiO_2$ content, % by wt. | $T_i$, °C | $\Delta T_M$, °C | $t_M$, sec |
|---|---|---|---|---|---|---|---|
| 0 (Glass 13) | 873 | 7 | 420 | 6 | 790 | 51 | 120 |
| 1 | 865 | 11 | 375 | 7 | 785 | 44 | 180 |
| 2 | 863 | 10 | 305 | 8 | 789 | 46 | 135 |
| 3 | 830 | 22 | 285 | 9 | 787 | 32 | 160 |
| 4 | 785 | 55 | 123 | 10 | 766 | 8 | 375 |
| 5 | 790 | 45 | 115 | 11 | 760 | 6 | 405 |

grams under our experimental conditions. We can say that glass 13 with 5% $TiO_2$ contains 70% of the main phase (spodumene) while the content of the second (high-temperature) phase does not exceed 30%.

For glass 13 with 5% $TiO_2$ thermograms were obtained at heating rates from 2 to 29 degrees/min (Table 1).

Although the results are somewhat scattered, it can be seen that as the heating rate is decreased, the heat effect begins at lower temperatures, proceeds more slowly, and the maximum temperature deviation gradually diminishes. However, at heating rates in excess of 20 degrees/min, both the duration of the effect and the maximum temperature deviation are virtually independent of the heating rate.

As the temperature is lowered a point is reached where the rate of heat evolution by the thermal effect becomes equal to the rate of accompanying heat loss, so that the readings of the differential thermocouple remain zero although crystallization may yet occur. Therefore we used the following procedure for investigations at such temperatures: the glass was first held for a certain time at the given temperature, and the heat effect was then determined in the usual manner. It was found that as the glass is held at the given temperature, the residual heat effect diminishes and commences at a lower temperature. After a certain holding time the effect disappears. This shows that the glass is completely crystallized after that time.

At 740° the heat effect disappears after 1 h; at 700° after 50 h; at 690° an appreciable change of the heat effect begins after 35 h, and it disappears completely after 98 h. At 660° a change of the heat effect begins to appear after 100 h, while at 630 and 600° there were no changes of the heat effect even after 600 h.

The time during which the heat effect disappears entirely at a given holding temperature determines the crystallization rate of the glass at that temperature.

The effect of additions of 1 to 11% of titanium dioxide by weight on the heat effect was studied in the case of glass 13. Thermograms were obtained at 7 degrees/min (Table 2).

Table 2 shows that the heat effect in the original glass and in glasses containing up to 3% $TiO_2$ begins at a high temperature and its maximum value does not exceed 22°. Starting with the glass containing 4% $TiO_2$, the temperature at which the effect begins drops appreciably, the effect is more rapid, and $\Delta T_M$ reaches 50°. Finally, in glasses with 10 and 11% $TiO_2$, the heat effect again diminishes somewhat, the temperature at which it begins drops still lower, and the effect becomes slower. However, the heat effect given by quenched glasses with 10 and 11% $TiO_2$ was of the same order as for glasses containing 4 to 9% $TiO_2$. This shows that the crystallization rate is higher in glasses with 10 and 11% $TiO_2$ than in the other glasses, and therefore these glass specimens crystallized to a greater extent during normal cooling and annealing.

The first four glasses (Table 2) give thermograms of type a (Fig. 2). In this case, crystallization occurs at the surface only. The other glasses undergo volume crystallization; the thermograms are of the form b (Fig. 2).

Thus it is possible to determine from the number and nature of the exothermic effects on the thermogram how much crystalline material separates out in a glass and which of the crystalline phases are predominant. Data on the temperature range of crystallization and on the relative change of the rate of crystal growth with temperature can be obtained from thermograms obtained for the same glass at different heating rates. The effect of heat treatment on glass crystallization can be assessed from changes in the heat effect as a result of such treatment. The relative amount of crystallized phase can be calculated from the thermograms by the calculation procedure of A. G. Vlasov and A. I. Sherstyuk.

LITERATURE CITED

1.    A. G. Vlasov and A. I. Sherstyuk, this collection, p.122.

# INVESTIGATION OF CHANGES IN THE PHYSICAL PROPERTIES OF GLASS IN THE $Li_2O-Al_2O_3-SiO_2$ SYSTEM DURING CRYSTALLIZATION

## I. D. Tykachinskii and E. S. Sorkin

In the formation of a glassceramic, all the physical properties of the original material change sharply. The course of the crystallization can be assessed from the nature of these property changes, reflecting structural changes in the glass.

The purpose of our investigation was to elucidate the variations of physical properties during crystallization of glass in the $Li_2O-Al_2O_3-SiO_2$ system.

We studied variations of density, refractive index $n_D$, the coefficient of linear thermal expansion, the coefficient of total light transmission, and isothermal static compressive deformation of the specimens in relation to the holding time of the glasses at constant temperature.

The specimens were heat-treated in the furnace of the DKV dilatometer [1] in the temperature range above and below the softening temperature of the original glass. Automatic control was used for raising the temperature to the required level and for maintaining it during the experiments [2].

Specimens cut from the same piece of annealed glass 1, in which $TiO_2$ was used as the catalyst, were kept for various times at constant temperatures of 690, 700, and 710°.* All the specimens were heated up from room temperature at the same rate, 1.5 degrees/min.

After heat treatment in the furnace at constant temperature, the specimens were sharply chilled in air at room temperature. Density of the chilled specimens was measured by the sink-float method [3,4] to a precision of $\pm 1 \cdot 10^{-4}$ g/cm³.

The density of the original glass was $\rho_{20°} = 2.4366$ g/cm³. At the initial stages of heat treatment the density is due to the fact that the glass, which is heated up from room temperature to the experimental temperature at a certain finite rate, passes into the equilibrium state during the initial treatment stage [5].

The density remains constant for some time with increase of the treatment time and only begins to rise sharply at a definite time instant, which depends on the treatment temperature (Fig. 1). For example, the density begins to increase after 3.5 h at 690°, after 2 h at 700°, and after 1 h 15 min at 710°.

Figure 1 shows that at any treatment temperature in the range studied the density values tend asymptotically to the same final value.

The standard IRF-22 refractometer was used for

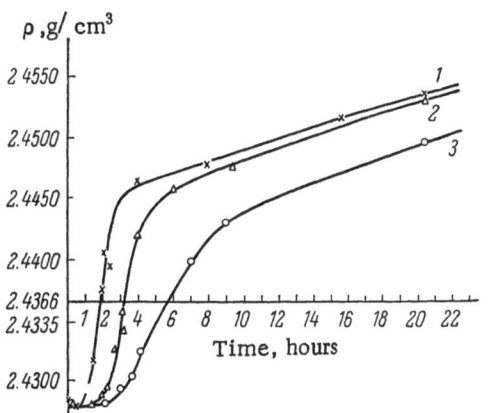

Fig. 1. Density variations during crystallization of glass 1 at different temperatures. 1) 710°; 2) 700°; 3) 690°.

*The transformation temperature $T_g$ of the original glass was 630°.

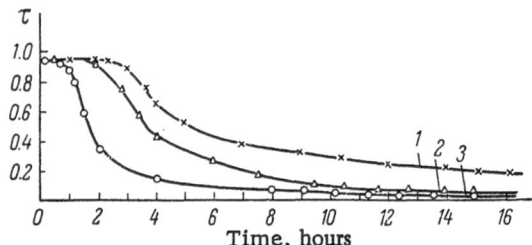

Fig. 2. Variations of the coefficient of total light transmission of the original glass 1 with the treatment time at various temperatures. 1) 690°; 2) 700°; 3) 710°.

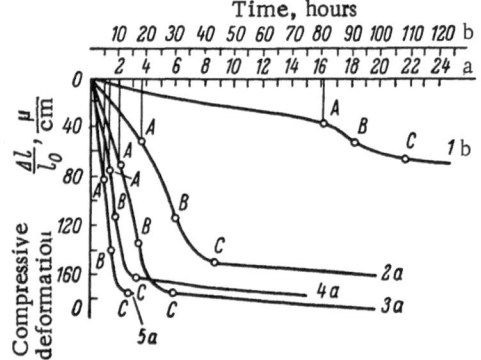

Fig. 3. Deformation under isothermal static compression of the original glass 1 heat-treated at various temperatures. 1b) 650°; 2a) 690°; 3a) 700°; 4a) 710°; 5a) 720°.

determination of the refractive indices of chilled glass 1 specimens after various heat treatments. Transmitted light was used for transparent specimens and reflected light for turbid glasses. The refractive index of the original annealed glass 1 was $n_D{}^{20}$ = 1.5300. At the first stage of heat treatment, $n_D$ remains unchanged. The variations of the refractive index $n_D$ of the original glass 1 during crystallization are analogous to the density variations (Fig. 4,4). At any treatment temperature in the range studied the values of $n_D$ tend asymptotically to the same final value.

The total light transmission of glass 1 specimens after various heat treatment was measured by means of a luxmeter with an F-102 photocell and a Yu-16 galvanometer. The transmission coefficient of the original glass 1 was $\tau_{or}$ = 0.934.

The coefficient of total light transmission remains unchanged during the initial treatment stage. The transmission coefficient begins to drop sharply only at a definite time, depending on the treatment temperature (Fig. 2).

Variations of the coefficient of total light transmission of the original glass 1 during crystallization lead to the conclusion that at any treatment temperature the transmission coefficient tends asymptotically to zero.

Specimens intended for measurement of the coefficient of linear thermal expansion were heat-treated for various times at 710°. After treatment at this temperature, the specimens were annealed under the same conditions. The measurements of the coefficients of linear thermal expansion were performed with the DKV dilatometer in the 20-400° range.

The coefficient of linear thermal expansion of the annealed original glass 1 in the 20-400° range was $\alpha_{or}$ = 58.1 · $10^{-7}$ degree$^{-1}$.

At the initial treatment stage at 710° heat treatment for 1 h does not result in any appreciable change of the coefficient of linear thermal expansion in the 20-400° range. The value of $\alpha$ begins to decrease only 1 h 15 min after the start of treatment; during the second and third hours of treatment, $\alpha$ falls from 58.1 · $10^{-7}$ degree$^{-1}$ to zero (Fig. 4, curve 2).

Variations of the physical properties of the original glass during heat treatment can be utilized for creation of a method for continuous control of the crystallization process. However, continuous measurement of physical constants of materials at high temperatures is very complicated. It is much simpler to achieve control of the crystallization process by continuous measurement of the compressive deformation of the original glass under constant load at constant temperature.

It is known [6] that the compressibility of silicate glass at constant temperature in the range of $10^8$-$10^{13}$ poises can be represented by the expression

$$\frac{1}{l}\Delta l = \left(\frac{1}{l}\Delta l\right)_\eta + \left(\frac{1}{l}\Delta l\right)_\rho + \left(\frac{1}{l}\Delta l\right)_\sigma, \tag{1}$$

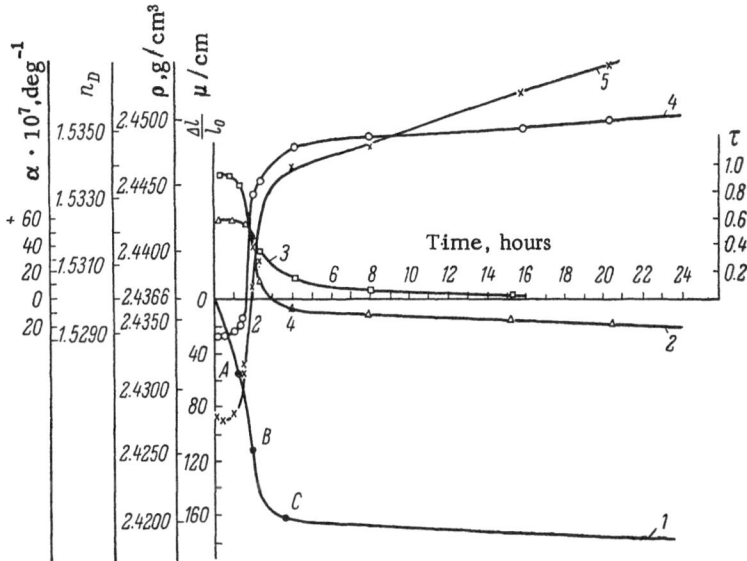

Fig. 4. Compressive deformation and variation of the physical properties of glass 1 in relation to the time of treatment at 710°. 1) Compressive deformation $\Delta l / l_0$; 2) variation of the coefficient of linear thermal expansion $\alpha$; 3) variation of the coefficient of total light transmission $\tau$; 4) variation of the refractive index $n_D$; 5) variation of the density $\rho$.

where $\frac{1}{l} \Delta l$ is the relative change in the length of the specimen in unit time; $\left(\frac{1}{l} \Delta l\right)_\eta$ is the relative change in the length of the specimen in unit time due to viscous flow; $\left(\frac{1}{l} \Delta l\right)_\rho$ is the relative change in the length of the specimen in unit time due to density changes; $\left(\frac{1}{l} \Delta l\right)_\sigma$ is the relative change in the length of the specimen in unit time due to surface tension.

Since the compression due to surface tension is relatively small, the third term in the right side of Eq. (1) may be omitted. The other two terms determine differences in the character of the compressibility curve. The relation between the density change and the resultant change in the length of the specimen is given by the expression

$$\frac{d\rho}{\rho} = -3 \frac{dl}{l}. \tag{2}$$

The deformation of the original glass specimens under isothermal static compression was measured with the DKV dilatometer at a load of 2.6 g/mm². The measurements were performed over a wide temperature range below and above the dilatometric softening point of the original glass.

Figure 3 shows the static compressive deformation curves of glass 1 treated at various temperatures.† The curves are all of the same character for the entire temperature range. The inflection points A and B and the point C are clearly seen on every curve.

For correlation between the form of the deformation curves and the variations of the physical constants, Fig. 4 represents, on the same scale, variations of the physical properties and static compressive deformation with the time of heat treatment at 710°.

It follows from Fig. 4 that at the first stage of heat treatment, corresponding to the region 0A of the deformation curve, there is a slight change of density. The refractive index $n_D$, the coefficient of linear thermal

_____

† The dilatometric softening point of the original glass 1 is 700°.

expansion, and the light transmission remain constant within the limits of experimental error. The slight convexity of the region OA of the deformation curve toward the negative ordinates is due mainly to the fact that the viscosity of the glassy component increases during treatment at constant temperature because the system passes into the equilibrium state [7].

The first inflection point A on the deformation curve corresponds to the instant during heat treatment when the physical properties of the material begin to alter rapidly. The second inflection point B corresponds to the instant when the rate of density increase falls. The point C on the deformation curve corresponds to the treatment time when viscous flow of the material virtually ceases and the subsequent deformation is due mainly to density changes (the deformation curve becomes almost linear).

If we plot deformation curves and variations of the physical properties for other temperatures on the same time scale, we find that the curves representing isothermal compressive static deformation reflect variations in the physical properties of the original glass during crystallization.

Investigations of the compressibility of original glasses in the $Li_2O$-$Al_2O_3$-$SiO_2$ system with other mineralizers showed that the deformation curves are analogous in character.

LITERATURE CITED

1. E. S. Sorkin, Steklo i Keramika No. 10, 40, 1958.
2. E. S. Sorkin, Byull. "Steklo," No. 1, 1962.
3. R. D. Duff, J. Am. Ceram. Soc. 30(1):12, 1947.
4. A. P. Galushkin, Legkaya Promyshlennost' 5:39, 1958.
5. A. I. Stozharov, Trudy GOI 4(39):1, 1928.
6. Watanabe, Chida, and Takizawa, J. Soc. Glass Techn. 202:295, 1957.
7. H. R. Zillie, J. Am. Ceram. Soc. 16:619, 1933.

# VARIATIONS OF THE PHYSICAL PROPERTIES DURING HEAT
# TREATMENT OF GLASSES IN THE $Li_2O - Al_2O_3 - SiO_2$ SYSTEM
# MINERALIZED BY TITANIUM DIOXIDE

## I. M. Buzhinskii, E. I. Sabaeva, and A. N. Khomyakov

Glasses of various compositions in the spodumene field were heat-treated in the temperature range from 550 to 900°; the treatment times at these temperatures were from 1 to 500 h. The following properties were measured: the optical constants $n_d$ ($n_F - n_c$) with the IRF-25 instrument, the light transmission $\tau_\lambda$ with the SF-4 spectrophotometer, the coefficient of linear expansion $\alpha$ and the elongation $\Delta l$ with the GOI interferometer, with a precision of $\pm 0.5 \cdot 10^{-7}$ degree$^{-1}$ and 0.05 $\mu$/cm, respectively, and density changes $\Delta d$ of 30-mm cubes to a precision of 1 $\mu$ with the aid of an optical meter. In addition, the following values were determined for some of the glasses: volume resistivity $\rho$, tangent of the dielectric loss angle $\tan \delta$, the glass transformation temperature $T_g$, the softening temperature $T_f$, and chemical durability.

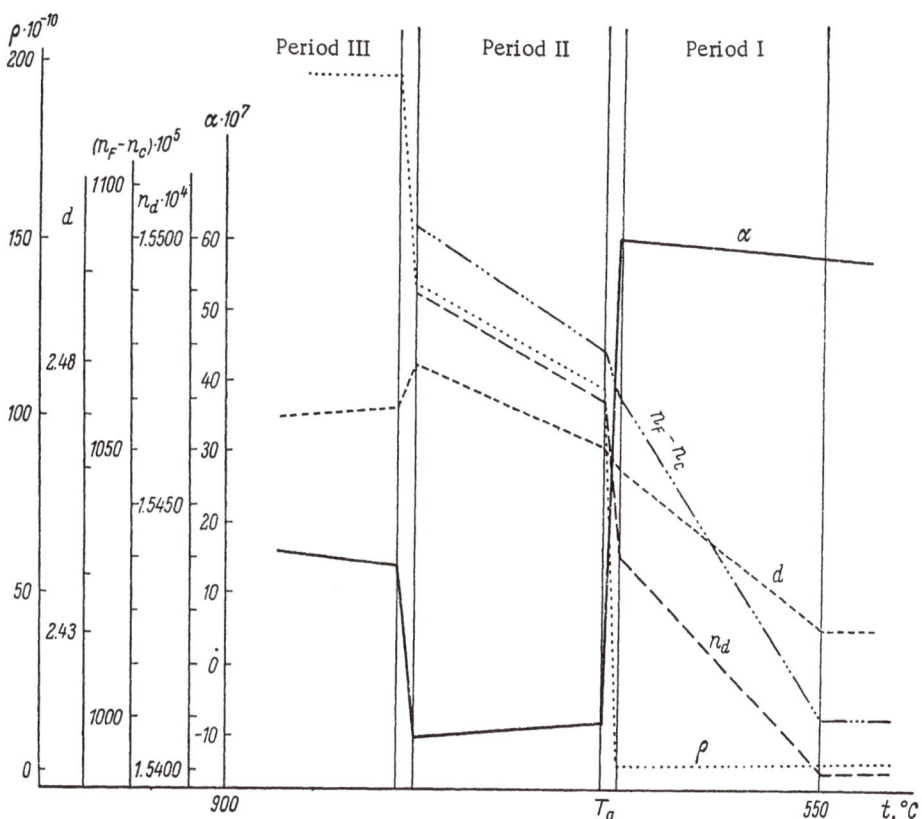

Fig. 1. Variations of glass properties during heat treatment.

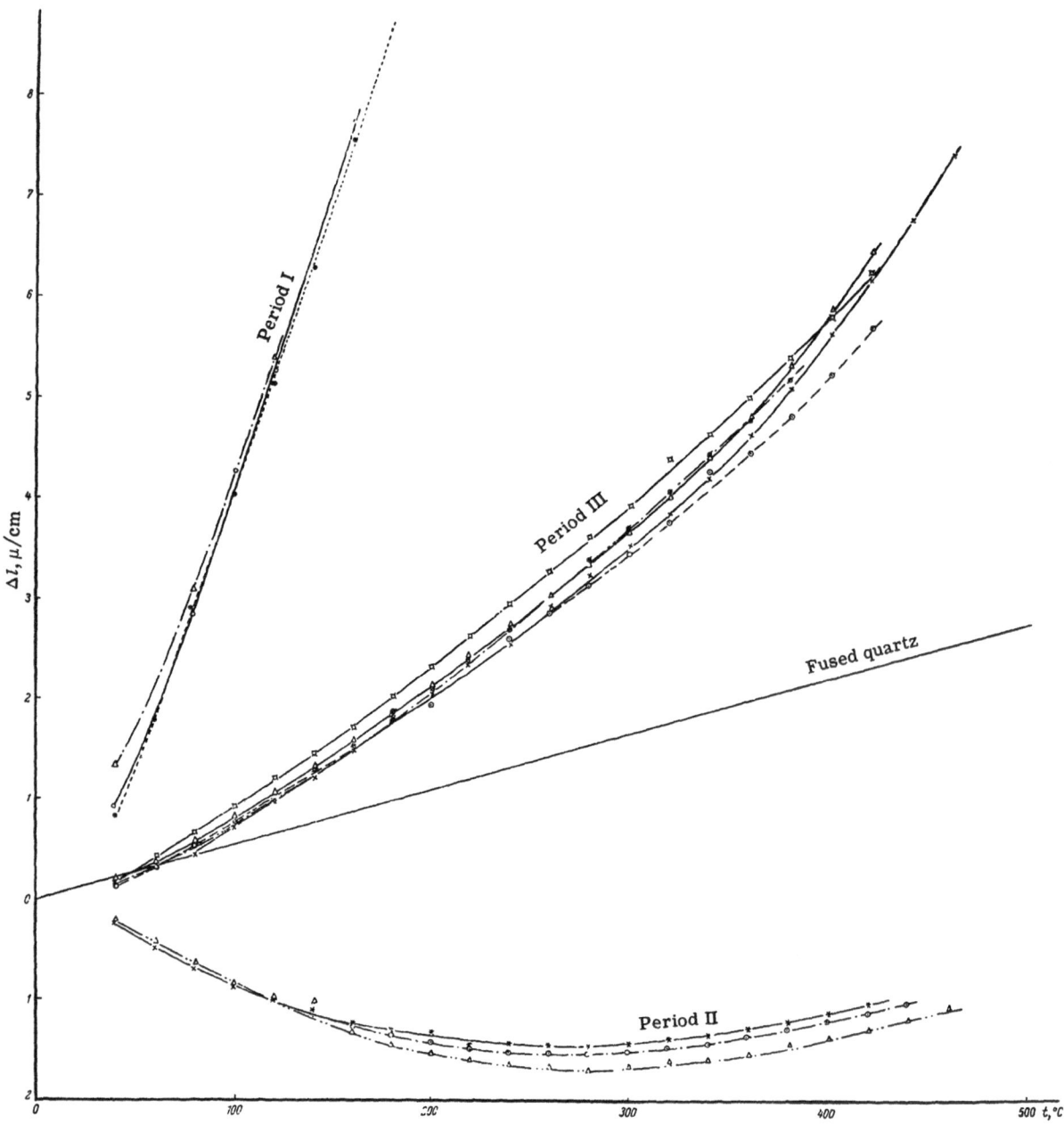

Fig. 2. Linear thermal expansion of glasses in the different periods.

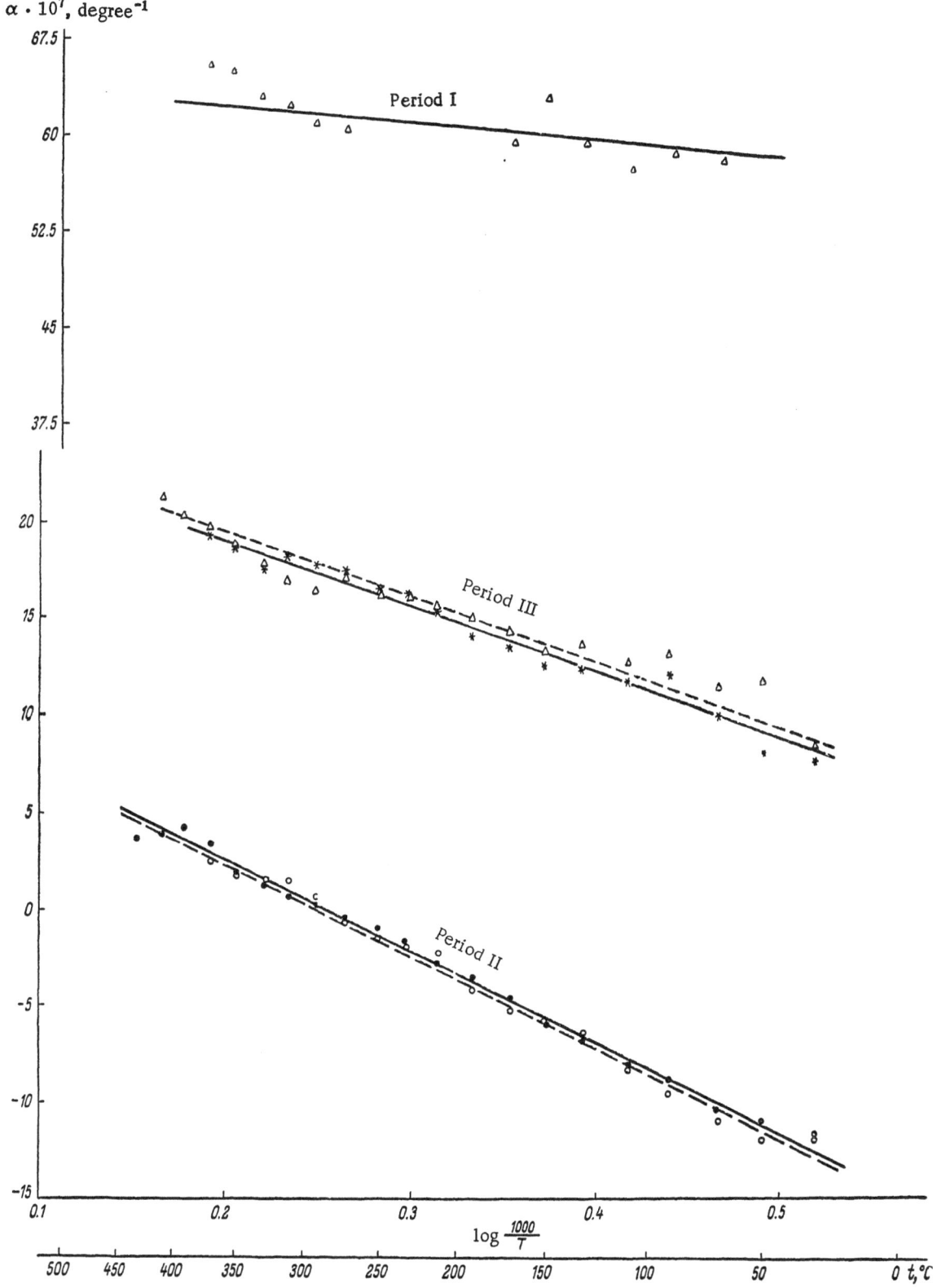

Fig. 3. Variations of the coefficient of linear expansion in the different treatment periods.

It was found that glasses of all compositions, when subjected to certain heat treatments, exhibit a fundamental change of $\alpha$ accompanied by more or less considerable variations of all the other properties as the crystalline phase appears in the form of $\beta$-eucryptite and then of spodumene (according to determinations kindly carried out by the staff of the Leningrad Art Glass Factory).

All the processes are separated sharply into three periods separated by three transition regions (Fig. 1). The temperature range of each period varies by tens of degrees in accordance with the temperature and duration of the preceding heat treatment. Variations of the coefficient of linear expansion in each transition region are independent of the conditions of heat treatment. It decreases sharply in the first transition region and increases rapidly in the second. The magnitude of these changes depends only on the composition of the original glass. Changes of the coefficient of linear expansion in each of the periods are negligible in comparison with the changes observed in the transition regions. Variations of chemical durability, volume resistivity, tangent of the dielectric loss angle, and the equilibrium between ferric and ferrous iron are similar in character. The refractive index, mean dispersion, density, boundary of transmission in the visible region of the spectrum, and light scattering vary during all the three periods and in the transition regions.

The observed changes of properties indicate that various structural changes occur continuously over the entire temperature range studied, but in the transition regions, covering several degrees, crystalline phases are formed; this is confirmed by the results of x-ray phase analysis.

In glasses which have passed only through period I of heat treatment, which we describe as preparation and which lies in the temperature range between the lower annealing temperature and $T_g$, and which may reach somewhat higher temperatures when the times are short, crystalline phases are not detected.

In all the glasses relating to period II, which we describe as the period of the first crystalline phase, the presence of $\beta$-eucryptite crystals is established by x-ray phase analysis.

Glasses of period III, termed the period of the second crystalline phase, were found to contain spodumene and $\beta$-eucryptite crystals.

The sharp increase of volume resistivity, viscosity, resistance to the action of hydrofluoric acid, and the change in the coefficient of linear expansion, all observed in the transitions from period I to period II and from II to III, are consistent with the formation of $\beta$-eucryptite crystals in the first case and of spodumene crystals in the second, as these are precisely the changes to be expected from the properties of these crystals.

The observed slight decrease of volume resistivity and the increase of viscosity during prolonged heat treatment in period I are easily explained on the assumption that, in this case, two phases form and separate in the glass; this is in agreement with the present views on the preparatory period [1,2].

For a fuller analysis and elucidation of the nature of the processes taking place in lithium aluminosilicate glasses, let us consider the changes of some of the most characteristic properties during all the periods of heat treatment.

Figure 2 shows the elongation of series I glass in the temperature range from 0 to 500°, and Fig. 3 gives the absolute coefficient of expansion in the same temperature range for glass after various heat treatments. It follows from Figs. 2 and 3 that materials corresponding to a particular period are characterized by a family of curves occupying a definite position in the diagram.

The curves of glasses and materials of period I are in the upper part of the diagram, those of period II in the lower, and those of III between them. The higher the $Li_2O \cdot Al_2O_3$ content in the glass, the greater is the difference between the elongations and expansion coefficients of materials of period I and materials of periods II and III. The elongation and coefficient of expansion of crystalline glass materials (periods II and III) differ from those of glass in character as well as in magnitude.

For all the investigated glasses and materials of period I, the elongation is a linear function of the temperature, and the coefficient of expansion increases slightly at high temperatures and is almost the same for the different compositions of the series $[(50-60) \cdot 10^{-7} \text{ degree}^{-1}]$.

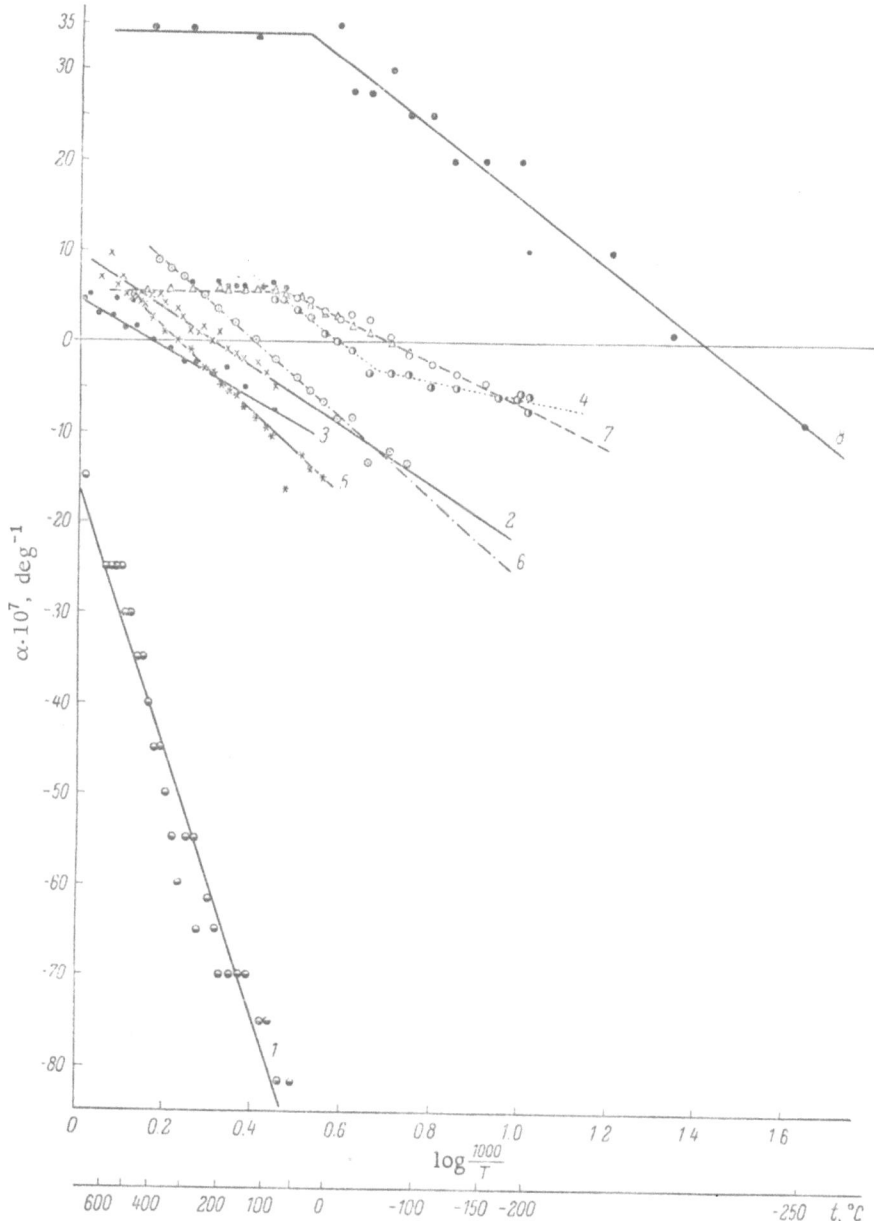

Fig. 4. Variations of the coefficients of expansion of various materials as functions of the temperature (data of various authors). 1) Eucryptite; 2) spodumene; 3) petalite; 4) Pyroceram; 5) glass of series I; 6) glass of series II; 7) fused quartz; 8) Pyrex.

The elongation of materials with a crystalline phase is represented by a second-order curve, and the coefficients of expansion increase continuously with the temperature over the entire range studied. The coefficient is a linear function of the logarithm of the reciprocal absolute temperature:

$$\alpha_T = K \log \frac{1}{T} + C,$$

where $\alpha_T$ is the true coefficient of expansion at temperature T; K is a proportionality factor, defined as the tangent of the angle formed by the straight line; C is a constant determining the position of the line on the diagram.

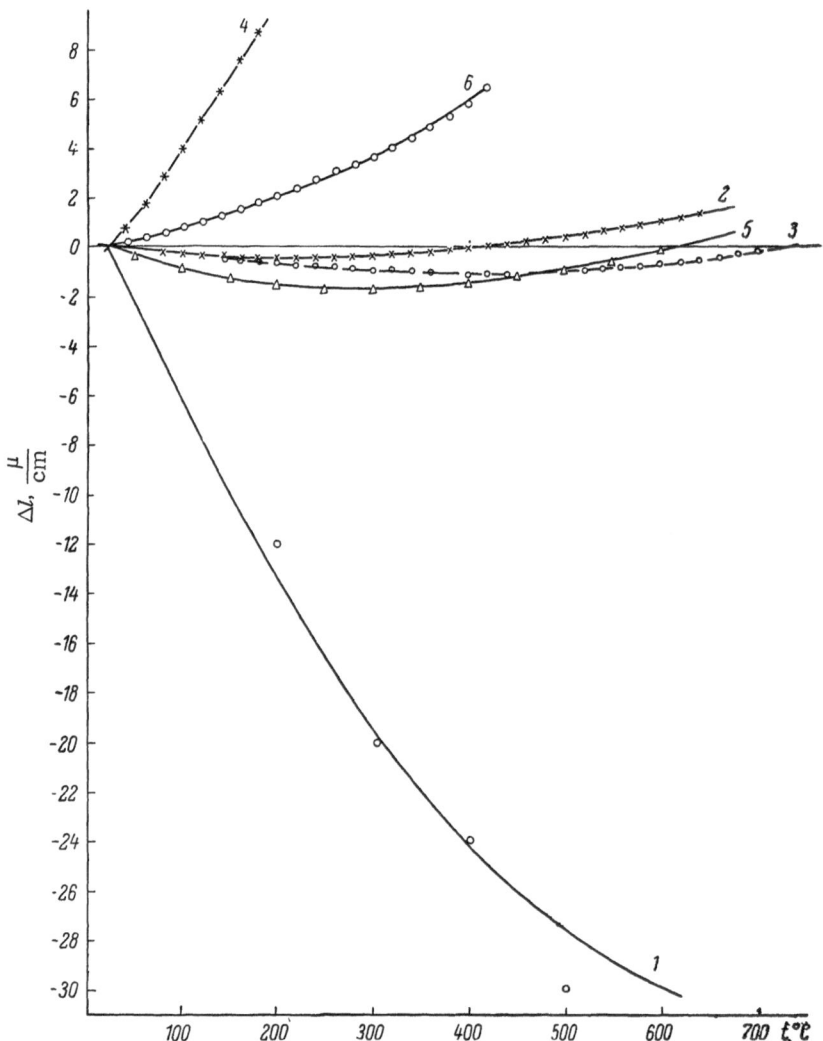

Fig. 5. Thermal expansion of certain materials. 1) Eucryptite; 2) spodumene;
3) petalite; 4) raw glass of period I; 5) glass of period II; 6) glass of period III.

It is relevant to note here that such variations of the coefficient of expansion are not usually observed in glasses at temperatures above 20°. At lower temperatures the course of the variations of the coefficient of expansion of quartz and Pyrex glass (Fig. 4) becomes of the nature described above [3]. The crystalline materials petalite, spodumene, and eucryptite exhibit similar variations of the coefficient of expansion at all temperatures up to +700° [4,5].

The similarities between the temperature dependence of elongation and the variations of the coefficients of expansion of materials in treatment periods II and III, petalite, spodumene, and eucryptite suggest that it might be possible to identify the crystalline phases in these materials by calculation of their elongation from additivity formulas [6]. It was assumed that $\alpha$ of the crystalline phase cannot differ considerably from $\alpha$ of the original glass.

The elongation curves of petalite, eucryptite, and spodumene are given in Fig. 5; the expansion coefficients are taken from [4,5]. The calculation was performed by solution of several systems of equations for different temperatures with the aid of the formulas

$$\Delta l_{t_1} \text{ (orig. mat.)} = \Delta l_{t_1} \text{ (spod.) a} + \Delta l_{t_1} \text{ (eucr.) b} + \Delta l_{t_1} \text{ (glass) c,}$$
$$\Delta l_{t_2} \text{ (orig. mat.)} = \Delta l_{t_2} \text{ (spod.) a} + \Delta l_{t_2} \text{ (eucr.) b} + \Delta l_{t_2} \text{ (glass) c,}$$
$$\Delta l_{t_3} \text{ (orig. mat.)} = \Delta l_{t_3} \text{ (spod.) a} + \Delta l_{t_3} \text{ (eucr.) b} + \Delta l_{t_3} \text{ (glass) c,}$$

138

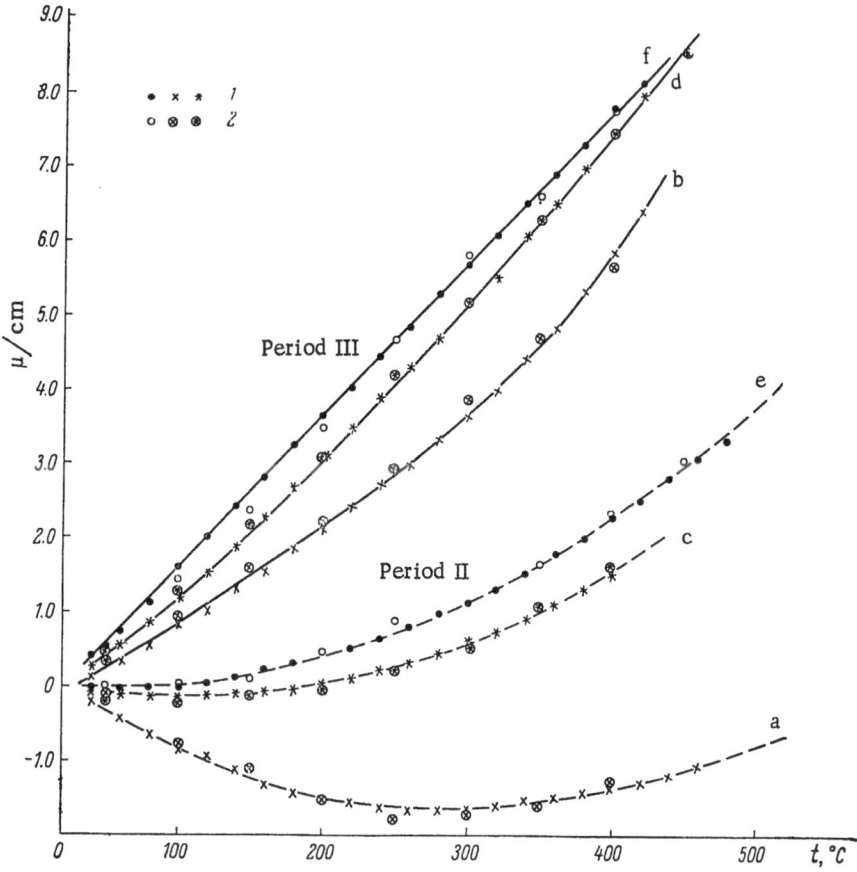

Fig. 6. Thermal expansion of crystalline glass materials of various compositions.
1) Experimental values; 2) calculated values. a,b) Glasses of series I; c,d) glasses
of series II; e,f) glasses of series III.

where a + b + c = 1, where a, b, and c are the fractions of spodumene, eucryptite, and glass, respectively. The
agreement between the calculated and experimental values for certain glasses is shown in Fig. 6.

Calculations showed that in materials of period II, the crystalline phase can be present only as $\beta$-eucrypt-
ite, the content of which falls from 49% in the first glass to 41.5% in the third. The calculated amount of $\beta$-
eucryptite is 4-10% less than is to be expected from the composition of the original glass; this is evidently because part of the $Li_2O$ and $Al_2O_3$ remains in the glassy phase.

Materials of period III are calculated to contain spodumene and a small amount (3-11%) of $\beta$-eucryptite.
The total content of the crystalline phase rises to 60 and 55%, which is 12-28% below the possible content
based on the composition of the original glass. This may be because the coefficient of expansion of the glassy
phase was assumed constant in the calculations.

On the other hand, the low content of the crystalline phase may be attributed to the formation of eucryptite in a compound enriched with $SiO_2$ rather than in the form $Li_2O \cdot Al_2O_3 \cdot 2SiO_2$, as indicated in [7]. When
spodumene forms, $\alpha$ should increase, because the glassy phase becomes poorer in silica. On the assumption that
spodumene is formed as the result of recrystallization of $\beta$-eucryptite, the amount of crystalline phase should
increase by a factor of 1.5 and become 73.5 and 62.5%, respectively.

The optical properties, density, light transmission, and light scattering alter considerably during all the
periods of heat treatment. Figures 7 and 8 show the variations of $n_d$, $n_F - n_c$ and $\Delta l$. The density change $\Delta d$
and $\Delta l$ are connected by the expression $\Delta d = 3d\Delta l \cdot 10^{-4}$. The optical constants increase continuously with

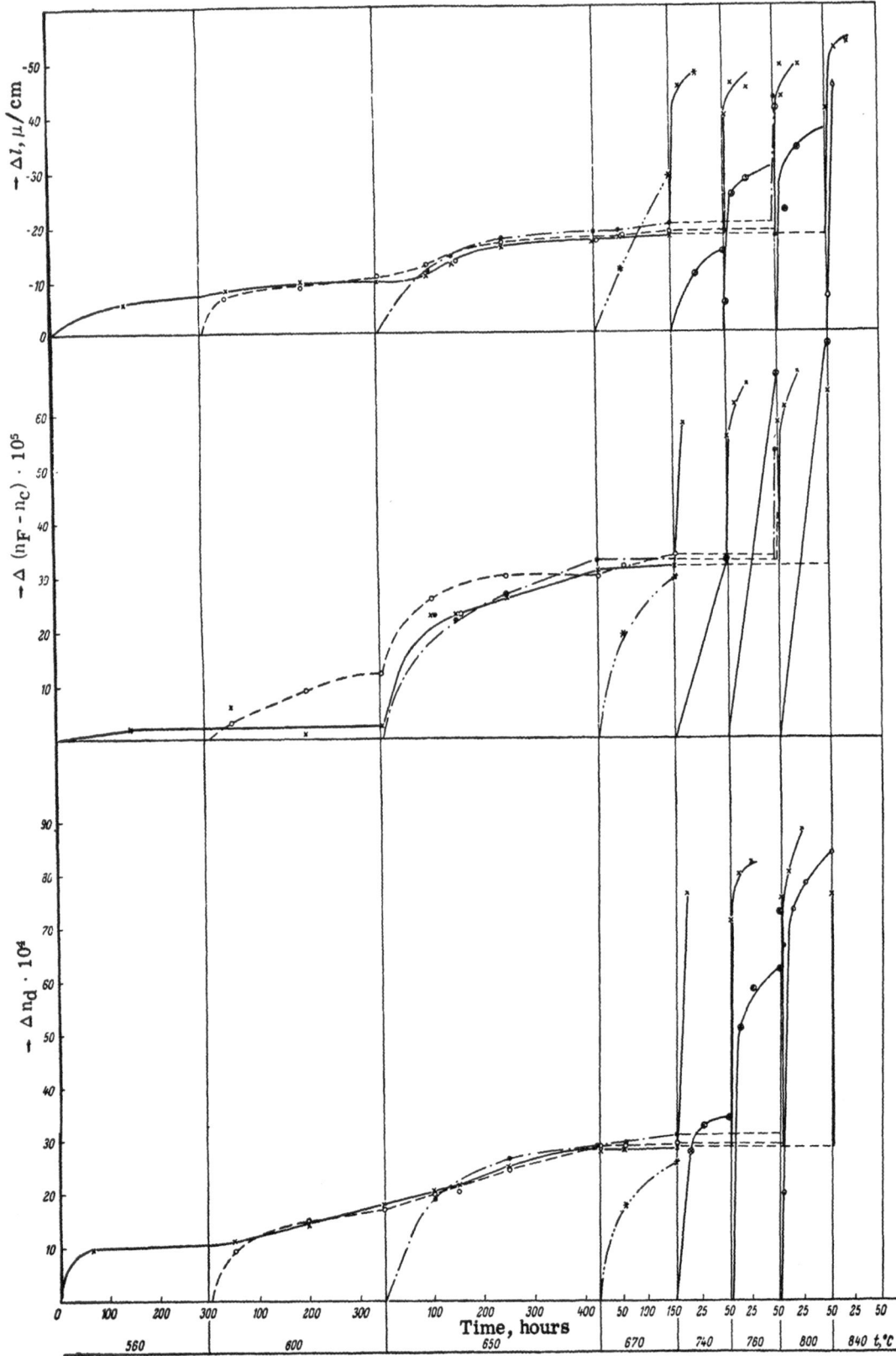

Fig. 7. Variations of optical constants and linear dimensions during different heat treatments.

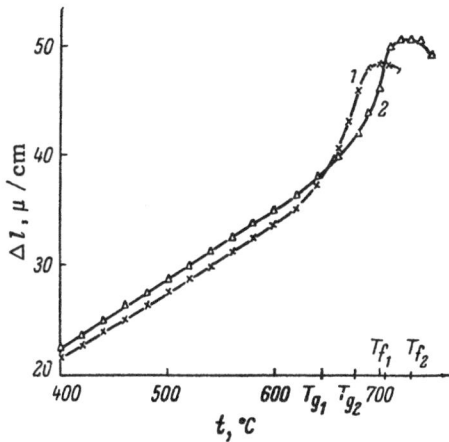

Fig. 8. Thermal expansion of period I glasses. 1) Raw glass of series II; 2) series III glass with prolonged preparation.

rise of the heat-treatment temperature, but the rate of increase grows less with temperature than is the case for ordinary glasses [8,9]. These changes are irreversible.

This anomalous change of the rate of growth with temperature can be attributed only to a continuous increase in the viscosity of the material (which was found by measurements of $T_g$ and $T_f$ of glasses prepared in different ways in period I) and especially in the transition from one period to another. The same applies to the density changes, except that, in the transition to period III, the density sometimes decreases.

In Fig. 9, $n_d$ is plotted as a function of $n_F - n_C$ for one series of glasses. Regardless of the heat-treatment conditions, all the glasses of periods I and II give "property-property" plots in the form of two straight lines with considerable relative displacement. The observed changes of $n_d$ and $n_F - n_C$ in period I and II may be attributed to a change in the structural form of titanium, as these high $\Delta(n_F - n_C)/\Delta n_d$ ratios are given by titanium-containing glasses and the change of $n_d$ during heat treatment is proportional to the amount of $TiO_2$ introduced into the glass (Fig. 10).

Our explanation of this process is that phase separation occurs continuously during heat treatment in period I [1,2,10], while in period II crystal growth takes place; this creates conditions for retention of the high-temperature structural forms of titanium owing to the increased viscosity.

The change in the first transition region (between periods I and II) on the "property-property" diagram differs sharply from that described above and is characterized, as already stated, by rapid growth of $n_d$ with slight changes of $n_F - n_C$. This cannot be associated with changes in the structure of $TiO_2$; we attribute it to an increase in the amount of aluminum in sixfold coordination in the glass during crystallization. Conversion of 1% $Al_2O_3$ from $AlO_4$ to $AlO_6$ configuration increases $n_d$ by $18 \cdot 10^{-4}$ and $(n_F - n_C)$ by $1.5 \cdot 10^{-5}$; i.e., this is the ratio observed in the transition from period I to period II. This hypothesis is supported by other facts. Calculation of the refractive indices of glasses in the $xLi_2O-Al_2O_3(1-x)SiO_2$ system [11] and of our glasses by the additivity formula [6] shows that the amount of aluminum in the $AlO_6$ form increases with decrease of the silica content (Fig. 11).

The data show that as we pass from compositions with low $Li_2O$ and $Al_2O_3$ contents to higher contents of these oxides, the amount of aluminum in the $AlO_6$ form increases and $n_d$ and $n_F - n_C$ vary as described above.

In the glasses studied, separation of the crystalline phase in the form of $\beta$-eucryptite results in all cases (by calculation) in changes of composition which increase the $AlO_6$ concentration, and the refractive index should therefore increase sharply. Glasses were made in which the ratio of $AlO_6$ and $SiO_2$ (according to Fig. 11) corresponds to the eucryptite composition. As was to be expected from the foregoing, these glasses did not exhibit a sharp change of $n_d$ in the transition region.

Comparison of the changes of density, refractive index, and dispersion in periods I and II shows that these changes cannot be attributed to increased density of the glass. Thus, in the preparation process, $n_d$ changes by 0.25%, $n_F - n_C$ by 6%, and the density by 1.2%; and in period II, $n_d$ changes by 0.18%, $n_F - n_C$ by 3.2%, and the density by 0.9%. It is very important to note that the magnitude of the changes of $n_d$ and $n_F - n_C$ and the density are not associated with the magnitude and the nature of the change of the coefficient of expansion and therefore with the amount of the crystalline phase. We obtained glassceramics corresponding to period II by rapid heat treatment without a preparation period, with the same coefficient of expansion, having $n_d$, $n_F - n_C$, and density values much lower than could be obtained during preparation without change of the coefficient of expansion and formation of a crystalline phase. On further heat treatment of such glassceramics, these properties reach the same values as for glassceramics subjected to prolonged preliminary heat treatment without change of the coefficient of expansion.

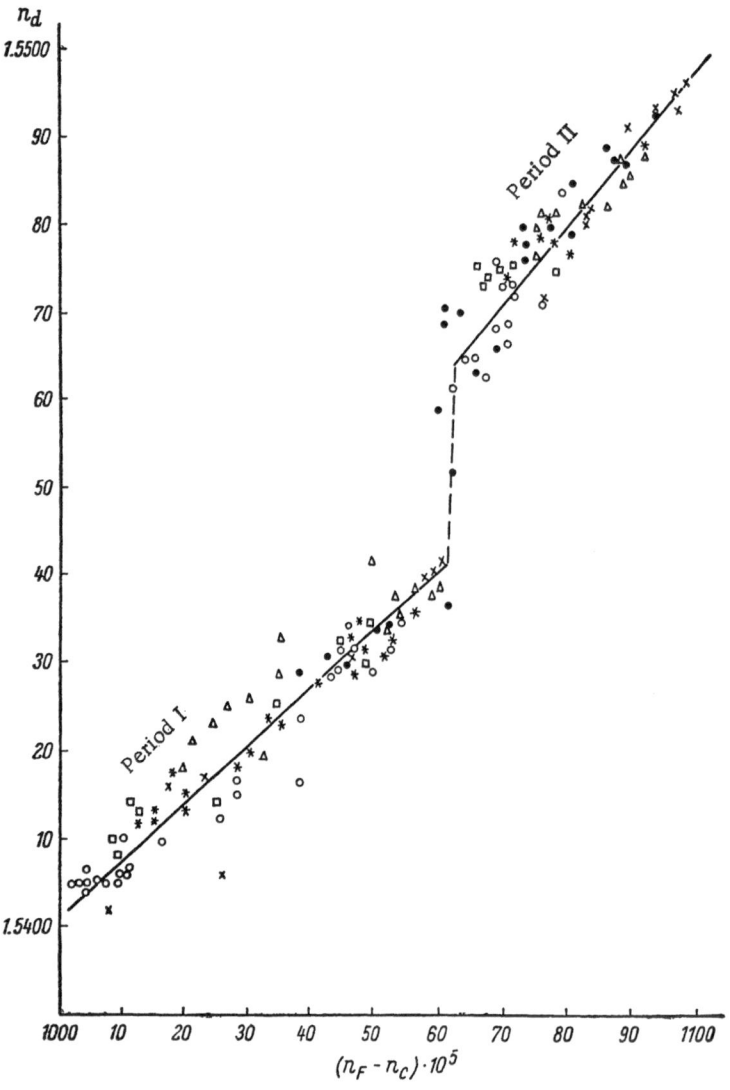

Fig. 9. Variations of optical constants at different stages of heat treatment.

This confirms our opinion that changes in the structural forms of titanium are associated with viscosity changes. In the last case there is no time for completion of the structural changes because of the short time during which the viscosity of the material is high.

Figure 12 represents the light transmission of one of the glasses in heat-treatment periods I and II. The specimens are numbered in accordance with rising temperature or increase of the time in heat treatment. The variation of $\tau$ with $\lambda$ is of the same character for all the specimens. The shift of the absorption curve into the long-wave region of the spectrum with increased duration of heat treatment in period I, attributable to the formation of complex iron and titanium compounds [12,13], is not accompanied by a decrease of the amount of ferrous iron ($\tau$ at 900-1200 m$\mu$ remains unchanged). The shift, although slight, of the absorption curve into the short-wave region of the spectrum after a certain maximum is inexplicable. This shift occurs even during period I of heat treatment, and the less it is in I the greater it is in II.

The transition from period I to II is always accompanied by increased absorption in the 900-1200 m$\mu$ region, which indicates considerable conversion of $Fe_2O_3$ into $FeO$. This change of equilibrium is in good agreement with the hypothesis that some $AlO_4$ is converted into $AlO_6$.

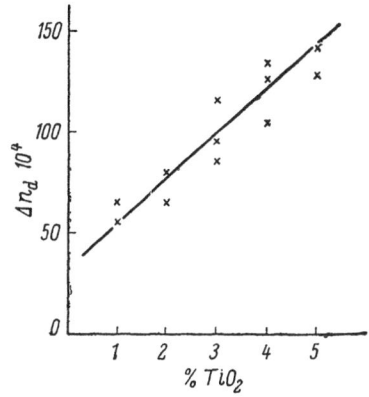

Fig. 10. Variations of refractive index
with the mineralizer content.

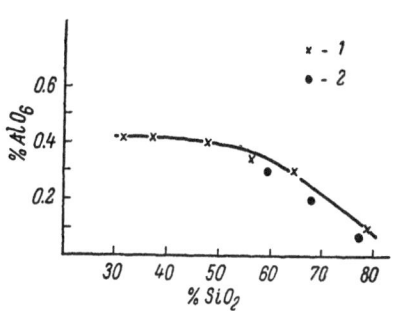

Fig. 11. Contents of $AlO_6$ in glasses of
different compositions. 1) Glasses $xLi_2O$
$\cdot Al_2O_3 \cdot (1-x)$ $SiO_2$; 2) glass of series II.

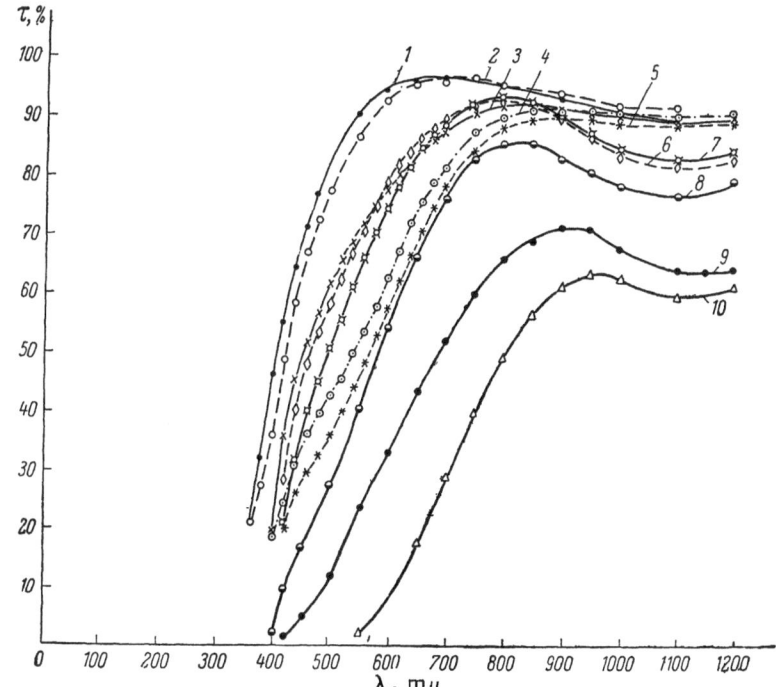

Fig. 12. Transmission curves of glasses after different heat treatments.
1-5) Glasses of period I; 6-10) glasses of period II.

The absorption changes in period II are entirely attributable to crystal growth, causing light scattering. As a rule, the scattering of period III materials is so great that their optical and spectral properties cannot be determined. The rate at which light scattering increases during period II is determined only by the conditions of heat treatment during period I, as the size and number of particles formed during crystallization are associated only with the nature of the heat treatment [14].

Conclusion

On the basis of our experimental data, the processes taking place in glasses of the $Li_2O$-$Al_2O_3$-$SiO_2$ system mineralized by titanium can be explained as follows.

1. When the original glass melt cools, it deposits minute particles of titanium compounds, most probably rutile, capable of spontaneous crystallization on cooling. At the usual cooling rates, the rutile particles are formed only at a certain titanium concentration, below which $TiO_2$ remains in the glass and does not yield crystallization nuclei. The particles are so small and so few that their formation does not produce any changes in the properties of the glass, and they are very difficult to detect.

During the subsequent heat treatment, starting with the lower annealing temperatures, the nuclei act as centers for deposition of titanium oxides with changes in their structural form, of ferrous and ferric oxide, and of the glass itself. This deposition results in the formation of minute droplets differing in composition from the original glass; separation of the glass into two phases occurs, leading to increased viscosity. The droplets are smaller and more numerous at lower temperatures of heat treatment.

At higher treatment temperatures, in the region near $T_g$ and above, droplet formation is accompanied by considerable growth, which leads to a decrease in the number of particles formed.

2. Structural changes with formation of a crystal lattice occur within the new phase only after two phases have been formed in the glass and a considerable amount of material has separated out as the new phase. According to our observations, this is always a spontaneous process. It cannot be assigned to any definite temperature (the temperature can vary by tens of degrees in accordance with the duration of the preceding heat treatment); neither is it associated with the extent of the changes in the structural forms of titanium. In our opinion this process is due only to supersaturation of the glass by the newly formed phase.

The composition of the crystalline product which separates out corresponds to the lowest compound of silica with lithium and aluminum, β-eucryptite. Formation of the crystalline phase leads to changes in all the properties of the material. Further heat treatment of a material in which the crystalline phase has separated out below the crystallization temperature does not lead to any processes, because the viscosity has increased sharply. Heat treatment at higher temperatures and for longer times leads to further changes in the structural forms of the titanium compounds (these changes are more pronounced if the process is less advanced in the initial stage) and to growth of crystals, primarily from the remaining glass, corresponding to eucryptite in composition and subsequently from the glassy silica and by aggregation of fine particles.

3. The composition of the crystals gradually changes from β-eucryptite to spodumene; at certain temperatures and treatment times, β-eucryptite recrystallizes into spodumene.

Thus all the processes taking place in these glasses during heat treatment can be subdivided into three periods separated by two narrow transition regions.

Period I is characterized by the fact that the glass does not contain a crystalline phase but separates into two phases with formation and growth of minute droplets of glass having the composition of the crystals which are subsequently deposited in the transition region. Structural changes also take place in titanium oxide and other compounds of variable-valence elements. The separated glass droplets crystallize in the first transition region.

In period II the material consists of glass and β-eucryptite crystals; crystal growth and structural changes of titanium compounds continue. In the second transition region, the β-eucryptite crystals are converted into spodumene crystals at the expense of glassy silica, with a corresponding increase of the total percentage content of the crystalline phase.

In period III the material consists of spodumene and a small amount of glass. The complex crystallization processes are virtually completed here.

LITERATURE CITED

1. W. Hinz and P.-O. Kunth, Glastechn. Ber. 34(9):431, 1961.
2. S. I. Sil'vestrovich and É. M. Rabinovich, Zhur. Vsesoyuzn. Khim. Obshch. im. Mendeleeva 5(2):186, 1960.
3. Pyroceram 9606 and 9608 [Russian translation], Gosudarstvennyi Nauchno-Tekhnicheskii Komitet Soveta Ministrov SSSR, Otdel Vneshnikh Snoshenii, 1961.
4. H. J. Smoke, J. Am. Ceram. Soc. 34(3):122, 1951.

5. F. H. Gillery and E. A. Buch, J. Am. Ceram. Soc. 42(4):175, 1959.

6. L. I. Demkina, Investigation of the Influence of Composition on the Properties of Glasses, Oborongiz, 1958.

7. H. Saalfeld, Ber. Dtsch. keram. Ges. No. 7, 38, 1961.

8. A. A. Lebedev, Trudy GOI, II(10):57, 1921; III(24):1, 1924.

9. A. N. Stozharov, Trudy GOI, IV(39):1, 1928.

10. A. I. Avgustinnik, "Formation of a crystalline phase from a silicate melt," in The Structure of Glass, Vol. 2, Consultants Bureau, New York, 1960, p. 95.

11. R. A. Hatch, Am. Mineralogist 28(9/10):477, 1943.

12. V. V. Vargin, Doklady Akad. Nauk SSSR 111(4):848, 1956.

13. V. V. Vargin, Doklady Akad. Nauk SSSR 103(1):105, 1955.

14. O. Knapp, Glass Ind. 40(6):307, 1959.

# INVESTIGATION OF THE PHASE COMPOSITION OF CRYSTALLINE GLASS MATERIALS OBTAINED FROM THE $Li_2O-Al_2O_3-SiO_2$ AND $Li_2O-MgO-Al_2O_3-SiO_2$ SYSTEMS

## I. I. Kitaigorodskii, L. S. Zevin, and M. V. Artamonova

Crystalline glass materials based on the $Li_2O-Al_2O_3-SiO_2$ and $Li_2O-MgO-Al_2O_3-SiO_2$ systems were investigated in the Department of Glass Technology, the D. I. Mendeleev Moscow Institute of Chemical Technology, with the URS-50I x-ray apparatus with ionization recording of the x-ray intensity.

Three ternary compounds are known in the $Li_2O-Al_2O_3-SiO_2$ system; these occur naturally as the minerals eucryptite (mole ratio of oxides $1:1:2$), spodumene ($1:1:4$), and petalite ($1:1:8$). On heating, eucryptite and spodumene are converted into high-temperature forms differing in structure from the native minerals. Native petalite gives a solid solution of $\beta$-spodumene with $SiO_2$ on heating. Syntheses of compounds with variable mole contents of $SiO_2$ from 1 to 10 in this system showed that $\beta$-spodumene forms solid solutions with $SiO_2$; the structure of the solid solutions is analogous to that of $\beta$-spodumene and their x-ray diagrams differ from that of $\beta$-spodumene by a small shift of the lines toward higher $\Theta$. When the oxide mole ratio is higher than $1:1:8$, crystalline forms of $SiO_2$ separate out together with the solid solution of $\beta$-spodumene.

The x-ray diagrams of most of the crystalline glass materials and synthesized compounds studied revealed characteristic changes in the intensity of the $\beta$-spodumene diffraction lines in accordance with the heat-treatment conditions. Figure 1 shows that rise of the heat-treatment temperature results in substantial increases of the line intensities corresponding to interplanar spacings 3.86, 3.13, and 1.93 A, and others, with a corresponding decrease in the intensity of the principal $\beta$-spodumene lines (d = 3.46, 1.87, 1.60 A, etc.). In a number of cases, such as in the crystallization of materials obtained from these systems (Fig. 2), the first group of lines is entirely lacking in low-temperature crystallization (700-850°) and appears at a higher crystallization temperature (900°).

Crystallization of certain materials directly from the melts yielded large crystals in the form of hexagonal bipyramids up to 10 mm in size (Fig. 3). It was found that lines of the first group are missing from the structure of these crystals, and that their x-ray diagrams are analogous to the diagram in Fig. 2a. These lines appear when the crystals are heated again at a higher temperature. Laue diffraction patterns led to the conclusion that the crystals formed are concretions of two or three single crystals and that breakdown of the original structure with formation of polycrystalline material occurs during the heating. Further investigation of the crystals is hindered by the presence of a glassy phase which permeates the crystal in the form of fine threads.

The character of the structural changes suggests that they are caused by one of the following phenomena.

1. Formation of a "second phase," the composition of which cannot be determined from the x-ray diffraction data available for compounds of this system. Separation of the "second phase" was observed in all cases at oxide ratios from $1:1:4$ to $1:1:10$, and this phase always separates out in presence of the main $\beta$-spodumene phase. If the lines of the "second phase" are excluded, the x-ray diffraction diagrams of compounds with oxide ratios from $1:1:2$ to $1:1:10$ become very similar and differ only by a shift of the lines toward larger $\Theta$ in the transition from $1:1:2$ to $1:1:10$ ratios. This effect is probably associated with the formation of a large series of solid solutions, including $\beta$-eucryptite, $\beta$-spodumene, and petalite.

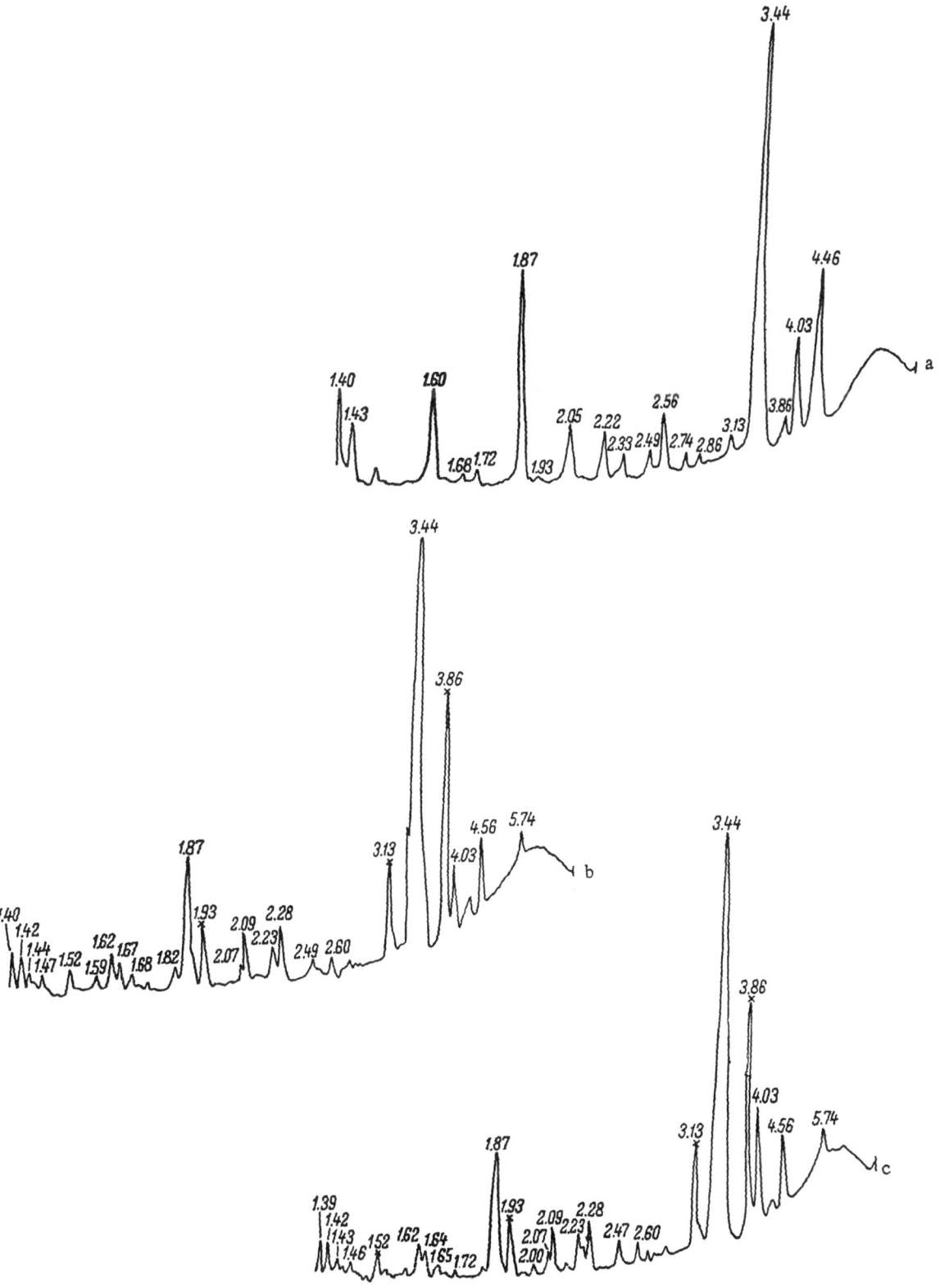

Fig. 1. Effect of the heat-treatment conditions on the phase composition of the compound with oxide mole ratio 1 : 1 : 10. a) Crystallization at 800°; b) at 900°; c) at 1000°; ×) lines of the "second phase."

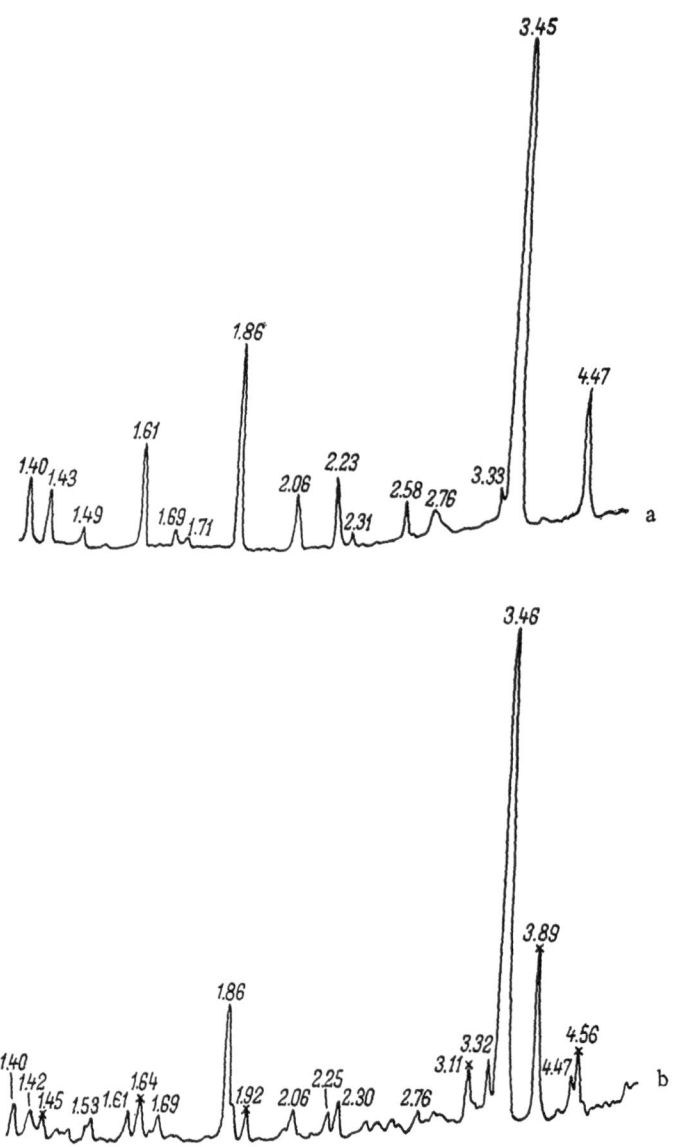

Fig. 2. Variation of the phase composition of material 87. a) Crystallization at 800°; b) repeated crystallization at 900°.

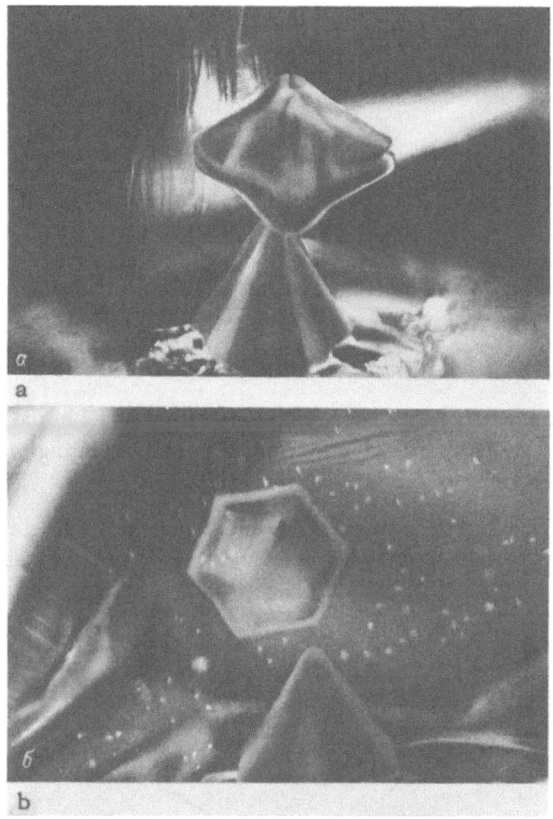

Fig. 3. Crystals (a and b) in glass.

2. Conversion of β-spodumene from the low-temperature form, stable in the 700-850° range, into a high-temperature form stable at temperatures above 900°. However, it cannot be decided which of these suggestions is correct until a single crystal of β-spodumene has been obtained.

# INVESTIGATION OF THE ELECTRICAL PROPERTIES OF CERTAIN GLASSES AND CRYSTALLINE GLASS MATERIALS BASED ON THE $Li_2O-Al_2O_3-SiO_2$ SYSTEM

## G. A. Pavlova, M. M. Skornyakov, and V. G. Chistoserdov

Investigations of the electrical properties of crystalline glass materials in the $Li_2O-Al_2O_3-SiO_2$ system in relation to composition are of practical significance for improving the insulating properties of photosensitive lithium materials, and are of theoretical interest.

The original glass chosen had the composition of a lithium photosensitive glass similar to that given in a patent [1]. We investigated the effects of varying the concentrations of $Li_2O$ and $Al_2O_3$ by replacement with $SiO_2$, and of replacement of a part of the $SiO_2$ by the oxides BaO, SrO, and CaO, with constant small amounts of other oxides.

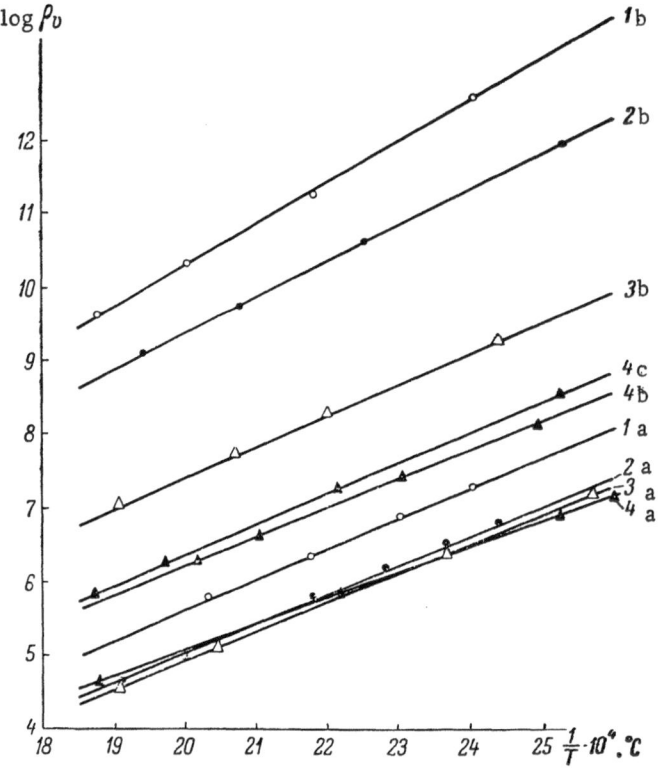

Fig. 1. Effect of temperature on the resistivity of certain glasses (a) and crystalline glass materials (b). 1) Composition 4; 2) composition 2; 3) composition 3; 4a) composition based on the $Li_2O-Al_2O_3-SiO_2-TiO_2$ system; 4b) specimen crystallized at 800°; 4c) specimen crystallized at 1150°.

TABLE 1

| Comp. No. | Oxide contents, % by wt. | | Original glass | | | | | Crystallized glass | | | | |
|---|---|---|---|---|---|---|---|---|---|---|---|---|
| | | | $\log \rho_V$ | | E, eV | $f = 10^6$ cps at 25° | | $\log \rho_V$ | | E, eV | $f = 10^6$ cps at 25° | |
| | $Al_2O_3$ | $Li_2O$ | 150° | 250° | | $\tan \delta \cdot 10^4$ | $\varepsilon$ | 150° | 250° | | $\tan \delta \cdot 10^4$ | $\varepsilon$ |
| 1 | – | 16 | – | – | – | – | – | 12.20 | 9.60 | 2.30 | 10 | 5.7 |
| 2 | 4 | 16 | 6.46 | 4.65 | 1.60 | 205 | 9.0 | 11.20 | 8.95 | 2.00 | 14 | 7.3 |
| 3 | 14 | 16 | 6.33 | 4.55 | 1.54 | 178 | 9.1 | 8.98 | 7.05 | 1.61 | 183 | 8.6 |
| 4 | 1 | 12 | 7.15 | 5.25 | 1.67 | – | – | 12.38 | 9.82 | 2.25 | 18 | 5.5 |
| 5 | 4 | 12 | 7.41 | 5.40 | 1.78 | – | – | 11.00 | 8.87 | 1.86 | 51 | 6.3 |
| 6 | 8 | 12 | 7.15 | 5.07 | 1.60 | 130 | 9.1 | 10.00 | 7.90 | 1.85 | 70 | 7.7 |
| 7 | 14 | 12 | – | – | – | – | – | 9.50 | 7.27 | 1.96 | 160 | 7.8 |
| 8 | 4 | 10.5 | 7.08 | 5.40 | 1.47 | 137 | 7.6 | 11.36 | 8.95 | 2.13 | 56 | 6.3 |
| 9 | 14 | 10 | 7.47 | 5.52 | 1.73 | 140 | 8.0 | 8.77 | 6.80 | 1.73 | 172 | 8.5 |

Note. E was calculated from the formula $\varkappa = A e^{\frac{-E}{2kT}}$

The instruments indicated in [2] were used for measurement of the volume resistivity $\rho_V$, the dielectric loss $\tan \delta$, and the dielectric constant $\varepsilon$. The test specimens were 25 × 25 × 0.5 mm in size, or 30 mm in diameter and 2-2.5 mm thick.

The results of the determinations (Table 1, Figs. 1 and 2) showed that crystallized materials containing less than 10% $Al_2O_3$ and free from $Al_2O_3$ have much higher insulation characteristics than glasses; $\rho_V$ increases after crystallization by 3-5 orders of magnitude at 150° (the activation energy of conductivity E increases from 1.5-1.6 to 1.85-2.3 eV), $\tan \delta$ decreases two- to threefold at 1 Mc/sec and 25°, and $\varepsilon$ decreases by 1-1.7.

Increase of the $Al_2O_3$ concentration leads to a sharp deterioration of the insulating properties of crystalline glass materials. It is interesting to note that for glasses with a high $Al_2O_3$ content (compositions 3 and 9), $\rho_V$ rises by 1-2.5 orders of magnitude after crystallization, while $\tan \delta$ does not decrease. The activation energy of conductivity is almost the same for these glasses as for the crystallized materials.

Increase of the $Li_2O$ concentration from 10.5 to 16% at a constant $Al_2O_3$ content of 4% (compositions 2, 5, and 8) has no adverse effect on the insulating properties of crystalline glass materials, whereas increase of alkali oxide concentration in glasses leads to the usual increases of conductivity, $\tan \delta$, and $\varepsilon$.

Replacement of a part of the $SiO_2$ by BaO, SrO, or CaO tends to improve the insulating properties of the crystallized materials. When 5-6% $SiO_2$ is replaced by BaO and SrO, the resistance increases by 2-2.5 orders of magnitude at 150°, $\tan \delta$ at 1 Mc/sec and 25° decreases by more than half, and the point at which $\tan \delta$ begins to increase sharply with temperature is shifted toward higher temperatures.

It is known from the results of x-ray diffraction and optical crystallographic studies that the predominant crystalline phase in crystalline materials which do not contain $Al_2O_3$ is a solid solution of lithium disilicate and silica. If the glass contains $Al_2O_3$, solid solutions of spodumene and eucryptite with silica are deposited as a result of crystallization. Either spodumene or eucryptite predominates, depending on the $Al_2O_3$ concentration. This variation of the crystalline phases determines the variations of the electrical properties of the crystallized materials.

It is noteworthy that in lithium crystalline glass materials with higher $Al_2O_3$ contents (Table 1), the dielectric loss does not decrease after crystallization, despite the fact that $Li_2O$ passes into the crystalline phase, and $\rho_V$ is higher for the crystallized materials than for the original glasses. It is likely that the crystalline phase has a significant influence on the dielectric loss in this case.

In investigations of the electrical properties of crystalline glass materials based on the $Li_2O$-$Al_2O_3$-$SiO_2$-$TiO_2$ system containing less than 10% $Li_2O$ (other properties of these materials are described in [3]), we observed that the dielectric loss depends sharply on the conditions of heat treatment (Fig. 3). Specimens crystallized at 800° have $\tan \delta$ approximately double and $\varepsilon$ 1.5 times the corresponding values for the original glass at 1 Mc

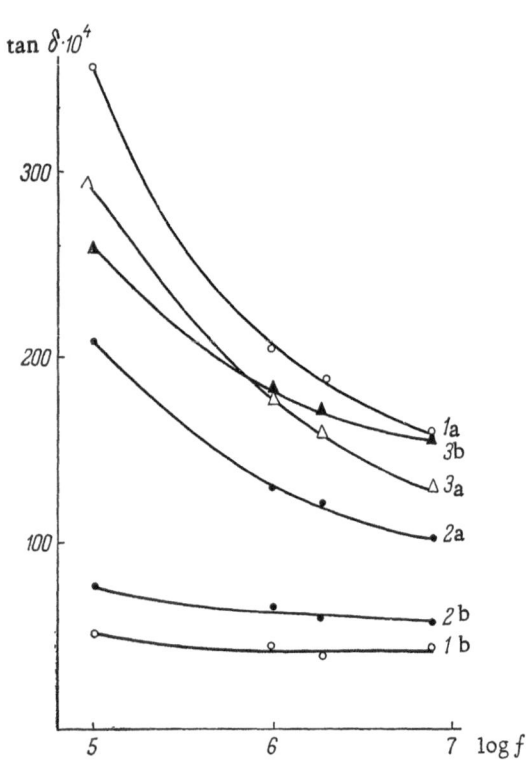

Fig. 2. Variations of $\tan \delta$ with frequency for certain glasses (a) and crystalline glass materials (b). 1) Composition 2; 2) composition 6; 3) composition 3.

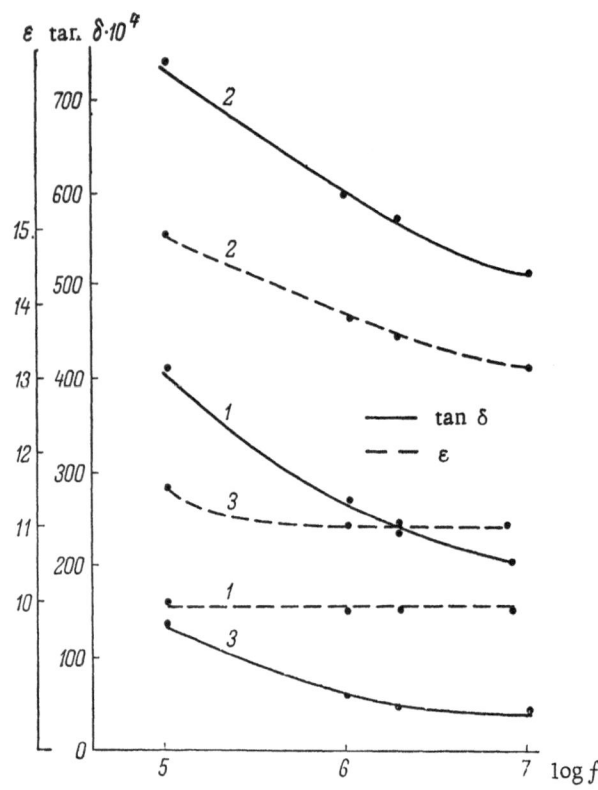

Fig. 3. Dependence of $\tan \delta$ and $\varepsilon$ on frequency for glass and crystalline glass materials based on the $Li_2O$-$Al_2O_3$-$SiO_2$-$TiO_2$ system. 1) Original glass; 2) specimen crystallized at 800°; 3) specimen crystallized at 1150°.

TABLE 2

| Maximum crystallization temperature, °C | Brazing temperature of silver paste, °C | $\log \rho_V$ | | $f = 10^6$ cps at 25° | |
|---|---|---|---|---|---|
| | | 150° | 250° | $\tan \delta \cdot 10^4$ | $\varepsilon$ |
| 750 | 500 | 9.68 | 7.73 | 90 | 8.0 |
| | 700 | 11.00 | 8.55 | 84 | 8.0 |
| | 800 | 11.33 | 8.99 | 68 | 7.6 |
| | 850 | 11.70 | 9.82 | 20 | 7.2 |
| 825 | 500 | 10.00 | 7.90 | 78 | 8.0 |
| | 700 | 10.44 | 8.00 | 72 | 8.3 |
| | 800 | 11.03 | 8.55 | 72 | 8.8 |
| | 850 | 10.32 | 8.02 | 36 | 7.3 |

Note. The table gives average values of $\log \rho_V$, $\tan \delta$, and $\varepsilon$ found from measurements on two specimens.

per sec and 25°. When the crystallization temperature is raised to 1150°, $\tan \delta$ decreases to below the value for the original glass. The resistivity of all the crystallized glasses is approximately the same, and is about ten times that of the original glass (Fig. 1). It is known that lithium aluminosilicates comprise the main crystalline phase deposited in these materials, but the phase composition varies in accordance with the heat treatment.

It therefore follows from these results that when alkali oxides (in particular $Li_2O$) enter the crystalline phase, $\tan \delta$ of crystalline glass materials does not always decrease, as was reported earlier [4], but may decrease, increase, or remain close to that of the original glass (Figs. 2 and 3).

In order to determine how the temperature of brazing of silver paste influences the electrical properties of crystalline glass materials obtained from photosensitive glasses, we formed silver electrodes by brazing at 500, 700, 800, and 850°. The results (Table 2) showed that the insulating properties of crystalline glass materials are even improved somewhat when the brazing temperature is raised. The greatest decrease of $\tan \delta$ was obtained when silver paste was brazed at 850°.

It may be that at higher temperatures (in the region of 850°), silver diffusing into the material tends to alter the crystalline phase by formation of quartz compounds.

The following conclusions may be drawn from the results.

1. The electrical insulating characteristics of crystalline glass materials obtained from photosensitive glasses based on the $Li_2O-Al_2O_3-SiO_2$ system may be enhanced considerably by decrease of the $Al_2O_3$ concentration and replacement of part of the $SiO_2$ by BaO, SrO, and CaO.

2. Dielectric losses of lithium aluminosilicate glasses may decrease, increase, or remain the same, after crystallization, as those of the original glasses, provided that lithium ions enter the crystalline phase. In this case the resistivity of crystallized materials is always higher than that of the glasses.

These data suggest that the change of the dielectric loss of glasses based on the $Li_2O-Al_2O_3-SiO_2$ system as the result of crystallization depends not only on the composition, but also on the structure of the crystalline phase formed, and on the crystalline glass material as a whole.

3. Increase of the temperature used for brazing of silver paste from 500 to 800° does not have an adverse effect on the insulating properties of photosensitive materials. If the silver paste is brazed at 850°, when the rate of silver diffusion into the material is highest, $\tan \delta$ decreases considerably.

## LITERATURE CITED

1. Federal German Patent 922,734, 1955; Steklo i Keramika No. 5, 41, 1958.
2. G. A. Pavlova and V. G. Chistoserdov, this collection, p. 200.
3. V. G. Chistoserdov and V. I. Novgorodtseva, this collection, p. 154.
4. G. I. Skanavi, Dielectric Polarization and Losses in Glasses and Crystalline Glass Materials with High Dielectric Constants, Gosénergoizdat, Moscow-Leningrad, 1952.

# MICROCRYSTALLINE MATERIALS WITH POSITIVE, NEGATIVE, AND ZERO COEFFICIENTS OF THERMAL EXPANSION

## V. G. Chistoserdov and V. I. Novgorodtseva

In order to obtain crystalline glass materials with low positive and negative or zero values of the thermal expansion coefficient, we synthesized glasses which deposit crystalline phases with small positive or negative values of the coefficient of thermal expansion.

By regulation of the amounts of these crystalline phases and of the residual glassy phase, which has a considerable positive coefficient of linear thermal expansion ($\alpha = 50$-$70 \cdot 10^{-7}$ degree$^{-1}$), it is possible to obtain crystalline glass materials with a wide range of coefficients of thermal expansion.

It is known [1] that a region of negative thermal expansion of ceramic materials exists in the lithium aluminosilicate system in the regions containing petalite ($Li_2O \cdot Al_2O_3 \cdot 8SiO_2$), spodumene ($Li_2O \cdot Al_2O_3 \cdot 4SiO_2$), and eucryptite ($Li_2O \cdot Al_2O_3 \cdot 2SiO_2$). When heated to 700-900°, these minerals are changed irreversibly into high-temperature $\beta$-forms; the structures of these forms are based on the tetrahedral framework of high-temperature quartz [2,3]. The structure of $\beta$-eucryptite is basically identical with that of high-temperature quartz. The symmetry of $\beta$-eucryptite is retained up to the composition $Li_2O \cdot Al_2O_3 \cdot 3.5SiO_2$.

Table 1 gives the coefficients of linear expansion of certain synthetic lithium minerals, taken from [4].

According to [5], the coefficient of thermal expansion of single $\beta$-eucryptite crystals is anisotropic; it is $+82 \cdot 10^{-7}$ degree$^{-1}$ along the a axis, and $-176 \cdot 10^{-7}$ degree$^{-1}$ along the c axis (Fig. 1). This pronounced anisotropy leads to the formation of cracks and cavities in polycrystalline eucryptite on heating; therefore, there is at present no information in the literature on the formation of fairly strong materials based on $\beta$-eucryptite with negative or zero coefficients ($\alpha$) of linear thermal expansion.

Glass compositions lying in the center of the spodumene–eucryptite regions were chosen for the investigation. Multicomponent glasses similar to those reported in [4] were used in order to lower the crystallization rate. The glass compositions were chosen so that eucryptite should be deposited first during crystallization

TABLE 1

| Ratio $Li_2O : Al_2O_3 : SiO_2$ | $\alpha \cdot 10^{-7}$ degree$^{-1}$ in temp. range | | | |
|---|---|---|---|---|
| | 20-400° | 20-600° | 20-800° | 20-1000° |
| 1 : 1 : 2 (eucryptite) | −80 | — | — | −65 |
| 1 : 1 : 3 | −19 | −14 | −11 | − 7 |
| 1 : 1 : 4 (spodumene) | + 4.5 | + 7 | + 9 | + 9 |
| 1 : 1 : 6 | 0 | + 1.6 | + 4.5 | + 5 |
| 1 : 1 : 8 (petalite) | 0 | + 1.2 | + 2.5 | + 3 |
| 1 : 1 : 10 | + 7.5 | + 7 | + 6 | + 4 |

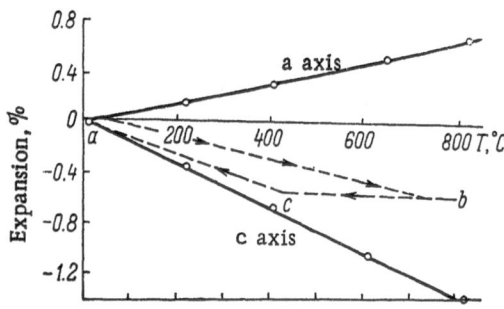

Fig. 1. Thermal expansion of a single crystal of $\beta$-eucryptite along the a and c axes, and of polycrystalline $\beta$-eucryptite (broken lines).

154

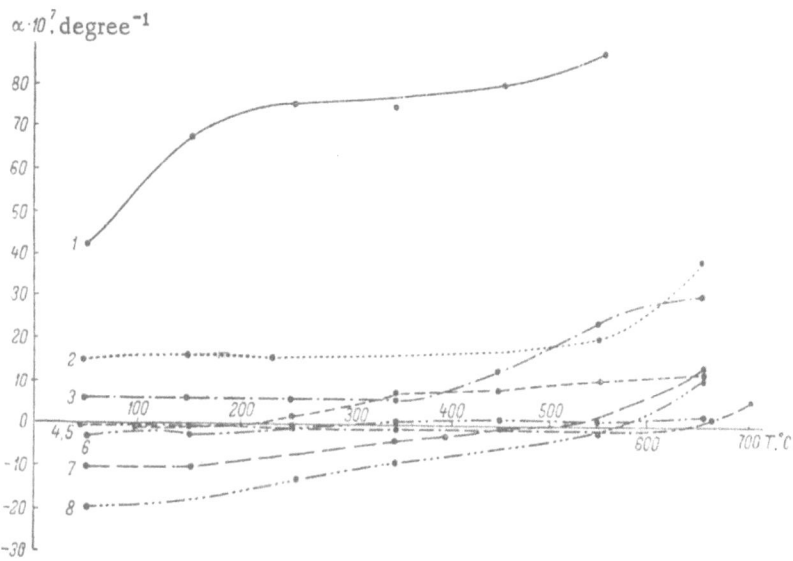

Fig. 2. Coefficients of linear thermal expansion of the original glass (1)
and of microcrystalline materials obtained from it (2-8).

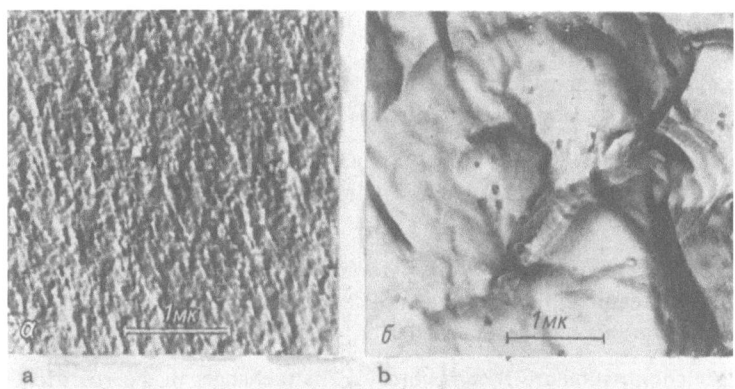

Fig. 3. Electron micrographs (a and b) of microcrystalline materials.

in the form of small crystalline formations. The linear shrinkage of the specimen then reaches 1%, the bending strength increases two- to threefold to 15-25 kg/mm², and the microhardness increases slightly, from 560 to 610-640 kg/mm².

Figure 2 represents the coefficients of linear thermal expansion $\alpha$ of specimens of the original glass and of crystalline glass materials obtained from it by heat treatment under various conditions. The precision in the determination of $\alpha$ is about 3%. Crystalline glass materials with zero and negative values of $\alpha$ (curves 5-8) were obtained by low-temperature crystallization procedures (up to 820°). Materials with positive (curves 2-4) values were obtained by high-temperature procedures (850-1150°). Specimens with zero and small positive or negative $\alpha$ withstand more than 70 heat cycles (from 800-850° to water at 5-6°) without change of properties.

The electron micrograph in Fig. 3a is of specimens represented by curves 6 and 7 (Fig. 2), and that in Fig. 3b, of the specimen represented by curve 3 (Fig. 2). The photographs show that the structure becomes coarser when the heat-treatment temperature is raised.

The results of an x-ray diffraction study of specimens of crystalline glass materials and comparison standards are presented in Table 2 in the form of a list of the principal lines.

TABLE 2

| β-Eucryptite standard | | Crystalline glass material, specimens 6 and 7 | | Spodumene standard | | Crystalline glass material, specimen 3 | |
|---|---|---|---|---|---|---|---|
| Interplanar spacings, A | I, % | Interplanar spacings, A | I, % | Interplanar spacings, A | I, % | Interplanar spacings, A | I, % |
| 4.60 | 30 | 4.57 | 12 | 4.61 | 10 | 4.576 | 11 |
| 3.52 | 100 | 3.50 | 100 | 3.97 | 40 | 3.89 | 33 |
| 2.63 | 25 | 2.627 | 6 | 3.48 | 100 | 3.46 | 100 |
| 2.29 | 25 | 2.265 | 5 | 3.17 | 12 | 3.148 | 12 |
| 2.13 | 25 | 2.09 | 5 | 1.94 | 4 | 1.925 | 7 |
| 1.92 | 50 | 1.892 | 22 | 1.89 | 10 | 1.867 | 13 |

The β-eucryptite and spodumene standards were taken from the ASTM index; they agree with the standards obtained in our laboratory. The crystalline glass materials corresponding to curves 6 and 7 (Fig. 2) are identical with β-eucryptite, and the material corresponding to curve 3 (Fig. 2) is identical with spodumene. It is evident that β-eucryptite is the main crystalline phase deposited on heating to 800-850°; at higher temperatures spodumene appears.

The sharp dependence of the electrical properties of the glass on the crystallization temperature is of great interest. The tangent of the loss angle at $f = 10^5$ cps and 25° is $400 \cdot 10^{-4}$ for the original glass, over $700 \cdot 10^{-4}$ for glass crystallized at 800°, and less than $100 \cdot 10^{-4}$ for glass crystallized at 1150°. The dielectric constant is 9.7, 23, and 13, respectively. At $f = 10^3$ cps and 160°, the dielectric constant alters from 44 to values exceeding 100, depending on the crystallization temperature. The value of $\varepsilon$ increases rapidly with rise of temperature.

As a result, we obtained strong nonporous crystalline glass materials in the lithium aluminosilicate system with coefficients of linear thermal expansion ranging from +15 to $-20 \cdot 10^{-7}$ degree$^{-1}$, including zero.

By different thermal treatments it is possible to obtain, from the same glass, microcrystalline materials with predetermined coefficients of expansion, which depend mainly on the contents of β-eucryptite or spodumene in the material.

Crystalline glass materials obtained by low-temperature heat treatment contain β-eucryptite as the main crystalline phase. If the heat-treatment temperature is raised, considerable amounts of spodumene and other crystalline phases appear and the structure becomes coarser.

Changes in the phase composition are accompanied by sharp changes in the electrical properties of the materials; this offers great potentialities for deeper study of problems associated with the structure of glass and crystalline glass materials.

## LITERATURE CITED

1. E. Smoke, J. Am. Ceram. Soc. 34(3):87, 1951.
2. F. Hummel, J. Am. Ceram. Soc. 34(8):235, 1951.
3. H. Saalfeld, Ber. Dtsch. keram. Ges. 38(7):281, 1961.
4. W. Hinz and P. Kunth, Silikattechnik No. 11, 506, 1960.
5. S. Gillery and E. Bush, J. Am. Ceram. Soc. 42(4):19, 1959.

# DISCUSSION

Yu. M. Vorona, replying to a question on the possibility of investigating the structure of glassceramics by electron microscopy without the use of replicas, pointed out the difficulties in satisfying the three main requirements in the specimens: 1) specimen thickness of several hundred angstroms; 2) corresponding orientation of the crystals in the glassceramic; 3) presence of identity periods not smaller than 5-10 A in the crystals.

N. S. Andreev, replying to a question on the causes of opalescence in the glasses investigated by him (microseparation of phases or presence of a small amount of a crystalline phase), asserted that there were no traces of crystallization in the cases under consideration, even when the heterogeneity regions become as large as 1000 A (the size may be varied by heat treatment) and about half the glass volume is occupied by these regions. The only explanation which can be offered for the effects, studied in detail (both by low-angle x-ray scattering and by other methods), is the presence of amorphous regions in these glasses.

E. V. Podushko spoke about the influence of $TiO_2$ on the mechanism of glass crystallization. He said that absence of volume crystallization in specimens with low titanium contents was checked by reference to the coefficient of expansion. At high temperatures the presence of titanium in the glass leads to phase separation, i.e., to breakdown of the structure with formation of new interfaces. These interfaces are the sites on which the crystal nuclei grow in the main phase during the precrystallization treatment of the glass. He pointed out that the precrystallization period should be regarded as the state of the glass when its viscosity is $10^{13}$ poises, i.e., the annealing range. In conclusion, he referred to the contradictory nature of data on glassceramics belonging to the same system, such as the $Li_2O$-$Al_2O_3$-$SiO_2$ system. In particular, this applies to interpretation of x-ray diffraction diagrams, infrared spectra, etc., which makes correlation of the data difficult.

I. D. Tykachinskii spoke on the question of the precrystallization period. In his opinion, this concept is not clear at this time, because it is not defined by any parameters. He also opposed N. A. Tudorovskaya's assertion that the exothermic effect is associated with crystal growth. He cited an example of an exothermic effect without growth of crystals in the glass. In his view, the exothermic effect is associated with deposition of primary crystalline phases.

E. A. Porai-Koshits was of the opinion that the argument as to which period should be termed the precrystallization period is terminological in character: it is simply the period of heat treatment which does not cause any crystallization in the glass. In contradicting the view of I. D. Tykachinskii that the precrystallization period is characterized by an absence of changes in the main properties, Porai-Koshits referred to the considerable changes of the submicrostructure and degree of chemical heterogeneity detected by the low-angle x-ray method in a number of glasses. Disagreeing with the view of E. V. Podushko that catalyzed crystallization of glass occurs on "defects," Porai-Koshits stated that a defect is a disturbance of the general structure, i.e., an exception in the general regularity, and therefore crystallization which begins on "defects" cannot have the dispersity and uniformity characteristic of glassceramics. The submicroscopic heterogeneity regions in glasses (i.e., metastable phase separation) are not "defects," but constitute the normal state of the glass, favoring simultaneous uniform and fine crystallization. By heat treatment it is possible to vary the glass structure in the precrystallization period in the direction required for the production of a "good glassceramic."

E. V. Podushko considered that titanium dioxide begins to act as a catalyst during the cooling period while still in the liquid phase. Crystallization centers are not formed during this period. It is impossible to obtain a crystalline glass material by holding the glass during cooling from a high temperature, as the formation of crystallization centers corresponds to the precrystallization period. There is a definite precrystallization period for each composition. In reply to I. D. Tykachinskii, Podushko emphasized that nucleation is an endothermic and not an exothermic process. It is followed by crystal growth, accompanied by exothermic effects. The crystallization centers are regarded as particles smaller by at least an order of magnitude than the crystals themselves, and their volume should not exceed 1% of the glass volume.

I. D. Tykachinskii insisted that the precrystallization period is characterized by absence of changes in the main properties. When the first crystals appear, the structure becomes consolidated and the properties change sharply. Replying to E. V. Podushko and discussing the effect of titanium dioxide, Tykachinskii associated himself with the opinion that crystals of the main phase grow on rutile, and emphasized the similarity between the lattice constants of the catalyst and the main crystalline phase.

É. M. Rabinovich discussed the use of thermal analysis for investigating crystallization of lithium aluminosilicate glasses. He noted that this method makes it possible to detect the appearance of metastable phases in the lithium aluminosilicate system, as was shown in a paper published in Voprosy radioélektroniki (ser. 4, No. 7, 69, 1961).

Yu. I. Kolesov, commenting on the paper by A. I. Sherstyuk and A. G. Vlasov, pointed out that if two crystalline phases with different heats of formation separate out simultaneously, it is impossible to calculate the amount of crystalline phase formed from the results of differential thermal analysis.

S. V. Nemilov, with reference to the same paper, noted that the sharp increase in the specific heat of the glassy phase in the softening range should affect the precision of calculations from the results of differential thermal analysis.

A. I.Sherstyuk , in his reply to Yu. I. Kolesov, said that the theory was developed for separation of the phases in different temperature ranges, which was always the case experimentally. Even similar crystalline phases are separated sharply at high heating rates, as they are deposited in narrow temperature ranges. The reply to the second comment was that the thermal constants of the glass before and after crystallization differed little.

*Other Three-Component and Multicomponent Systems*

# CERTAIN PROPERTIES OF THE LATENT IMAGE
# IN PHOTOSENSITIVE GLASSES

## V. A. Borgman and V. G. Chistoserdov

It has been shown [1,2] that the effectiveness of the action of ultraviolet radiation on photosensitive glasses greatly depends on the irradiation temperature. It was suggested [2] that heating of the glass increases the number of structural defects which become centers of the latent image of capturing photoelectrons.

Let us denote by N the concentration of free electrons created by irradiation in the glass. The loss of electrons captured by defects or recombining with ions is made up by fresh photoelectrons; equilibrium is established and N remains constant during the irradiation.

Before the start of irradiation, the concentration of free traps in the glass was $n_0$. Formation of the latent image during irradiation is essentially filling of the free traps with electrons; as a result, at time t there are n free and $n_1$ filled traps. Evidently the electron — trap interaction is represented by the equation

$$\frac{dn}{dt} = -\alpha n N, \tag{1}$$

where $\alpha$ is the probability of capture of an electron.

Hence it is easy to derive the expression for increase of the density of the latent image with time:

$$n_1 = n_0 (1 - e^{-\alpha N t}). \tag{2}$$

As stated in [2], the latent image centers have an absorption band in the ultraviolet region. The optical density at the maximum of the band increase during irradiation in accordance with the expression

$$D_m = \alpha n_0 (1 - e^{-\alpha N t}). \tag{3}$$

Taking logarithms of the derivative of (3) with respect to t and replacing the differentials by finite differences, we have

$$\ln \frac{\Delta D_m}{\Delta t} = \ln (\alpha n_0 \alpha N) - \alpha N t. \tag{4}$$

This expression makes it possible to check the agreement between (3) and experimental data, and to determine the values of $\alpha N$ and $\alpha n_0$. By determining the relationship (3) at different temperatures, we can follow the variations of these quantities with temperatures and thus verify the hypothesis put forward in [2].

Experiments were carried out for this purpose with the lithium photosensitive glass described in [1,2], in which irradiation creates an absorption band with a maximum at 270 m$\mu$.

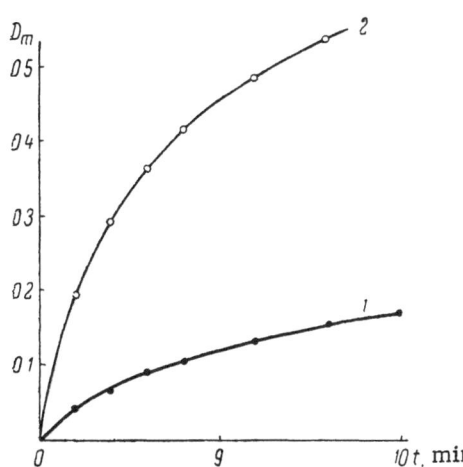

Fig. 1. Effect of the irradiation time on the optical density at the maximum of the absorption band of the latent image centers. 1) Irradiation at 20°; 2) at 180°.

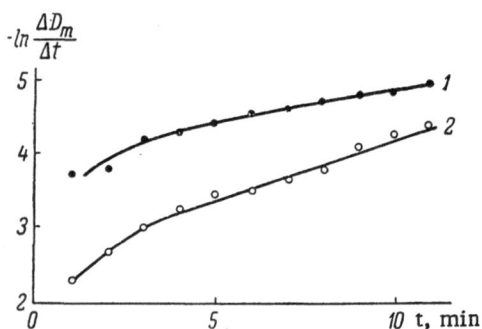

Fig. 2. Variations of $-\ln(\Delta D_m/\Delta t)$ with the irradiation time. 1) Irradiation at 20°; 2) 180°.

In Fig. 1, $D_m$ is plotted against the irradiation time (the optical density was not reduced to unit thickness, as the effect of active radiation weakens with increasing penetration into the glass; the specimens were 0.4 mm thick).

In Fig. 2, $-\ln(\Delta D_m/\Delta t)$ is plotted against the irradiation time for the same specimens. The linear character of these plots shows that Eq. (1) is in agreement with experimental data. During the first 3 min, the increase of $D_m$ is slower than is indicated by Eq. (3). Evidently, nonstationary processes occur during the first stage of irradiation.

Investigation of the plots in Fig. 2 gave the following results: at 20°, $\alpha N = 0.1$ and $an_0 = 1.17 \cdot 10^{-16}$; at 180°, $\alpha N = 0.17$ and $an_0 = 2.6 \cdot 10^{-6}$ (for the given specimen thickness and irradiation conditions, as N depends on the irradiation intensity). In experiments with other specimens the values of $an_0$ showed a scatter over two orders of magnitude, as the properties of the glass specimens were somewhat different owing to heterogeneity. There is no doubt, however, that in the temperature range studied $\alpha N$ varies little at the same irradiation intensity, whereas the trap concentration $n_0$ increases by 9-10 orders of magnitude with rise of temperature. This accounts for the increase of sensitivity on heating within the temperature range where a latent image is formed.

After the glass has been heated and cooled, $n_0$ and sensitivity return to the initial values; therefore, the increase of $n_0$ on heating must be attributed to reversible changes of the glass structure.

## LITERATURE CITED

1.  V. A. Borgman, V. M. Petrov, and V. G. Chistoserdov, Zhur. Fiz. Khim. 35(6):1383, 1961.
2.  V. A. Borgman and V. G. Chistoserdov, Optika i Spektroskopiya 12(1):141, 1962.

# IMAGE FORMATION IN PHOTOSENSITIVE GLASS
# CONTAINING GOLD AND SILVER

## E. V. Gurkovskii

Crystallization centers are formed in photosensitive glass under the influence of radiation. Image formation in photosensitive glass was studied with glasses in the $Na_2O$-$CaO$-$SiO_2$ system with additions of cerium dioxide, gold, and silver.

Figures 1-3 represent variations of light absorption under the influence of ultraviolet radiation and subsequent thermal deexcitation for glass with no addition (Fig. 1), glass containing cerium (Fig. 2), and glass containing cerium and gold (Fig. 3). In the first two cases, the absorption returns to the original level as a result of deexcitation, but in the last case an absorption band (Figs. 3 and 4) is formed with a maximum at about 280 m$\mu$, which does not disappear after thermal deexcitation. The intensity of this band increases to a certain limit with the irradiation time.

Preliminary experiments with potash glass showed that it also gives rise to an absorption band which persists after thermal deexcitation, but no image is formed after subsequent heat treatment. The formation of an absorption band is a necessary but not a sufficient condition for image formation.

The sensitivity of glass containing gold to ultraviolet irradiation depends on the cerium dioxide content. There is an optimum amount of cerium dioxide which gives the maximum image density (Fig. 5). The role of an electron donor in exposure to ultraviolet radiation is attributed to $Ce^{3+}$. An absorption band at about 320 m$\mu$ is produced by $Ce_2O_3$ in glass. The action of cerium as an electron donor is due to the $CeO_2 \rightleftharpoons Ce_2O_3$ equilibrium in the glass.

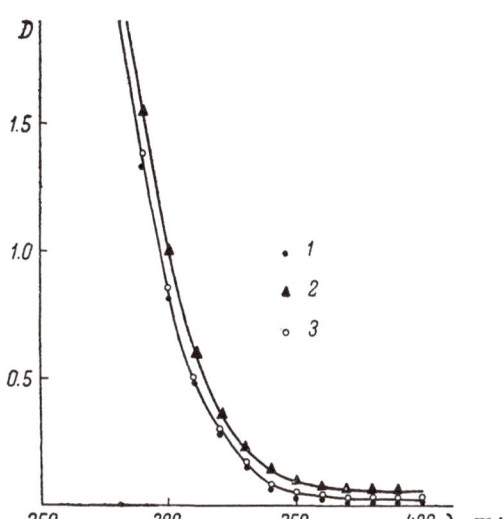

Fig. 1. Absorption curves of glass without gold and cerium in the ultraviolet region. 1) Unexposed glass; 2) irradiated; 3) deexcited. d = 0.8 mm.

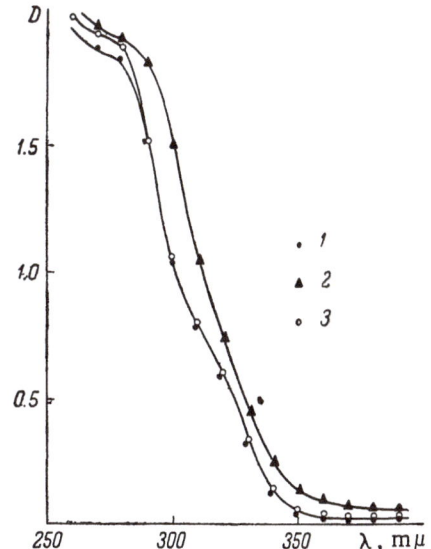

Fig. 2. Absorption curves of glass with 0.05% $CeO_2$ in the ultraviolet region. 1) Unexposed glass; 2) irradiated; 3) deexcited. d = 1.1 mm.

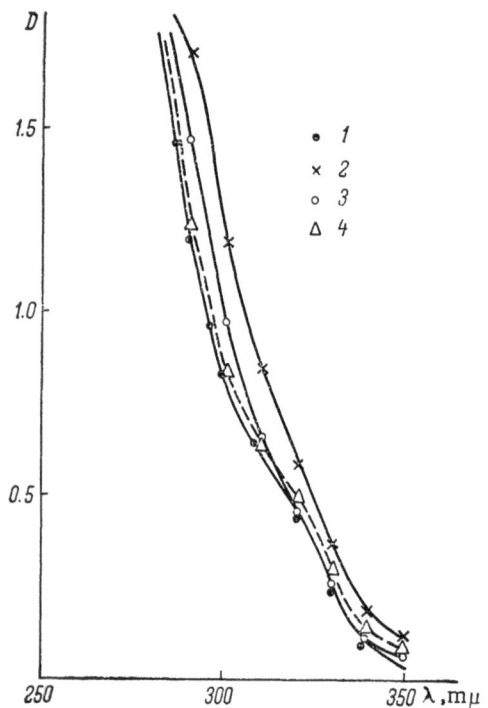

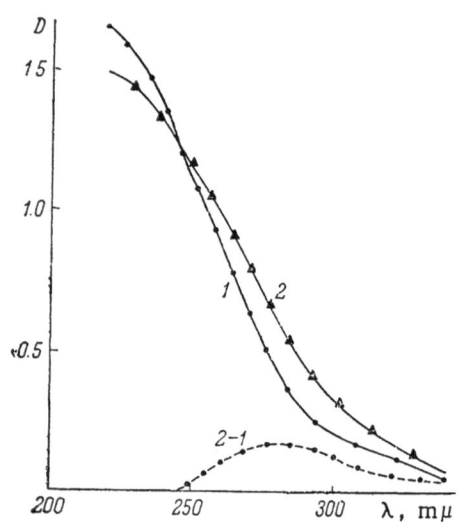

Fig. 3. Absorption curves of glass with 0.05% CeO$_2$ and 0.01% Au in the ultraviolet region. 1) Unexposed glass; 2) irradiated; 3) deexcited; 4) developed. d = 0.85 mm.

Fig. 4. Formation of an absorption band (2−1) in photosensitive glass with gold. 1) Unexposed glass; 2) irradiated; 2−1) difference. d = 0.11 mm.

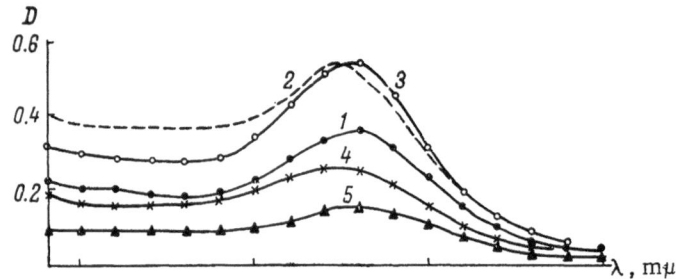

Fig. 5. Effect of the amount of CeO$_2$ on the spectral absorption curves of photosensitive glasses with 0.02% Au. 1) 0.05%; 2) 0.075%; 3) 0.010%; 4) 0.125%; 5) 0.150%.

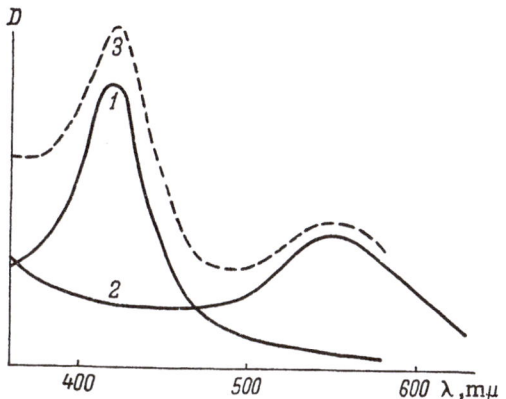

Fig. 6. Addition of spectral absorption curves of glass resulting from the presence of silver (1) and gold (2); result of addition of the ordinates of curves 1 and 2 (3).

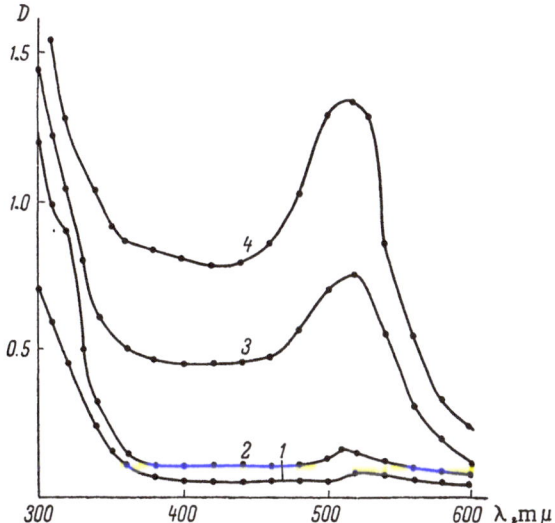

Fig. 7. Effect of the amount of gold on the spectral absorption curves of glasses with 0.025% Ag and 1% CeO₂. 1) 0.005%; 2) 0.010%; 3) 0.020%; 4) 0.040%. d = 0.8 mm.

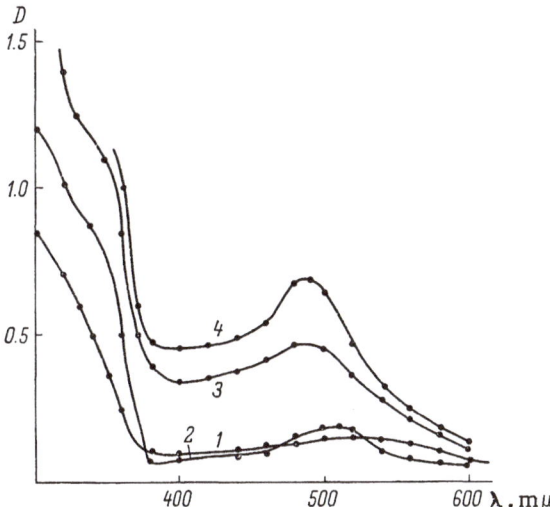

Fig. 8. Effect of the amount of silver on the spectral absorption curves of glasses with 0.010% Au and 0.1% CeO₂. 1) 0.012%; 2) 0.025%; 3) 0.050%; 4) 0.100%. d = 0.8 mm.

The optical density spectral curves of photosensitive glasses containing gold and silver are not the result of superposition of optical density curves resulting from the presence of gold and silver separately (Fig. 6), but have a single maximum, the position of which depends on the relative amounts of gold and silver; variations of the amount of silver have the greater influence on the shift of the maximum (Figs. 7 and 8). The shift of the maximum of the optical density curves was observed between 420 and 520 mμ. These results show that a continuous series of solid solutions of gold and silver is formed in glass containing gold and silver simultaneously.

A pale pink or yellow color is formed as a result of prolonged heat treatment of nonirradiated glasses. It may be concluded from this that photosensitive glass contains sensitivity centers — neutral metal atoms. However, experiments with potash glass showed that the presence of sensitivity centers does not always lead to image formation.

Study of the conditions of image formation in photosensitive glasses containing gold and silver may be one method of studying the nature of the glassy state.

# OPTICAL INVESTIGATIONS OF PHOTOSENSITIVE GLASSES

## A. A. Gorbachev, Yu. M. Polukhin, A. M. Ravich and L. M. Yusim

The use of crystalline glass materials based on photosensitive glasses in electronics makes it necessary to study processes of image formation and crystallization in such glasses. Various properties of photosensitive glasses have been studied by numerous workers [1-5].

The present paper is concerned with an investigation of the mechanism of image formation in photosensitive glasses of the lithium aluminosilicate system, and of nucleation kinetics. The following main optical characteristics were studied for this purpose: 1) absorption spectra of irradiated, nonirradiated, and heat-treated glasses; 2) the dependence of absorption on temperature during continuous heating of the glass specimens; 3) thermoluminescence; 4) luminescence spectra of irradiated and nonirradiated glasses in relation to the temperature of heat treatment.

The principal investigations were carried out with photosensitive glass 2L of the lihtium aluminosilicate system, made in a 1.5-ton tank furnace. The specimens were irradiated by a PRK-7 lamp at a distance of 400 mm. The absorption spectra were recorded with the aid of the SF-4 quartz spectrophotometer. The results of the determinations are given in Fig. 1.

Irradiation of the glass gives rise to a band in the 275 m$\mu$ region, the intensity of which depends on the irradiation intensity and is independent of the silver concentration. The intensity of this band decreases with rise

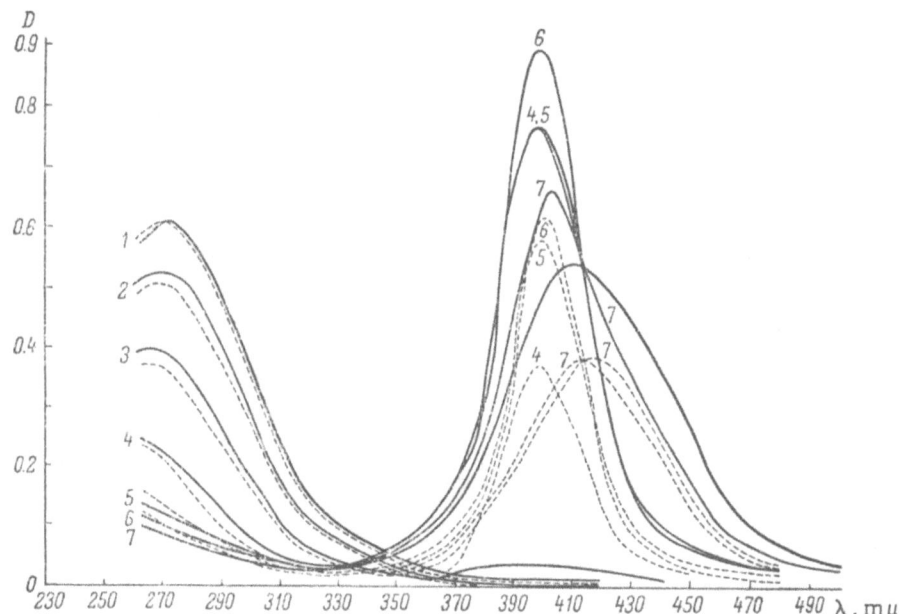

Fig. 1. Absorption spectrum of 2L glass in relation to heat treatment and silver concentration (relative to nonirradiated glass). 1) Heat treatment at 20°; 2) at 250°; 3) at 300°; 4) at 350°; 5) at 400°; 6) at 450°; 7) at 500°. Continuous lines, 0.08 mole% Ag; broken lines, 0.04 mole% Ag.

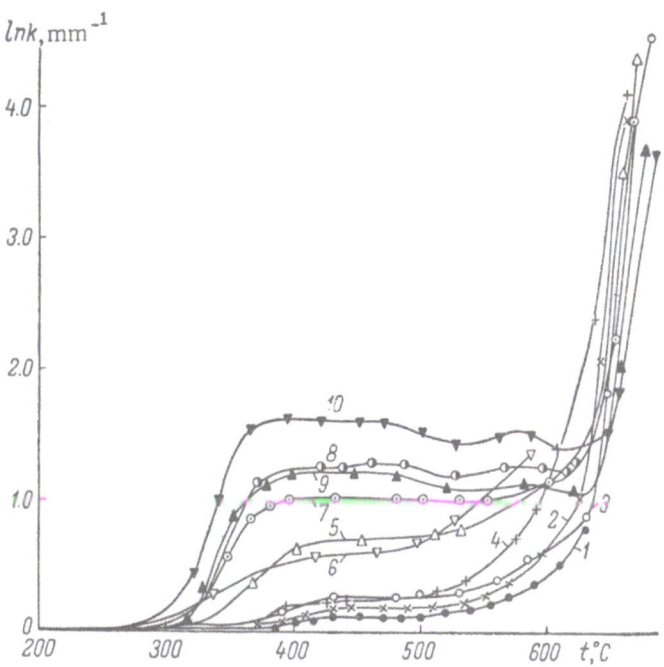

Fig. 2. Dependence of the absorption of glass 2L on the temperature with different exposures (ultraviolet irradiation). 1) 1 min;
2) 2 min; 3) 4 min; 4) 8 min; 5) 16 min; 6) 32 min; 7) 1 h; 8) 2 h;
9) 3 h; 10) 4 h.

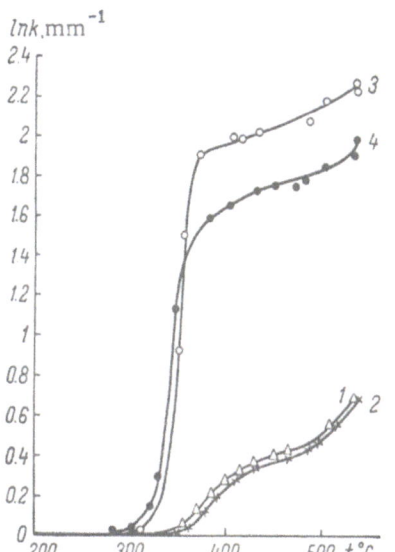

Fig. 3. Absorption curves of glass 2L
irradiated for 8 min (1,2) and 4 h (3,4).
1,3) Without heat treatment; 2,4) held
at 540°.

of the temperature of heat treatment; starting with 300°, the decrease is accompanied by the appearance and intensification of an
absorption band in the 405 m$\mu$ region. The intensity of the latter
is a linear function of the Ag concentration, to within 0.01-0.1
mole%, which is in agreement with the Lambert-Beer law. On
further heat treatment the linear relationship between the absorption and the Ag concentration breaks down, which indicates the
start of association of the absorbing centers [6]. The intensity of
the band in the visible region reaches a maximum at 450°, corresponding to a sharp decrease in the rate of decrease of the intensity of the band in the ultraviolet region.

After treatment at 500°, the maximum begins to shift toward
longer waves. Increase of the treatment time results in a decrease
of the absorption intensity, a further shift of the maximum to 414
m$\mu$, and asymmetry of the band, indicating coagulation of the particles. The dependence of the shift of the maximum on the Ag
concentration is also characteristic of the coagulation process.

Special apparatus was used for recording the curves for the
dependence of absorption on temperature during continuous heating of the specimens (Fig. 2). At short irradiation exposures (up to
16 min) the rate of absorption increase is at first relatively low,
and development begins at about 350°. However, at about 500° the
absorption again begins to increase sharply. The absorption increases at a much higher rate with long exposures
(over 16 min) than with short. Development begins at 300°, saturation is reached at 370°, and a further increase
of absorption is observed only at temperatures in the region of 600°, which is associated with the start of crystallization. As will be shown later, the decrease of absorption observed at 520-560° is of considerable importance.

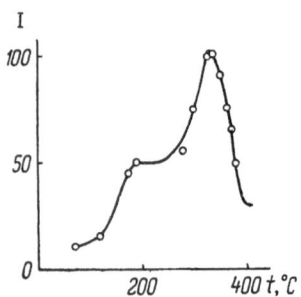

Fig. 4. Thermoluminescence
ceruve of glass 2L irradiated for
4 h.

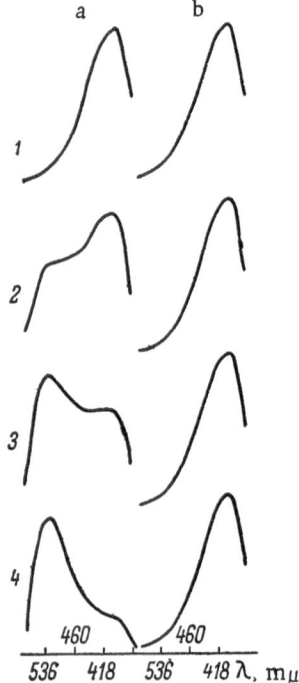

Fig. 5. Luminescence spectra
of irradiated (a) and nonirradiated (b) glass 2L heat-
treated at various temperatures.
1) 250°; 2) 290°; 3) 320°; 4) 350°.

We consider that the sharp increase of absorption at 500-520° in specimens irradiated for less than 16 min is due to deposition of silver on existing centers. Increased absorption during treatment in this temperature range
is characteristic of this process. This view is supported by the curves in
Fig. 3. At exposures longer than 16 min there is a decrease of absorption
in the 540-560° range (in agreement with Fig. 1); this effect increases with
the Ag concentration, which may be attributed to growth of the colloidal
particles as a result of coagulation.

The thermoluminescence curve (Fig. 4) shows that most electrons are
released from the level characterized by the temperature 340°, corresponding to the highest rate of development. Moreover, some electrons are released at a lower energy level (200°), which coincides with the decrease in
the intensity of the absorption band in the ultraviolet (Fig. 1) at that temperature.

The glasses studied contain cerium as the sensitizer. We determined
the luminescence spectra of irradiated and nonirradiated glasses heat-
treated at various temperatures. The luminescence spectrum of nonirradiated glass is not altered by heat treatment (Fig. 5). A maximum appears
in the long-wave region of the luminescence spectra of irradiated glasses
at definite temperatures preceding the increase of absorption. The intensity of this maximum increases in accordance with the absorption. The
luminescence maximum in the long-wave region of the spectrum disappears
completely when saturation absorption is reached.

Hence the appearance and growth of a luminescence maximum in
the long-wave region of the spectrum after heat treatment of irradiated
glasses may be attributed to formation of luminescence centers. These
centers are apparently associated with atomic silver centers, as their luminescence maximum appears at the temperature of the start of development.

On the basis of these investigations, the following mechanism of
image formation in photosensitive glasses may be postulated. Photoelectrons, the release of which is associated with the presence of a sensitizer in
the glass, pass to metastable levels as a result of irradiation and become
localized there because of the high viscosity of the glass. When the glass
is heated, the electrons can move and become localized near the silver ions,
forming so-called "atomic centers," the formation of which is accompanied
by increased absorption in the visible region of the spectrum. At higher
temperatures either the centers increase in size as a result of deposition
of silver on them (with short exposures), or the particles grow in size as a
result of coagulation (with long exposures). After reaching the critical size,
these particles can become crystallization nuclei in the glass.

LITERATURE CITED

1.    S. D. Stookey, Ind. Eng. Chem. 41(4), 1949.
2.    R. D. Maurer, J. Appl. Phys. 29(1), 1958.
3.    D. Barth, Silikattechnik No. 3, 1960.
4.    A. I. Berezhnoi, Photosensitive Glasses and Glassceramics of the Pyroceram Type, Moscow, 1960.
5.    V. A. Borgman, V. I. Petrov, and V. G. Chistoserdov, Zhur. Fiz. Khim. 35(6), 1961.
6.    M. V. Savost'yanova, Uspekhi Fiz. Nauk 22(1), 1939.

# ACTION OF DILUTE HYDROFLUORIC ACID SOLUTIONS
# ON LITHIUM SILICATES

N. A. Shmeleva, V. G. Chistoserdov, and A. I. Gerasimova

It is known that the solubility of certain photosensitive glasses in hydrofluoric acid after treatment greatly depends on whether they had previously been subjected to ultraviolet irradiation. This is a matter of great practical importance. It is of interest to determine what crystalline phases increase the solubility in HF.

Photosensitive glass containing 8% $Al_2O_3$ was used for the investigation in the form of 20 × 20 × 0.5 mm plates. Solubility, expressed as the difference between the weights before and after treatment for 30 min in 10% hydrofluoric acid, was investigated with specimens crystallized for 2 h at temperatures in the range of 500-800° at intervals of 50°. The results are given in Figs. 1 and 2.

In the case of glass 137, the only factor determining the difference between the curves 1 and 2 (Fig. 1) was that the specimen was irradiated in the first case but not in the second. The highest solubility (0.58 g) was found for irradiated specimens crystallized at 550 and 600°. At 600° there was considerable scattering of the experimental points, but they all lay on the downward branch of the curve, indicating a nonequilibrium state in the system.

Glass 133 (same composition as glass 137, but from a different melting) gives a solubility curve of the same character, but the temperature range in which a phase of higher solubility is formed is somewhat narrower (Fig. 2, curve 1). The solubility at 600° was 0.15 g, or approximately the same as for nonirradiated specimens. The course of curves 1-4 (Fig. 2) shows that the solubility maximum shifts with increase of the storage time toward higher temperatures, with a simultaneous decrease of solubility, especially at low crystallization temperatures. It also follows from these curves that crystallized glass of this composition is in a nonequilibrium state

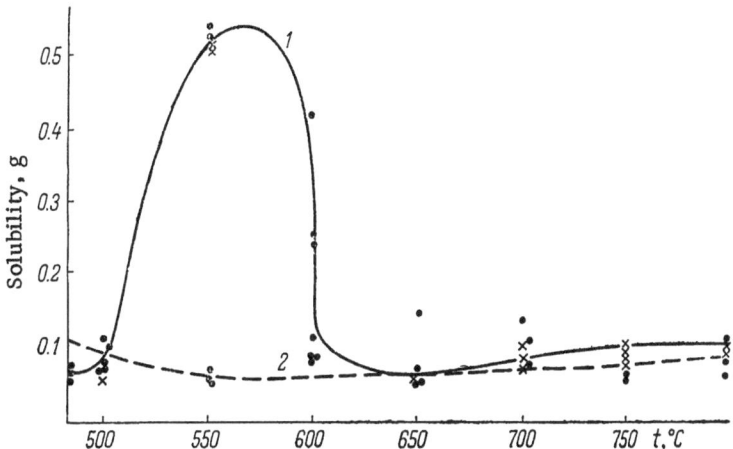

Fig. 1. Solubility of glass 137 in 10% HF solution as a function of the crystallization temperature. 1) Irradiated glass; 2) nonirradiated glass.

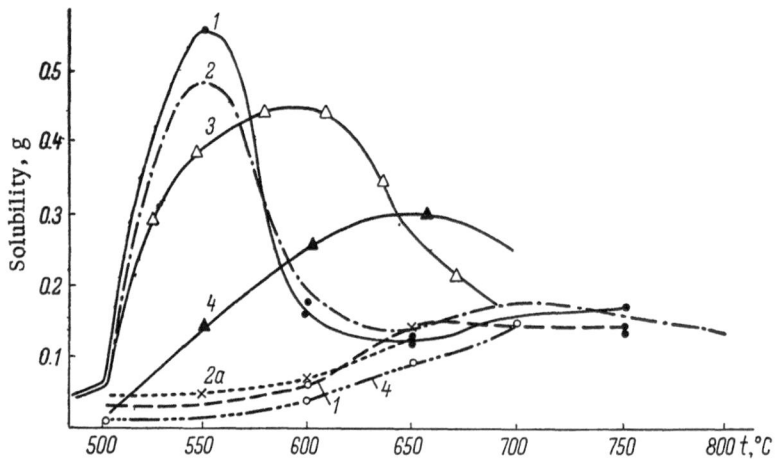

Fig. 2. Solubility of glass 133 in 10% HF solution as a function of the crystallization temperature. Curves 1-4 were determined on the 2nd, 7th, 14th, and 31st day after crystallization of irradiated glass; curves 1a, 2a, and 4a correspond to the 2nd, 7th, and 31st day after crystallization of nonirradiated glass.

| Standard | Solubility in 1% HF solution | |
|---|---|---|
| | Sample wt., % | K* |
| High-temperature eucryptite | 94.3 | 2.5 |
| Lithium metasilicate | 60.8 | 1.62 |
| Lithium disilicate (>700°) | 58.1 | 1.54 |
| Spodumene (>700°) | 56.1 | 1.50 |
| Lithium disilicate (700°) | 43.3 | 1.15 |
| $\alpha$-Cristobalite | 17.2 | 0.46 |
| Quartz (20°) | 0.3 | 0.17 |

* The solubility of lithium glass with 8% $Al_2O_3$ was taken as the arbitrary unit of solubility.

and undergoes changes in time even at room temperature, leading to decreased solubility in hydrofluoric acid.

The table gives the relative solubilities of the most probable crystalline phases in our glasses. Standard specimens, previously checked by x-ray diffraction and petrographic analysis, were used for this purpose. The following method was used for the solubility determinations: a finely divided sample (0.5 g) was treated for 1 h with 50 $cm^3$ of 1% hydrofluoric acid solution at 20° with stirring.

It is seen that eucryptite has the highest solubility (K = 2.5) and quartz and $\alpha$-cristobalite have the lowest. As is evident from Fig. 3, essentially only one phase crystallizes in glasses 133 and 137. Our investigations showed that irradiated crystallized glasses dissolve at 6-10 times the rate of nonirradiated glasses.

The region of increased solubility coincides with the initial stages of the formation of solid solutions of spodumene — eucryptite mixtures with $SiO_2$. As the crystallization temperature increases, the composition gradually alters in the same series of solid solutions, with separation of $SiO_2$.

In the initial stage in the transformation of the solid solutions, strictly parallel striation appears within each crystal, with a yellow-brown coloration, in consequence of volume shrinkage of the microcrystals (different in different axial directions) leading to the formation of microcracks.

Electron micrographs of specimens crystallized at 600° revealed (at 18,000 magnification) pronounced striation even in crystals 1 $\mu$ in size (Fig. 4). The striation revealed by prolonged etching of a freshly fractured surface in 0.1% HCl solution gave a positive relief; i.e., it was due to a substance difficultly soluble in HCl. This substance was undoubtedly cristobalite, in accordance with the following characteristics: low refractive index, low solubility in HCl, and even faceting with growth forms characteristic of cristobalite; i.e., in this case additional crystallization took place in the microcracks.

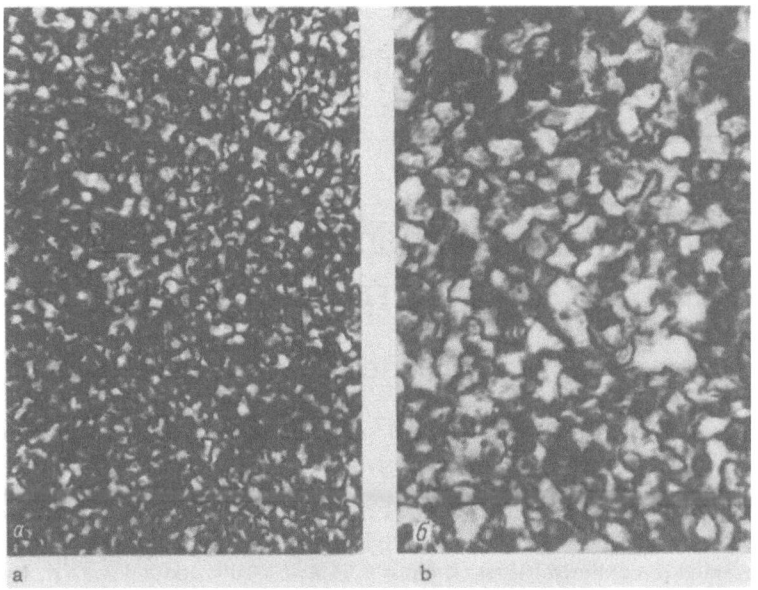

Fig. 3. Glass 133 crystallized for 2 h. a) At 600°; b) at 750°.
(CrossedNicols, magnification 640.)

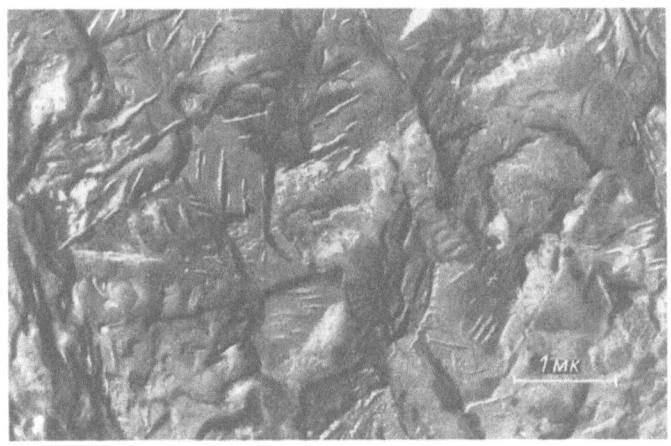

Fig. 4. Glass 137 crystallized at 600° (magnification 18,000).

Thus, the restriction of the region of higher solubility of irradiated and crystallized glasses (in 10% HF solution) should be attributed to transformation in the solid solution, condensation of its structure, and concentration of cristobalite at the shrinkage cracks of the crystals.

# STUDY OF CRYSTALLIZATION PRODUCTS IN THE MAGNESIUM ALUMINOSILICATE SYSTEM WITH ADDITIONS OF TiO$_2$

## V. G. Chistoserdov, N. A. Shmeleva, and A. M. Serdyuk

Study of the crystallization products of glasses in the MgO-Al$_2$O$_3$-SiO$_2$-TiO$_2$ system presents considerable difficulties because of the possible formation of numerous crystalline phases.

For elucidation of the influence of added TiO$_2$ we first studied the crystallization of titanium-free glasses in the cordierite field or near the boundary of that field. In glasses crystallized at 1200°, petrographic and x-ray diffraction methods revealed $\alpha$-cordierite (lines with d = 8.54 A, I = 100 and d = 3.37 A, I = 30) as the main crystalline phase; at 1000° there was a small amount of $\mu$-cordierite (d = 3.45 A, I = 100), an extremely small amount of $\alpha$-cordierite, and a main phase with d = 3.69 A. This last crystalline phase, which disappears on increase of temperature, was not identified.

The dynamics of crystalline phase formation were studied by means of x-ray diffraction diagrams recorded at high temperatures. One of the glasses investigated was composition 59, corresponding to the triple eutectic: 61.4% SiO$_2$, 18.3% Al$_2$O$_3$, and 20.3% MgO. To obtain a microcrystalline mass throughout the volume, 0.25% of elemental silicon was added to glass 59.

The resultant ionization tracings (Fig. 1) show that the unidentified phase (3.69 A) and $\mu$-cordierite are formed at 960° (curve 5). On increase of temperature the first phase disappears and $\mu$-cordierite passes into the $\alpha$-form; the $\mu$-cordierite line with d = 3.40 A (I = 100) becomes one of the principal $\alpha$-cordierite lines with d = 3.37 A (I = 77). At 1150°, $\alpha$-cordierite is present in a small amount, and at 1200° it is the only phase present (curves 8-10).

For studying the influence of TiO$_2$ on crystallization, TiO$_2$ was added to glass 59 in amounts from 2 to 18% by weight above 100%. With up to 8% TiO$_2$ in the glass, the structure remains coarse-grained. Crystallization at 1000° proceeds with formation of $\mu$-cordierite, which is in the stage of transformation into $\alpha$-cordierite. The amount of the phase with d = 3.69 A in titanium-containing glasses rapidly diminishes with increase of the TiO$_2$ content. With more than 10% TiO$_2$ microcrystalline structures were obtained; the principal phase was $\mu$-cordierite in the stage of transformation into the $\alpha$-form, and a small amount of clinoenstatite.

All the glasses containing from 2 to 18% TiO$_2$ and crystallized at 1100° contained $\alpha$-cordierite as the principal phase, and glasses with 10% TiO$_2$ and over also contained geikielite (MgTiO$_3$).

Figure 2 shows high-temperature diffraction diagrams of glass with 10% TiO$_2$; these are almost the same as the curves in Fig. 1, differing from the latter by some delay in the transformation of $\mu$-cordierite into the $\alpha$-form.

Figure 3 shows thermograms of glass with 10% TiO$_2$ and of a number of specimens of this glass after preliminary heat treatment. The strongest exothermic effect II at 960-1000° is attributed to the formation of $\mu$-cordierite, and effects IV and III (1150-1200°) to the transformation of $\mu$-cordierite into the $\alpha$-form and the formation of geikielite. Effect I is apparently associated with the unidentified phase with d = 3.69 A. Preliminary heat treatment eliminates effects I and II almost entirely and, at the same time, intensifies effect IV, which increases with rise of the heat-treatment temperature. The diffuse nature of the peak on the thermogram of glass 59 (Fig. 3, curve 5) shows that the formation of $\mu$-cordierite and its transformation into the $\alpha$-form are gradual processes. In the case of glass containing titanium, the formation of $\mu$-cordierite is abrupt (960-1000°), while its transformation into the $\alpha$-form occurs abruptly only after preliminary heat treatment.

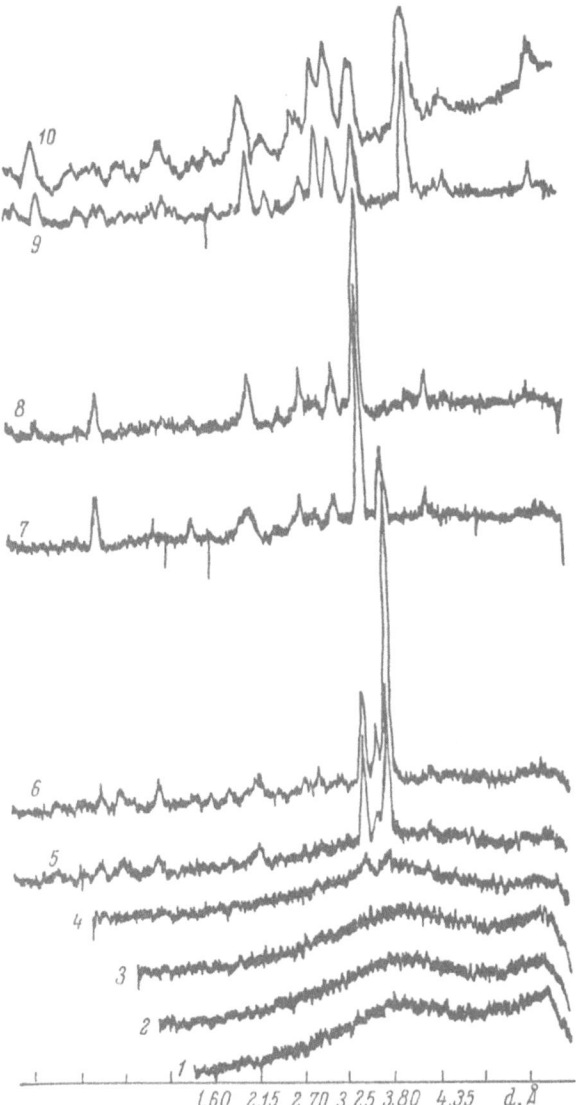

Fig. 1. Ionization tracings of glass 59 with addition of 0.25% silicon, recorded at high temperatures. 1) Uncrystallized glass; 2) crystallized at 600°; 3) at 800°; 4) at 900°; 5) at 960°; 6) at 1020°; 7) at 1100°; 8) at 1150°; 9) at 1200°; 10) cooled specimen.

Fig. 2. Ionization tracings of glass 59 with addition of 10% $TiO_2$, recorded at high temperatures. 1) Crystallized at 1000°; 2) instantly at 1100°; 3) at 1100° with delay; 4) at 1150°; 5) instantly at 1200°; 6) at 1200° with delay.

Thus, crystallization of titanium-containing glasses involves formation of $\alpha$-cordierite by way of the $\mu$-form (a metastable and insufficiently individualized compound showing variations from 1.540 to 1.610 in the average refractive index). In glasses with 10-18% $TiO_2$ with a microcrystalline structure, geikielite and sometimes clinoenstatite is formed.

These data do not agree with the results of Hinz and Baiburt [1], who reported the formation of magnesium titanate as the main phase at the initial crystallization stages (900°), and at 1000-1100° found spinel, sapphirine, and sometimes clinoenstatite in addition to the titanate. Such a result is indeed possible in studies of glass with the composition of cordierite with addition of 13% $TiO_2$, when the principal line with d = 3.37 A may be assigned to a titanium − magnesium compound. However, our analogous x-ray diffraction patterns of glasses without titanium contradict this.

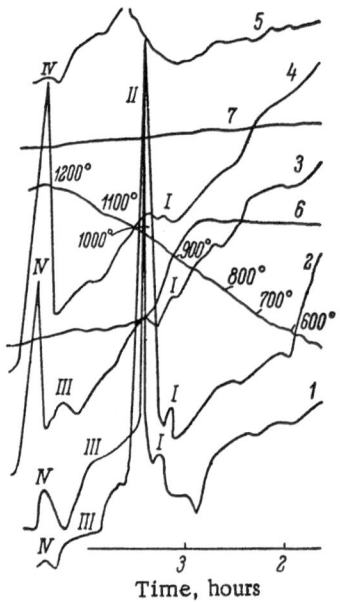

Fig. 3. Thermograms of glass 59 with addition of 10% $TiO_2$.
1) Without preliminary heat treatment; 2-4) preliminary heat
treatment for 3 h at 600-820, 900, and 960°, respectively; 5)
glass 59 without preliminary heat treatment; 6, 7) volume
shrinkage curves recorded simultaneously with thermograms
2 and 3.

The phase composition and microstructure of the resultant glassceramics determine their physicochemical properties. Because of its low coefficient of expansion (about $10 \cdot 10^{-7}$ degree$^{-1}$), $\alpha$-cordierite confers high temperature endurance on them, but the cellular — granular structure limits the possibility of obtaining high strengths.

LITERATURE CITED

1.    W. Hinz and L. Baiburt, Silikattechnik 11(10):455, 1960.

# CRYSTALLIZED GLASSES BASED ON CORDIERITE

## M. I. Kalinin and E. V. Podushko

We studied the catalyzed crystallization of glasses in the $MgO$-$Al_2O_3$-$SiO_2$ system, corresponding to cordierite in composition or containing not less than 70% of that compound, with 8-20% by weight of titanium dioxide as catalyst.

We investigated variations in the character of crystallization in relation to the catalyst content, and the influence of preliminary heat treatment on subsequent crystallization of the glasses. Crystallization of the glasses was studied by the polythermal method in the 700-1200° range; the exposure time was the same, 24 h, for all the glasses.

Glasses containing up to 8% of the catalyst exhibit surface crystallization. When the catalyst content is increased to 10%, crystallization begins at a lower temperature and is of the volume type. At a constant ratio of cordierite to catalyst in glasses with different cordierite contents, the character of the crystallization remains the same. Further increase of the catalyst content leads to even greater changes in the character of the crystallization, which become very considerable in glasses containing 1.5-2 times the minimum amount of catalyst needed to produce volume crystallization. We shall therefore consider separately the crystallization of glasses containing small and large amounts of catalyst.

The character of crystallization of glasses with small amounts of catalyst varies in accordance with the conditions of preliminary heat treatment; the crystallized glass may be transparent or opaque. The curve showing the dependence of the temperature at which transparency is lost on the conditions of heat treatment has a maximum, indicating that there are optimum conditions of preliminary heat treatment corresponding to the formation of the maximum amount of crystallization centers. Glasses subjected to such heat treatment lose transparency at temperatures higher by 100-150° than glasses not subjected to heat treatment.

Transparent crystallized glasses with high $TiO_2$ contents have the same temperature of transparency loss, without previous heat treatment, as glasses with small amounts of catalyst after preliminary heat treatment. Thus, preliminary heat treatment of glasses with high $TiO_2$ contents is ineffective with regard to displacement of the temperature of transparency loss. Apparently, at high $TiO_2$ contents the formation of the necessary amount of crystallization centers already occurs during the cooling and subsequent heating of the original glass to the crystallization temperature.

Differential thermal analysis showed that glasses with small amounts of $TiO_2$, not subjected to preliminary heat treatment, give one endothermic and two exothermic effects (Fig. 1). The differential heating curve of glass after preliminary heat treatment has three exothermic effects. The appearance of an additional effect, probably corresponding to separation of $\mu$-cordierite [1], is associated with the formation of numerous crystallization centers, on which this crystalline phase separates out, during the preliminary heat treatment.

The second exothermic effect, in our opinion, is associated with separation of a new crystalline phase from the residual glass. It appears that in untreated glass both crystalline phases separate out simultaneously owing to the lack of sufficient crystallization centers, which arise only as a result of preliminary heat treatment; this is indicated by the height of the peak on the heating curve and by the cracking of the material, as the two crystalline phases which separate out simultaneously have different coefficients of thermal expansion. The two crystalline phases can be produced separately by modification of the preliminary heat treatment.

The differential heating curve of glass with a high $TiO_2$ content is similar in nature to the heating curve

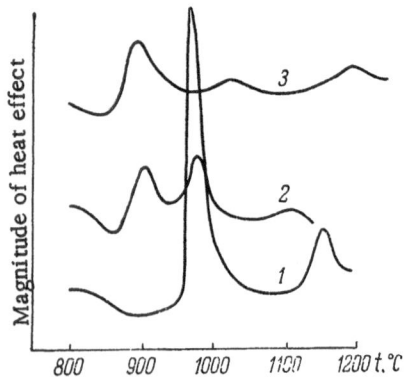

Fig. 1. Differential heating curves. 1) Glasses with low $TiO_2$ contents, without heat treatment; 2) low $TiO_2$ content, with preliminary heat treatment; 3) high $TiO_2$ content, without heat treatment.

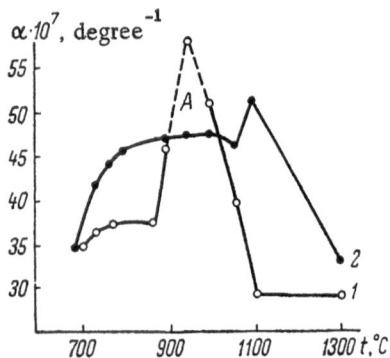

Fig. 2. Variation of the coefficient of linear expansion of crystallized glasses with the crystallization temperature. 1) Glasses with low $TiO_2$ contents; 2) glasses with high $TiO_2$ contents. A) Cracking region.

of glass with a low $TiO_2$ content subjected to preliminary heat treatment, although the heat effects are displaced: the first exothermic effect is shifted somewhat toward lower temperatures and the other two toward higher temperatures. This indicates that with high $TiO_2$ contents crystallization centers arise during cooling and subsequent heating of the glass to the crystallization temperature in a very short time without preliminary heat treatment. The reason why the second heat effect is smaller is probably because a larger amount of crystalline phase separates out in the temperature region corresponding to the first exothermic effect than in the case of glasses with a small amount of catalyst.

The third (high-temperature) effect on the heating curves of all the glasses is very weak; it apparently corresponds to inversion of the low-temperature form of cordierite, which is the primary crystalline phase formed in the crystallization of all the glasses studied.

The variations of the coefficient of linear expansion of glasses crystallized at different temperatures (Fig. 2) give an indication of the thermal expansion of the crystalline phases which separate out in the various temperature regions. Figure 2 shows that glass with a high $TiO_2$ content exhibits a regular increase of the coefficient of linear expansion up to a certain constant value corresponding to the thermal expansion of $\mu$-cordierite. Further increase of the crystallization temperature above the range of separation of the secondary crystalline phase leads to an increase of the coefficient of linear expansion; it follows that the secondary phase has greater thermal expansion than the primary. If the crystallization temperature is raised still further, the thermal expansion of the resultant material decreases considerably owing to the formation of $\alpha$-cordierite.

Glass with a small amount of catalyst, subjected to preliminary heat treatment, exhibits similar variations of the coefficient of linear expansion with the crystallization temperature.

In the case of glass with a low $TiO_2$ content, not heat-treated, the thermal expansion curve is entirely different. The thermal expansion of such glass increases slightly with rise of the crystallization temperature. Crystallization of this glass in the temperature region corresponding to separation of two crystalline phases gives a material the thermal expansion of which is greater than that of $\mu$-cordierite. In this case we may presume the formation of a fairly large amount of the secondary crystalline phase with high thermal expansion. Treatment of the glass at a higher temperature lowers the coefficient of linear expansion, owing to formation of $\alpha$-cordierite.

LITERATURE CITED

1. M. D. Karkhanavala and F. A. Hummel, J. Am. Ceram. Soc. 36(12):389, 1953.

# CATALYZED CRYSTALLIZATION OF GLASSES
# IN THE LITHIUM GALLOSILICATE SYSTEM

## G. T. Petrovskii, E. N. Krestnikova, and N. I. Grebenshchikova

In accordance with the concepts put forward by Petrovskii and Nemilov [1], the determining process in the formation of glassceramics in the lithium aluminosilicate system is switching of covalent bonds and change of the coordination of titanium and aluminum with respect to oxygen. Therefore, the most suitable systems for production of transparent glassceramics are those containing two oxides, one of which increases and the other decreases its coordination number when the temperature changes. It has been shown [2] that, in the high-temperature form of gallium oxide, gallium has two different coordination numbers with respect to oxygen: four and six. Several investigations [1,4] have shown that, in sodium gallosilicate glasses, gallium oxide passes from tetrahedral into octahedral coordination on decrease of the sodium oxide concentration in the glass. It is emphasized that fourfold coordination is less characteristic for gallium than for aluminum [4].

We investigated glasses of the lithium gallosilicate system with certain additions. The phase diagram of the system has never been studied and, despite the similarity between the properties of aluminum and gallium, there is no reason to expect it to be analogous to the corresponding system with alumina. Study of the system gallium oxide — silica [5,6] shows that it differs appreciably from the alumina — silica system by the existence of a phase-separation region and by the absence of mullite formation.

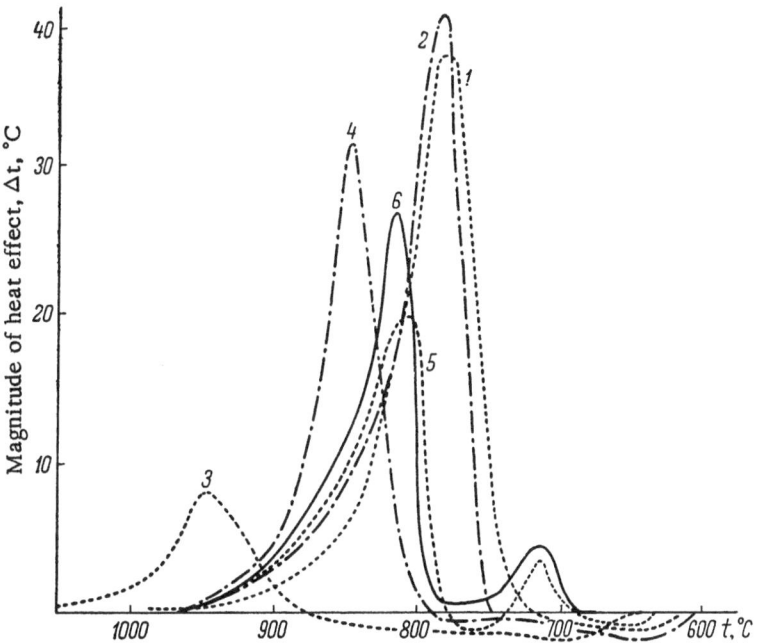

Thermograms of glasses of different compositions. 1) $Li_2O \cdot Ga_2O_3 \cdot 2SiO_2$; 2) $Li_2O \cdot Ga_2O_3 \cdot 4SiO_2$; 3) $Li_2O \cdot Ga_2O_3 \cdot 6SiO_2$; 4) $Li_2O \cdot Ga_2O_3 \cdot 4SiO_2$ + 5% $TiO_2$; 5) titanium-containing gallate glass with 2% $Li_2O$; 6) titanium-containing gallate glass with 5% $Li_2O$.

TABLE 1. Coefficients of Expansion of Aluminate and Gallate Glasses and Polycrystalline Specimens ($\alpha \cdot 10^{-7}$ degree$^{-1}$)

| Glass composition | Data, [9] | Data, [8] | Our data | |
|---|---|---|---|---|
| | | | glasses | polycrystals |
| $Li_2O \cdot Al_2O_3 \cdot 2SiO_2$ | 76 | — | — | — |
| $Li_2O \cdot Ga_2O_3 \cdot 2SiO_2$ | — | — | 80 | 68 |
| $Li_2O \cdot Al_2O_3 \cdot 4SiO_2$ | 66.6 | — | — | — |
| $Li_2O \cdot Ga_2O_3 \cdot 4SiO_2$ | — | 65.8 | 69 | 40 |
| $Li_2O \cdot Al_2O_3 \cdot 6SiO_2$ | 52.5 | — | — | — |
| $Li_2O \cdot Ga_2O_3 \cdot 6SiO_2$ | — | 54.5 | 52 | — |

TABLE 2. Changes in the Properties of Complex Gallium Glass on Crystallization

| Property | Glass | Glass-ceramic |
|---|---|---|
| Refractive index | 1.6023 | 1.6147 |
| Total dispersion | 0.01288 | 0.01350 |
| Density, g/cm$^3$ | 3.0666 | 3.1492 |
| Microhardness, kg/mm$^2$ | 660 | 824 |

In the $Li_2O$-$Al_2O_3$-$SiO_2$ system the region of compositions with negative and zero coefficients of expansion lie [7] mainly along the $SiO_2$-$Li_2O \cdot Al_2O_3$ section. We synthesized glasses corresponding to the formulas $Li_2O \cdot Ga_2O_3 \cdot 2SiO_2$, $Li_2O \cdot Ga_2O_3 \cdot 4SiO_2$, and $Li_2O \cdot Ga_2O_3 \cdot 6SiO_2$ in composition.

The coefficients of expansion of the glasses, according to our results and to Dubrovo's data [8], are close to the coefficients of expansion of the corresponding aluminate glasses (Table 1). However, in the crystalline state eucryptite, spodumene, and lithium orthoclase have low coefficients of expansion which vary in a very peculiar manner with the temperature. Table 1 shows that the coefficients of expansion of gallium glasses are changed very little as a result of crystallization.

The glasses were kept for 8 h at 950° for crystallization. Glass of the composition $Li_2O \cdot Ga_2O_3 \cdot 6SiO_2$ contained a large amount of glassy phase even after 10 h at 1000°. Thermal analysis of the glasses was carried out. The heating rate was 20° per minute. The curves were recorded with the aid of an electronic potentiometer with a fairly long time (8 sec) of carriage run over the scale. The figure shows that the heating curves (curves 1-4) contain only one effect associated with liberation of heat. Therefore, only one phase separates out in each glass in the temperature range studied. The evolution of considerable heat in a narrow temperature range during crystallization of $Li_2O \cdot Ga_2O_3 \cdot 2SiO_2$ and $Li_2O \cdot Ga_2O_3 \cdot 4SiO_2$ glasses indicates that in these cases crystallization spreads rapidly throughout the volume. In the case of $Li_2O \cdot Ga_2O_3 \cdot 6SiO_2$ glass, the liberation of heat is somewhat sluggish; it is probable that here crystallization develops slowly inward from the grain surfaces. Thus, if we postulate the existence of the compounds $Li_2O \cdot Ga_2O_3 \cdot 2SiO_2$ and $Li_2O \cdot Ga_2O_3 \cdot 4SiO_2$, the crystals of these compounds do not have especially low coefficients of expansion. Therefore, gallium glassceramics of complex composition also exhibit little change of the coefficient of thermal expansion on heat treatment.

Introduction of titanium dioxide into lithium gallosilicate glass does not lower the temperature of this heat effect, as might be expected on the assumption that titanium dioxide merely creates crystallization nuclei, but even raises it considerably (see figure, curve 4).

The position of the peak heat effect for complex glasses is intermediate (curves 5 and 6). Heating curves 5 and 6 are close together, although the glass contains 2% of lithium oxide by weight in the first case and 5% in the second. There is also a very weak low-temperature maximum at 710°, the nature of which is not clear.

It should be pointed out that because of the high rate of heating used, the crystallization temperatures determined by the thermal analysis method are higher than the temperatures at which conversion to glassceramics is actually effected. The increase of the temperature at which crystallization begins when the rate of heating is raised has been demonstrated very clearly [10] in a study of the crystallization of chromium oxide. Heat treatment of lithium gallosilicate glasses containing titanium dioxide at about 650° results in characteristic coloration of the glasses, the intensity of which can be weakened by addition of certain oxides. Changes produced in complex gallium glass as a result of crystallization are shown in Table 2.

As Myuller points out [11], specific-heat data indicate that "unfreezing" of valence vibrations occurs at lower temperatures in gallium oxide than in aluminum oxide. Accordingly, gallium glasses are more easily fusible than aluminum glasses, but gallium glassceramics lose transparency at lower temperatures.

## LITERATURE CITED

1. G. T. Petrovskii and S. V. Nemilov, this collection, p.118.
2. S. Geller, J. Chem. Phys. 33(3):676, 1960.
3. E. I. Galant, Doklady Akad. Nauk SSSR 141(2), 1961.
4. S. K. Dubrovo and A. A. Kefeli, Zhur. Priklad. Khim. 35(2):441, 1962.
5. F. P. Glasser, J. Phys. Chem. 63(12):2085, 1959.
6. N. A. Toropov and Lin Tsu Hsiang, Zhur. Neorg. Khim. 5(11):2462, 1960.
7. E. J. Smoke, J. Am. Ceram. Soc. 34(3), 1951.
8. S. K. Dubrovo, Zhur. Priklad. Khim. 35(1), 1962.
9. C. E. Brackbill and F. A. Hummel, J. Am. Ceram. Soc. 34(4):107, 1951.
10. T. V. Rode, Proceedings of the First Conference on Thermography, Moscow-Leningrad, 1955.
11. R. L. Myuller, Zhur. Fiz. Khim. 28(10):1834, 1956.

# CRYSTALLIZATION OF GLASSES
# OF THE Na$_2$O-Fe$_2$O$_3$-SiO$_2$ SYSTEM

## V. A. Sharai, N. N. Ermolenko, and I. G. Lukinskaya

An investigation of ways of utilizing aegirite minerals was carried out in the Belorussian Polytechnic Institute at the request of the Institute of Rare-Element Chemistry and Technology of the Kola Branch of the Academy of Sciences of the USSR.

Aegirite, NaFe[Si$_2$O$_6$], an alkali mineral of the pyroxene group, is a rock-forming component of volcanic alkali rocks — nepheline syenites, phonolites, and leucitophyres. Native aegirite contains various impurities — K$_2$O, MgO, Al$_2$O$_3$, TiO$_2$, etc. It can form isomorphous mixtures with diopside CaMg[Si$_2$O$_6$] and with augite Ca(Mg, Fe)[Si$_2$O$_6$].

The chemical composition of pure aegirite (in % by weight) is: SiO$_2$, 52; Fe$_2$O$_3$, 34.6; Na$_2$O, 13.4. The aegirite received from the Kola Branch of the Academy of Sciences of the USSR differed considerably from the theoretical composition and contained (% by weight): SiO$_2$, 44.66; Al$_2$O$_3$, 7.40; Fe$_2$O$_3$, 31.45; CaO, 4.24; MgO, 1.84; SO$_3$, 2.59; Na$_2$O, 8.59.

Although aegirite has a high alkali content, is readily fusible (m.p. 990°), and chemically stable, it has no industrial uses as a raw material.

We investigated glass formation in the Na$_2$O-Fe$_2$O$_3$-SiO$_2$ system and crystallization of glasses with the composition of aegirite.

Studies of this system and certain properties of glasses belonging to it have been reported in the literature. The phase diagram of glasses in this system was studied by Bowen, Schairer, and Willems [1]. It includes four three-component compounds, including aegirite.

We chose the composition range 5-45% Na$_2$O, 0-25% Fe$_2$O$_3$, and 55-85% SiO$_2$ for the investigation. The glasses were melted in quartz crucibles in a laboratory electric furnace for 1 h at 1350°. The results are shown in the figure. The figure shows that the composition corresponding to aegirite lies in the glass region. This means that glass can be obtained from aegirite without any additions at not very high production temperatures.

In addition to the glass-making properties, the chemical stability of the glasses was studied. The glasses were found to be resistant to water and sodium carbonate, but were easily attacked by acid.

Glass corresponding in chemical composition to aegirite was synthesized from chemically pure materials for a study of the crystallization properties. The following were added separately, 1 g per sample, to this glass: TiO$_2$, Cr$_2$O$_3$, CuO, cryolite, and zircon. The experimental glasses were crystallized in two stages: the specimens were kept first at the primary heat-treatment temperature (500 and 600°) and then at the secondary heat-treatment temperature (700, 800, and 900°). The duration of heat treatment was the same: 1, 3, and 6 h.

The first heat treatment did not result in any visible crystallization. There was only slight turbidity and a very thin crystalline film on the specimen surfaces.

All the glasses crystallized at the second stage of heat treatment. A compact black, flintlike mass was formed, slightly fused on the surface, with a wrinkled film of dark gray glass with a cherry-red tinge. Sometimes the main black crystallized mass also had a cherry-red tinge or contained dark red streaks and spots. At the second treatment stage crystallization proceeded from within rather than from the surface, which indicated that

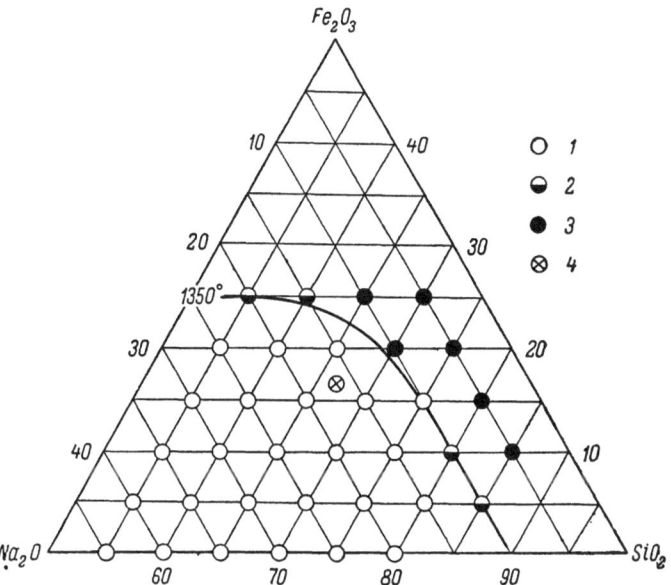

Region of glass formation in the $Na_2O$-$Fe_2O_3$-$SiO_2$ system at 1350°
(mole %). 1) Glass; 2) glass, incomplete melting; 3) incomplete
melting (sinter); 4) composition corresponding to aegirite mineral.

numerous crystallization centers had been formed in the glass. Specimens subjected to longer (6 h) primary treatment at 600° showed less surface fusion.

Aegirite is a ternary compound with an incongruent melting point. The point corresponding to this composition is in the hematite crystallization field. When an aegirite melt crystallizes, hematite separates out first and the composition changes until the maximum point is reached; the reaction then proceeds at a constant composition of the liquid phase: liquid + hematite = aegirite. The liquid and hematite are consumed simultaneously, and the whole mass solidifies in the form of aegirite. In our experiments all the crystallized glasses contained hematite and aegirite as the crystalline phases.

Hematite. In most of the specimens this mineral was in the form of minute spots smaller than 1 μ. Prolongation of the treatment had little effect on the growth of hematite. In the continuous crystallized mass its crystals were hardly distinguishable at 320 magnification. The hematite particles were larger in the glassy fused surface regions of the specimens. Here it formed plates up to 0.01 mm in size, and the abundant separation of these led to decolorization of the glass and lowering of its refraction.

Crystallization of hematite could be detected even in the thin crystalline film on specimens which had been subjected only to the first stage of heat treatment. The growth rate of the hematite crystals is very low (thousandths of a micron per minute) in regions where no surface fusion occurs during heat treatment. The hematite is often distributed irregularly in the crystallized glass, as is indicated by the spotted and layered character of the coloration in certain specimens.

Aegirite. In specimens subjected to two-stage treatment this mineral forms the main bulk of the crystallized material. It is present in the form of dark green, almost black prisms with vertical striation, a high refractive index, and high birefringence. The crystals are intertwined in dense feltlike masses with glass included in the interstices. The size of the aergirite crystals varies widely, from less than 1 μ to 0.3-0.4 mm.

The crystal size increases with the temperature at the second stage of heat treatment. The growth rate becomes especially high at 900°, reaching 0.8-0.9 μ/min. No skeletal forms or dendritic or spherulitic aggregates characteristic of devitrification of viscous media are observed here.

Under the optimum conditions of primary (6 h at 600°) and secondary (6 h at 800°) heat treatment, materials with a very uniform microgranular structure are formed; the crystal size reaches 1-3 μ.

The amount of glass in all the specimens decreases with increase of the treatment time. Under the experimental conditions the reaction of aergirite formation was not completed in practice; an incompletely crystalline material was formed with aegirite as the predominant solid phase. The course of the process was not greatly influenced by additives. The same two crystalline phases, hematite and aegirite, were formed in all specimens with different additives, and the resultant products were incompletely crystalline substances. Cryolite and $TiO_2$ had some effect. Specimens with these additives gave, under otherwise the same conditions, products free from surface fusion, with more uniform crystallization and small grains. It may be that these agents favor the formation of numerous crystallization centers, which confers high thermal stability to the glass (it fuses at higher temperatures) and gives rise to a more uniform granular structure.

Glass made from native aegirite and supplied by the Institute of Rare-Element Chemistry and Technology of the Kola Branch of the Academy of Sciences of the USSR was subjected to two-step crystallization at 600 and 800° for 6 h at each temperature. Under these conditions the glass crystallized with formation of an incompletely crystalline structure with very uniform crystals 1-3 $\mu$ in size. The crystallization was of the same nature as in glasses made from chemically pure substances and corresponding to the theoretical aegirite composition.

It is thus seen that glass resistant to the action of water and alkaline solutions can be obtained from native aegirite, $NaFe[Si_2O_6]$, without fluxes at 1250-1350°. Two-step heat treatment converts the glass into an incompletely crystalline material with a microgranular structure, consisting of aegirite, hematite, and an amorphous phase. Hematite crystallizes first.

These experiments on the crystallization of aegirite glasses lead to the conclusion that crystallization centers are formed at the first stage of heat treatment (500 and 600°), and crystal growth occurs at the second (700, 800, and 900°). Numerous crystallization centers are formed and the rate of crystal growth is low. Activating agents, 1% of $TiO_2$, $SnO_2$, $ZrSiO_4$, $Na_3AlF_6$, $Cr_2O_3$, and $CuO$, do not have any significant effect on the nature of the process.

## LITERATURE CITED

1. N. L. Bowen, J. F. Schairer, and H. W. Willems, Am. J. Sci. 5(20):405, 1930.

# ELECTRON MICROSCOPE INVESTIGATION OF THE STRUCTURE
# OF CERTAIN CRYSTALLINE GLASS MATERIALS

## I. I. Kitaigorodskii and M. D. Il'inichnina

This paper contains a report on the results of electron microscope investigations of the structure of certain crystalline glass materials obtained from metallurgical slag glasses. The investigation was carried out with the Tesla BS 242A electron microscope at 60 kV accelerating voltage and magnifications of about 6000, by means of carbon replicas previously shadowed with chromium. Fracture surfaces were investigated. To reveal the structure of the material, the freshly fractured surface was etched with dilute (2%) hydrofluoric acid. The etching time was chosen experimentally for the different specimens and ranged from 5 to 30 seconds in accordance with the composition of the material and the surface relief of the fracture.

The prepared surface was coated with a film of metallic chromium by evaporation under high vacuum at an angle of 20-30° to the surface, followed by an amorphous carbon film by direct contact of the rods at right angles to the surface. The carbon particles form a coherent film which retains the chromium on the specimen surface. The replica was separated from the surface with the aid of gelatin.

Electron microscope investigation of the crystallization of glass synthesized in the cordierite field of the $MgO-SiO_2-Al_2O_3$ system, a surface micrograph of which is shown in Fig. 1a, showed that the structure of the material greatly depends on the temperature and duration of heat treatment (negative electron micrographs were obtained, and therefore the shadowing appears black as in ordinary photographs).

The low-temperature stage of heat treatment leads to an increase in size of the microheterogeneities present in the original glass (Fig. 1b).

The closest packing of the structure is formed in the material after heat treatment at 950° for 2 h (Fig. 1c). The deposited crystals are rounded plates tenths of a micron in size. Two-stage heat treatment with preliminary treatment for 2 h near the softening point leads to the formation of transparent glassceramics with an extremely fine structure (Fig. 1f, g). The influence of the temperature at the first stage of heat treatment on the structure is shown in Fig. 1h, i, j.

The structure of crystalline glass materials in the $Li_2O-MgO-Al_2O_3-SiO_2$ system was investigated. The size of the crystals formed increases with rise of temperature from tenths of a micron up to 1 $\mu$ across (Fig. 2c, d, e). The glassceramic of which a surface micrograph is shown in Fig. 2d has the best properties. Two crystalline phases can be clearly seen; one, in the form of plates 0.5 $\mu$ in size, is identified as β-spodumene.

At 1200° with a treatment time of 1 h, the size of the β-spodumene platelets increases to 1 $\mu$. The other crystalline phase, concentrated in the center of each such platelet, remains unchanged (Fig. 2e).

Figure 3 shows micrographs of fracture surfaces of glassceramics obtained from glasses synthesized in the $Li_2O-MgO-Al_2O_3-SiO_2$ system with lower lithium contents (Fig. 3a, b) and in the forsterite region of the $MgO-Al_2O_3-SiO_2$ system (Fig. 3c, d, e). All these materials have good mechanical and electrical properties.

Electron microscope investigation of crystalline glass materials obtained from metallurgical slags showed that these slag glassceramics have a microgranular structure with crystals from tenths of a micron to 2 $\mu$ in length. Figure 4 shows electron micrographs of fracture surfaces of slag glassceramics obtained with various additives in the original composition under the same conditions of heat treatment (950°, 3 h). Introduction of 25%

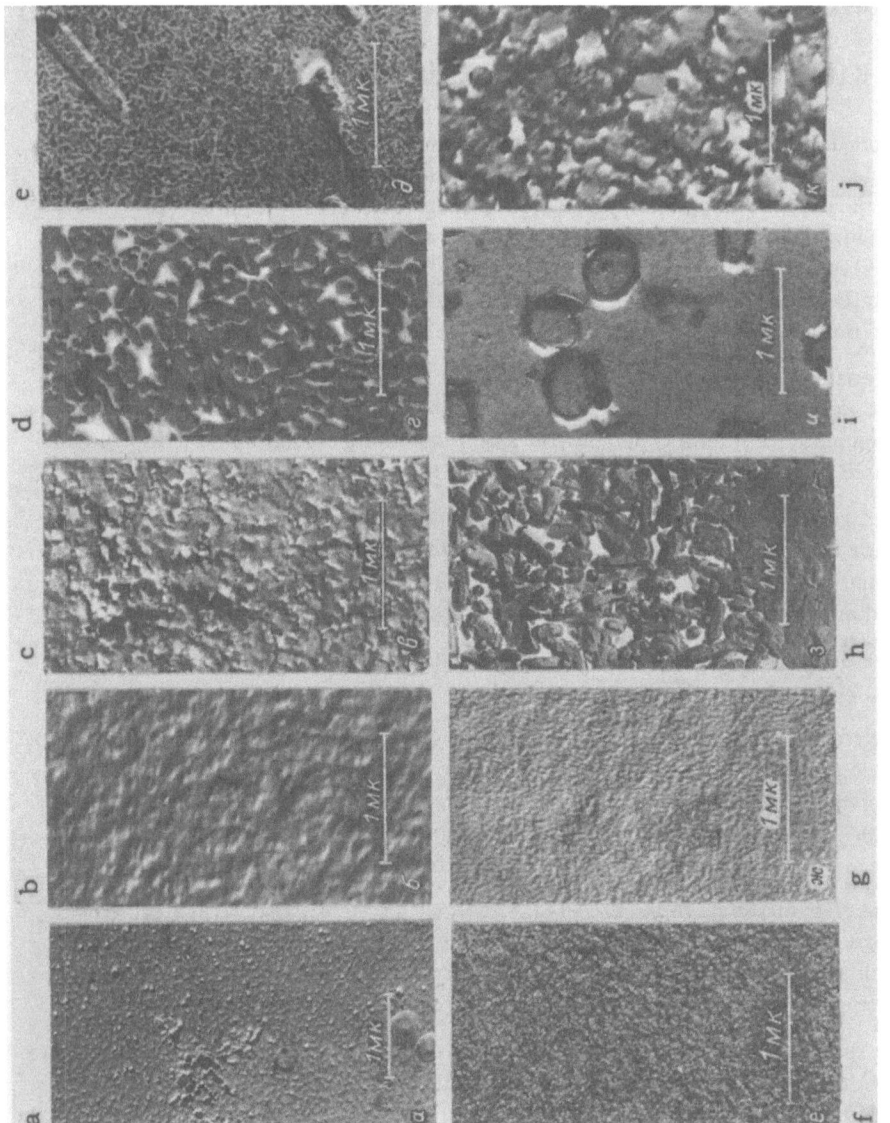

Fig. 1. Effects of the temperature and duration of heat treatment on the structure of the material. a) Original glass (MgO–Al$_2$O$_3$–SiO$_2$ system); b) heat treatment for 2 h near the softening point; c) heat treatment for 2 h at 950°; d) heat treatment for 2 h at 1000°; e) heat treatment for 2 h at 1200°; f,g) heat treatment at 850° (2 h) and 950° (2 h), respectively, with preliminary treatment for 2 h near the softening temperature; h, i, j) heat treatment at 1000° (2 h) with preliminary treatment for 2 h at 700, 750, and 800°, respectively.

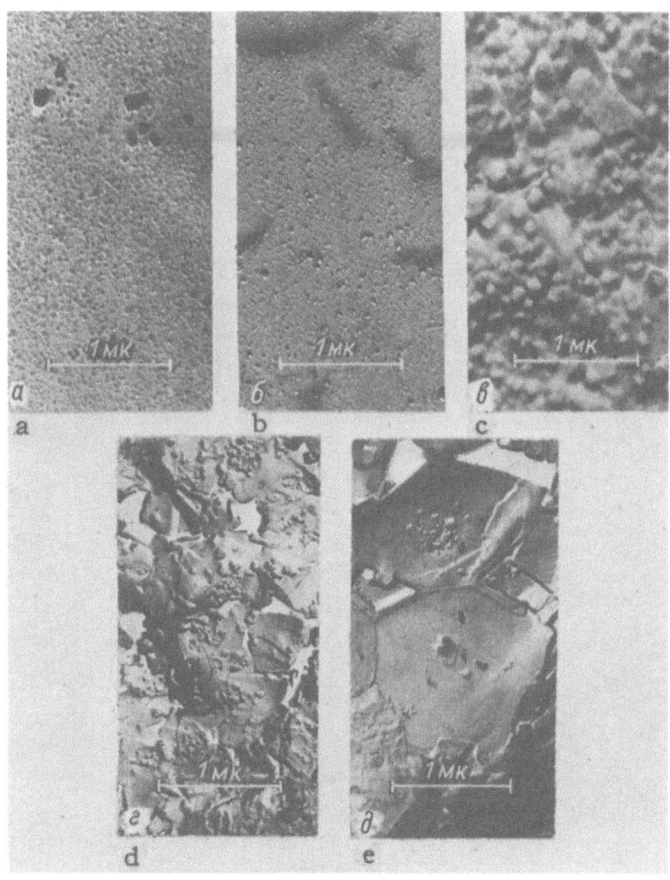

Fig. 2. Structure of the original glass and of crystalline glass material in the Li$_2$O-MgO-Al$_2$O$_3$-SiO$_2$ system. a) Original glass; b) heat treatment for 1 h at 550°; c) 3 h at 900°; d) 1 h at 600° and 2 h at 1000°; e) 1 h at 1200°.

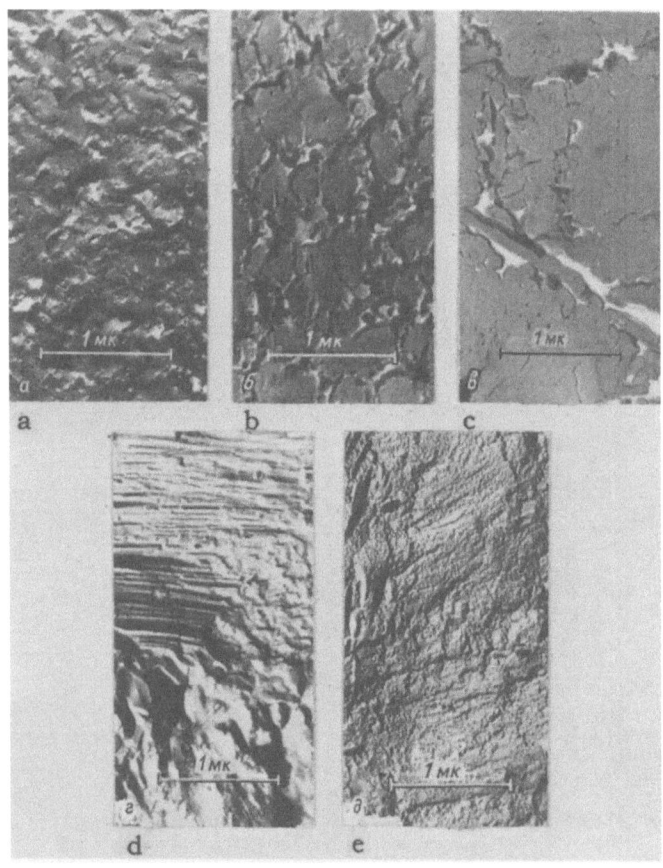

Fig. 3. Structure of crystalline glass materials. a,b) In the $Li_2O$-MgO-
$Al_2O_3$-$SiO_2$ system; c, d, e) in the forsterite region of the MgO-$Al_2O_3$-
$SiO_2$ system.

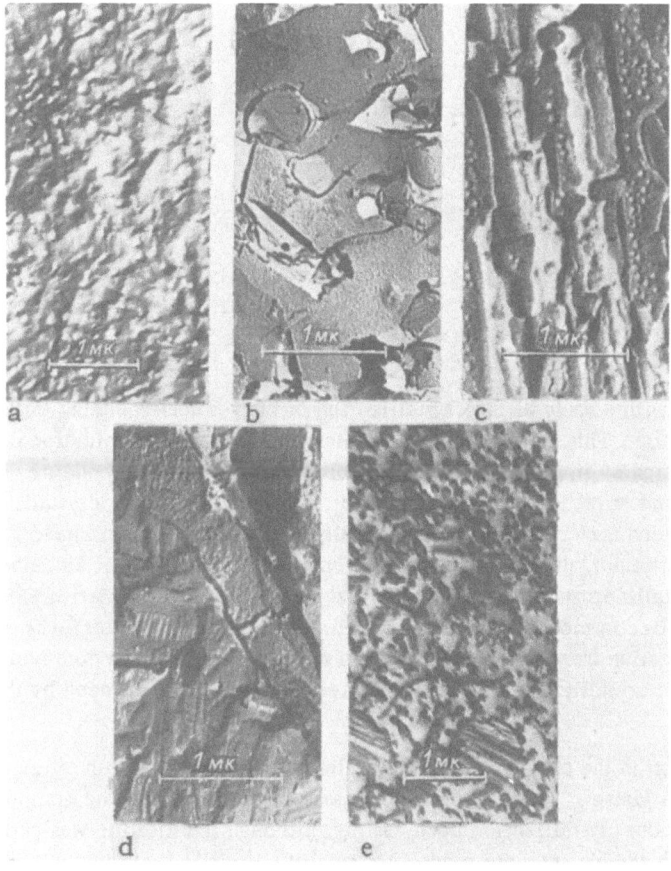

Fig. 4. Electron micrographs of fractured surfaces of several
slag glassceramics.

sodium fluosilicate, $Na_2SiF_6$, leads to the formation of a microcrystalline structure with grains tenths of a micron in size (Fig. 4a). The main crystalline phases are calcium fluoride, $CaF_2$, and anorthite, $CaO \cdot Al_2O_3 \cdot 2SiO_2$. Decrease of the sodium fluosilicate content leads to increase in the size of the anorthite platelets up to 2 $\mu$ in length (Fig. 4b, c, d).

One of the slag glassceramics studied has an entirely different structure, which is due to addition of $TiO_2$ to the original composition. The crystalline phase is in the form of rods and drops (apparently, undeveloped rods). According to the results of x-ray phase analysis, this crystalline phase is sphene, $CaO \cdot TiO_2 \cdot SiO_2$ (see Fig. 4e).

The slag glassceramic, the surface micrograph of which is shown in Fig. 4a, has the best mechanical properties. It has a closely packed microcrystalline structure with grains 0.1-0.3 $\mu$ in size.

# INVESTIGATION OF THE CATALYZED CRYSTALLIZATION
# OF ALKALI-FREE GLASSES

## V. S. Nikandrov

This paper is concerned with a study of the crystallization of glasses corresponding to the native mineral diopside in chemical composition. This composition was chosen because, first, elucidation of the conditions and peculiarities of the crystallization of diopside glasses would solve the problem of producing cheap and readily fusible glassceramics of great industrial importance from them; second, study of the crystallization of these glasses makes it possible to extend the crystallization procedure to other compositions having crystal structures close to that of diopside, in particular, the amphiboles. It is known that a number of amphiboles, including tremolite, having a cryptocrystalline structure, exhibit remarkable mechanical properties. Since both minerals, diopside and tremolite, crystallize in monoclinic cells with almost identical constants, it is to be expected that the mechanism of their regeneration from suitable chosen compositions has much in common. The common feature is probably the fact that crystallization of the respective glasses may be induced by the same impurities in certain cases.

Glass close in composition to the native mineral baikalite [1] was chosen for investigation of the crystallization mechanism of diopside glasses. This was because glasses corresponding to the composition $CaO \cdot MgO \cdot 2SiO_2$ could not be made without crystallization upon casting, although the melting was performed at 1400°. It must be pointed out here that the glasses were made from analytical grade materials in quartz crucibles. Noncrystallizing glasses in the diopside system were obtained only when up to 5% aluminum oxide was introduced into the glass composition. This action of aluminum oxide on glass formation in diopside glasses is apparently associated with the part played by the $Al^{3+}$ ion on entering fourfold coordination.

The diopside glasses investigated had the following composition (% by weight): CaO, 24.4; MgO, 17.5; $SiO_2$, 52.5; $Al_2O_3$, 5.6. To induce crystallization, Ag, Cu, $Cr_2O_3$, $TiO_2$, and apatite were added to the glasses in amounts from 0.01 to 5%.

The character of the crystallization and the temperature ranges in which these glasses crystallize were investigated by the gradient method: pieces of the glass in a nickel boat were placed in a tubular gradient muffle furnace.

The results of the investigation may be summarized as follows. In absence of the additives named above, crystallization of the glass always begins at the surface of the specimen and proceeds with great difficulty (taking 8-10 h at 960°); after crystallization, the specimen exhibits very large contraction and deformation and the deposited phase does not contain pyroxene. If additives are introduced into the glass, the nature and composition of the crystallized phase in the glass greatly depends on the kind of additive. Glasses containing dissolved silver crystallize in the same manner as glasses without additives. After crystallization, such glasses usually contain two or three silver particles up to 0.1 mm in size. This indicates that in this case silver ions cannot form stable bonds with the glass ions, and therefore the silver particles coalesce during crystallization into larger formations which constitute purely mechanical inclusions.

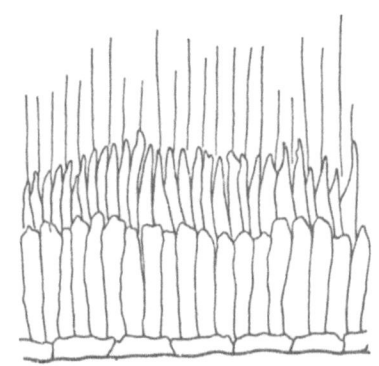

Fig. 1. Schematic representation of the crystallization of diopside glass with added $TiO_2$.

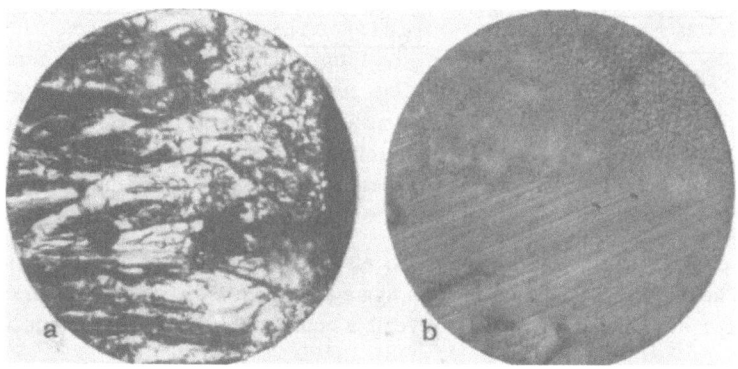

Fig. 2. Photomicrograph of a polished section of crystallized diopside glass with added TiO$_2$ (magnification 180). a) Surface layer; b) region of long prismatic and acicular formations.

Fig. 3. X-ray diagram of diopside specimens with added TiO$_2$.

Glasses containing Cu and Cr$_2$O$_3$ crystallize throughout their volume, forming numerous spherulites within the volume at the initial stage of crystallization. However, after crystallization these glass specimens also undergo considerable shrinkage and deformation, and the crystallized phase contains a small amount of diopside. Additions of TiO$_2$ and fluorapatite have the most favorable action in this respect. As the influence of these additives on the nature of glass crystallization is highly specific, we shall consider their action in greater detail.

Crystallization of Diopside Glasses with Additions of TiO$_2$

When titanium dioxide is introduced into the glass, it is found that during the melting process the walls of the quartz crucibles are rapidly attacked by the melt even at 1200-1260°. To eliminate this effect, CaO and SiO$_2$ have to be added to the glass.

The most favorable temperature conditions for crystallization of these glasses are in the range of 920-940°. Crystallization of diopside glasses containing TiO$_2$ also begins at the surface, but in this case it does not lead to shrinkage or deformation. The character of crystallization of these glasses is shown in Fig. 1. The figure shows that three regions can be distinguished in the crystallized material: 1) a region adjacent to the surface, where the crystals are oriented along the surface (Fig. 2a); 2) a region where the crystals are in the form of long prismatic formations oriented perpendicularly to the surface; 3) a region where the crystals are in the form of acicular formations without transverse faces (Fig. 2b). Microscopic examination of these crystals gave an average refractive index of 1.667 with an extinction angle of 30°. These crystals are therefore identified as diopside. X-ray investigation of the phase deposited after crystallization (Fig. 3) shows that it consists mainly of diopside.

Thus, introduction of TiO$_2$ into the glass composition leads to deposition of the mineral diopside, but even in that case crystallization of the glass begins at the surface. In order to obtain crystalline glass materials with a fine-grained structure, it is necessary to induce crystallization throughout the volume of the glass. We therefore investigated the influence of added fluorapatite on the character of crystallization of diopside glasses.

| Diopside | | Apatite | |
|---|---|---|---|
| Series | Constants, A | Series | Constants, A |
| [001] | 5.24 × 4 = 20.96 | [0001] | 6.85 × 3 = 20.55 |
| [100] | 9.78 × 2 = 19.56 | [2241] | 19.93 |

## Crystallization of Diopside Glasses with Additions of Fluorapatite

Fluorapatite was added to the batch in the form of a fine powder. The crystallization range of these glasses is considerably wider. The crystallization rate is at a maximum at 960° and at a minimum at 820°. Treatment at 960° leads to complete crystallization in 1 h, whereas treatment at 820° for the same length of time makes the glass opaque. Complete crystallization at 820° occurs after 10 h of treatment.

The crystals deposited differ in structure from the crystals which appear after crystallization of diopside glasses with added $TiO_2$. The difference is that they form aggregates with a divergent structure up to 2 mm in size. However, despite this, x-ray diffraction analysis reveals a considerable diopside content in this case also.

The presence of crystallization nuclei is necessary to induce crystallization of supercooled liquids such as glasses. The nuclei may be mechanical impurities, electric charges, or regions of the liquid having a higher degree of order than the rest of the volume. It follows from thermodynamic considerations that there is a relationship between the dimensions of such nuclei, their surface tension, the melting point, and the degree of supercooling [2], represented by the expression

$$r = \frac{2\sigma V_\mu T_0}{\lambda (T_0 - T)} , \qquad (1)$$

where r is the critical size of a crystallization nucleus; $\sigma$ is the interfacial surface energy; $T_0$ is the melting point; $T_0 - T$ is the supercooling of the nucleus; $V_\mu$ is the molar volume of the nucleus; $\lambda$ is the heat of fusion of the nucleus.

It has been noted recently [3] that for production of good crystalline glass materials with a fine-grained structure it is necessary to chose glass compositions with a wide demixing region. If a batch, the composition of which lies within the region of separation into two phases A and B, is melted at a temperature above the temperature of the demixing region, two phases separate out of the melt when it is cooled; one of these phases, distributed throughout the volume in the form of minute droplets, constitutes the nuclei for crystallization of the other phase. Naturally, the droplets of one phase, say A, dispersed in the other phase B must be of a size satisfying the relationship (1). In that case the droplets of phase A become nuclei of phase B. This view explains the action of added apatite in diopside glass.

There are extensive demixing regions both in the $CaO-CaF_2-Al_2O_3-SiO_2$ system and in the $CaO-SiO_2-P_2O_5$ system [4,5]. Thus, if we have a composition in the demixing region of the $CaO-CaF_2-Al_2O_3-SiO_2$ system and impurities from the $CaO-SiO_2-P_2O_5$ system are added to it, the glass undergoes phase separation during cooling.

However, phase separation alone is not sufficient to induce volume crystallization in glasses. Cases are known where glasses having a tendency to phase separation exhibit opalescence after heat treatment, which does not pass into volume crystallization. An example of this is provided by glasses of the $CaF_2-MgO-ZnO-Al_2O_3-SiO_2$ system.

As was pointed out above, droplets of phase A act as nuclei of phase B if the size of the phase A droplets satisfies Eq. (1). Here it must be taken into account that the physicochemical nature of the impurity has a significant influence on the growth of a given crystalline species. The greater the change produced by the impurity in the surface energy of the faces of the growing crystal in relation to the medium, the greater is this influence. The degree of this influence is determined by the relation between the lattice constants of the impurity and of the growing crystal.

If apatite is present as the impurity in diopside glass, the diopside and apatite structures coincide well if the (100) diopside face grows onto the (1010) prism face of apatite. The resulting relationships between the diopside and apatite constants are given in the table.

The good agreement between the diopside and apatite constants on these faces results in sharp lowering of the surface energy of the growing crystals, which favors their crystallization from the diopside melt.

With regard to the influence of $TiO_2$, it is difficult to envision agreement between the $TiO_2$ and diopside structures even in one direction. However, it is possible that the strong polarizing action of the rutile structure [6] sharply lowers the surface energy of the diopside crystals, but this lowering is not enough to induce volume crystallization; it is probable that only the presence of surface forces acting in the surface layer of the melt takes the process to the desired result.

## LITERATURE CITED

1. D. P. Grigor'ev and V. V. Guretskaya, Zap. Vserossiisk. Miner. Obshch. 68(4):4, 1939.
2. Ya. I. Frenkel', Statistical Physics, Moscow-Leningrad, 1948.
3. W. Hinz and P.-O. Kunth, Glastechn. Ber. 34(9):431, 1961.
4. Ya. I. Ol'shanskii, Doklady Akad. Nauk SSSR 114(6):1246, 1957.
5. D. S. Belyankin, V. V. Lapin, and N. A. Toropov, Physicochemical Systems of Silicate Technology, Promstroiizdat, Moscow, 1954.
6. F. A. Grant, Rev. Mod. Phys. 31(3):646, 1959.

# CRYSTALLIZATION OF GLASSES OF COMPOSITION
## $CaO - MgO - Al_2O_3 - SiO_2$ IN THE PRESENCE OF $Cr_2O_3$
## WITH FORMATION OF A STABLE PYROXENE PHASE

## L. A. Zhunina, V. N. Sharai, V. F. Tsitko, and N. N. Khripkova

Some aspects of the catalyzed crystallization of glasses in the diopside field of the $CaO-MgO-Al_2O_3-SiO_2$ system [1] were studied in the Glass Problems Laboratory of the Belorussian Polytechnic Institute.

The experimental specimens were glasses based on fusible calcareous clay of the following composition (% by weight): $SiO_2$, 49.8; $Al_2O_3$, 16.5; $Fe_2O_3$, 3.00; CaO, 9.6; MgO, 6.00; $R_2O_3$, 1.00; calcination loss 14.

This raw material was chosen for the glass because the clay is in abundant supply and could perhaps be utilized for production of crystalline glass materials for structural purposes. The purpose of the investigation was to elucidate the influence of the nature and amount of the catalyst on the course of crystallization of the glass and on the mineral phases formed. The catalysts tested were $SnO_2$, $P_2O_5$, ZnO, $ZrO_2$, $CaF_2$, NiO, CaO, $TiO_2$, and $Cr_2O_3$.

Most of these additives did not produce the desired results; $TiO_2$ and especially $Cr_2O_3$ were exceptions. In the presence of $TiO_2$ an enormous amount of crystallization centers was formed, but the rate of crystal growth was very low. Under the experimental conditions (temperature range in the gradient furnace 650-1000°; treatment time 2, 4, and 6 h) extremely microgranular, incompletely crystallized masses were obtained with crystals smaller than 1 $\mu$. The crystallized glasses were stratified and had heterogeneous coloration ranging from chocolate to violet-gray. The nature of the structure indicated the occurrence of phase separation, clearly visible in polished sections. The layers in the glass differed both in degree of crystallinity and in the nature of the minerals formed. Specimens obtained under the optimum conditions of crystal growth were found to contain the following minerals: a silicate of the pyroxene group (the predominant phase), ferruginous rutile, reduced rutile, and titanite [2,3].

The action of $TiO_2$ as a crystallization stimulator can be attributed to its limited solubility in alkali-free melts. The fine $TiO_2$ particles form mechanical centers for crystallization of the main mineral phase, pyroxene. This positive role of $TiO_2$ as crystallization stimulator is complicated by phase-separation effects which lead to heterogeneity of the material [4]. $Ti^{4+}$ has a fairly considerable field strength and favors phase separation in glass. In addition, $TiO_2$ becomes partially reduced to $Ti_2O_3$; this leads to a change in the coordination number of Ti and a simultaneous change in the color of the substance. Because of all this, $TiO_2$ cannot be used as a stimulator in a controlled process. The role of $Cr_2O_3$ must be regarded differently.

To simplify the crystallization mechanism we chose glass of the pyroxene composition, ensuring the formation of a monomineral phase. It is known that the pyroxenes are a group of minerals of fairly complex variable composition that crystallize fairly readily. If the glass composition is adjusted by addition of dolomite, all the oxides can be made to enter the pyroxene formed during the crystallization [5,6]. The required composition was calculated, and the presumed crystalline phase corresponded to the formula $mCaMgSi_2O_6 \cdot nCaAl_2SiO_6 \cdot pCaFe_2SiO_6$.

Pyroxenes melt congruently; if our system is regarded, for simplicity, as $MgO-CaO-Al_2O_3-SiO_2$, deposition of one solid phase can be expected at the point representing the composition in the pyroxene field. Glasses without $Cr_2O_3$ and with $Cr_2O_3$ contents ranging from 0.1 to 5% were made on the basis of the calculated composition.

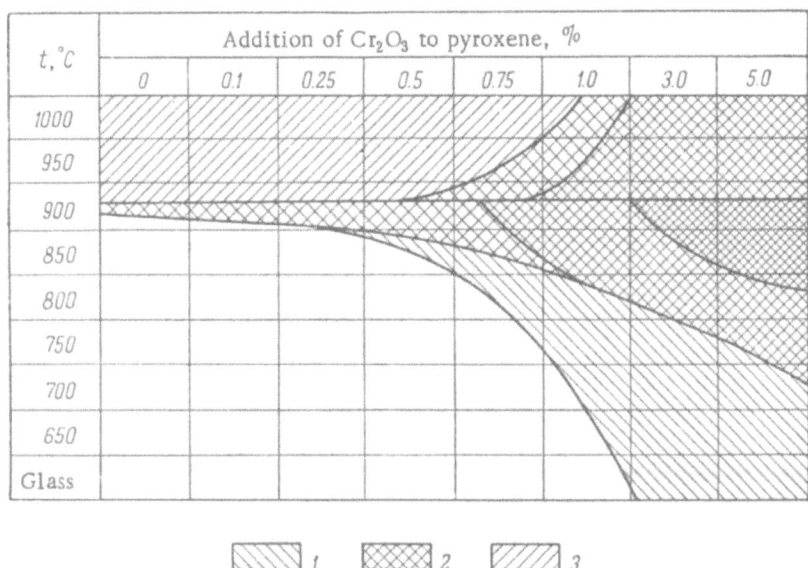

| $t$, °C | Addition of $Cr_2O_3$ to pyroxene, % | | | | | | | |
|---|---|---|---|---|---|---|---|---|
| | 0 | 0.1 | 0.25 | 0.5 | 0.75 | 1.0 | 3.0 | 5.0 |
| 1000 | | | | | | | | |
| 950 | | | | | | | | |
| 900 | | | | | | | | |
| 850 | | | | | | | | |
| 800 | | | | | | | | |
| 750 | | | | | | | | |
| 700 | | | | | | | | |
| 650 | | | | | | | | |
| Glass | | | | | | | | |

⬚ 1     ⬚ 2     ⬚ 3

Crystallization diagram of glasses in the $SiO_2$-$R_2O_3$-RO system (treatment time 4 h). 1) Spinels; 2) spinels and pyroxenes; 3) pyroxenes.

Investigations of the crystallization products gave the following results. Crystallization proceeds in two stages, with formation of two mineral phases, spinellide and pyroxene. At the first crystallization stage, at 650-680°, octahedral reddish-brown isotropic crystals with a high refractive index appear; these are spinellides. They occur as single forms or in clusters. The amount of this phase depends on the amount of $Cr_2O_3$ added, increasing with the latter. Glass made without $Cr_2O_3$ has the lowest spinellide content.

The formation of spinellides is followed by crystallization of the main mineral phase, pyroxene. The spinellides act as organizing centers or crystallization nuclei for pyroxene, and become gradually surrounded by edges, rosettes, and spherulites of that mineral. The growth continues until the aggregates join and form a continuous crystalline mass. The structure becomes increasingly microgranular with increase of the number of spinellide nuclei formed. The length of the crystalline needles and threads composing the spherulites ranges from 2-4 to 1-1.5 mm in accordance with the amount of catalyst, the treatment temperature, and the time of treatment.

The amount of the spinellide centers formed depends on the temperature as well as on the $Cr_2O_3$ content. These crystallization nuclei become less numerous at 950 and 1000° and in petrographic sections one may see disintegration of the nuclei and consequent brown coloration of the adjacent pyroxene zones. The nature of the crystallization is shown in the diagram.

The principal mineral phase, forming spherulitic and rosettelike aggregates, can be assigned to pyroxene by its optical properties. The average refractive index is in the range of 1.647-1.662, i.e., the lowest values for the representatives of this group. In a polished section the mineral is greenish at temperatures up to 950° and brownish at higher temperature. The extinction angle is oblique and ranges from 10 to 22°. X-ray diffraction analysis confirmed that the mineral belongs to the pyroxene group.

The part played by $Cr_2O_3$ in the crystallization of our glasses may be explained as follows. During crystallization of glass the composition of which is somewhat more complex than the MgO-CaO-$Al_2O_3$-$SiO_2$ system, a mineral of the spinellide group is deposited as the primary phase. Introduction of $Cr_2O_3$ creates conditions for formation of a chrome spinellide, more stable in silicate media than chromium-free varieties. The amount of chrome spinellide crystals increases with the amount of chromium in the glass; these crystals act as crystallization nuclei for the main phase, pyroxene. A possible composition of the spinellide is Mg(Fe)(Al, Fe, Cr)$_2$O$_4$.

The initial composition of the glass alters when the spinellides are formed, and the deposited pyroxene deviates from the predetermined composition, becoming richer in the Ca component. As the amount of spinellides depends on the crystallization temperature, the composition of the pyroxene phase must also alter, approaching the calculated composition at the minimum pyroxene content.

This variation of the pyroxene composition is confirmed by the fact that the chemical resistance of the glasses varies within certain limits in accordance with the $Cr_2O_3$ content and the treatment temperature. Increase of the duration of crystallization shifts all the effects toward lower temperatures, makes it possible to increase the number of nuclei, and favors the formation of a more fine-grained structure.

## LITERATURE CITED

1. D. S. Belyankin, V. V. Lapin, and N. A. Toropov, Physicochemical Systems of Silicate Technology, Promstroiizdat, Moscow, 1954.
2. V. N. Sharai, Sb. Nauchn. Trudov Belorussk. Politekhn. Inst. No. 82, 86, 1960.
3. L. A. Zhinina, A. M. Kripskii, and E. Z. Novikova, Sb. Nauchn. Trudov Belorussk. Politekhn. Inst. No. 82, 79, 1960.
4. A. Petzold, Enamels [Russian translation], Metallurgizdat, Moscow, 1958.
5. A. I. Tsvetkov, Trudy Inst. Geol. Nauk AN SSSR, No. 138, 1, 1951.
6. A. G. Kotlova, Trudy Inst. Rudnykh Mestorozhdenii AN SSSR No. 30, 56, 1958.

# DEPENDENCE OF THE TEMPERATURE RANGE OF CRYSTALLIZATION OF GLASSES IN THE $Na_2O-CaO-MgO-Al_2O_3-SiO_2$ SYSTEM ON THE COMPOSITION

## N. P. Danilova

In order to make it possible to obtain articles of complex configuration from glassceramics, the crystallization of the glass must begin at lower temperatures than the start of deformation under load. This can be achieved provided that crystallization centers are formed in the original glass (from which the glassceramic is obtained) at high viscosities of the order of $10^{10}$-$10^{11}$ poises, i.e., in the transformation range, somewhat above the annealing temperature.

According to existing literature data, fluorides are the most effective crystallization agents at high viscosities [1,2]. Glass compositions with good melting and working properties and of high crystallizability were obtained in the $Na_2O$-$CaO$-$MgO$-$Al_2O_3$-$SiO_2$ system with fluoride as mineralizer. These glasses crystallize rapidly and throughout the entire volume simultaneously, over a wide temperature range.

It was found that there is a certain optimum ratio of $SiO_2$ to $Al_2O_3$ at which the glass begins to crystallize at a lower temperature than that of the start of deformation under load. Thus, in glasses containing 60.5% $SiO_2$ by weight with addition of 4% fluoride, the optimum $Al_2O_3$ content is 10-11%; in glasses containing 57% $SiO_2$ with addition of 5% fluoride, the optimum $Al_2O_3$ content is 15-16%. If the alumina content is lower or higher than the optimum value, the glasses begin to crystallize in a narrow temperature range and the start of crystallization begins at the surface, i.e., it is nonuniform, and the crystalline structure is coarse and heterogeneous. In some instances cavities are formed within the specimens and the surfaces become wrinkled; as a result, the mechanical strength of the product deteriorates and is sometimes lower than that of the original glass.

The crystallizability of glasses is altered considerably when calcium oxide is replaced by magnesium oxide.

Figure 1 shows how the initial temperatures of weak and intensive crystallization vary for glasses containing variable amounts of MgO and CaO, ranging from 6 to 14% by weight. The temperature range of crystallization becomes wider and the process is much more intensive when the MgO content is increased. Glass containing 12% MgO begins to crystallize at a temperature below the temperature of initial deformation under load.

The formation, owing to the presence of MgO, of numerous crystallization centers at low temperatures ensures the formation of a microcrystalline and homogeneous structure with closer packing; this is mainly due to the fact that the ionic radius of magnesium is smaller than that of calcium and the former is therefore more mobile at high viscosities.

It was found that for the best melting properties and crystallizability the optimum contents are 7-8% CaO and 12% MgO by weight.

Increase of the fluoride content in glasses containing the optimum amounts of $SiO_2$, $Al_2O_3$, CaO, and MgO shifts the start of crystallization toward lower temperatures, below the temperature of deformation under load, and widens the crystallization range (Fig. 2). However, the optimum amount of fluoride added to the batch is 4-5% by weight; this gives a uniform and fairly dense microcrystalline structure. The amount of fluoride remaining in the glass is 2-3%.

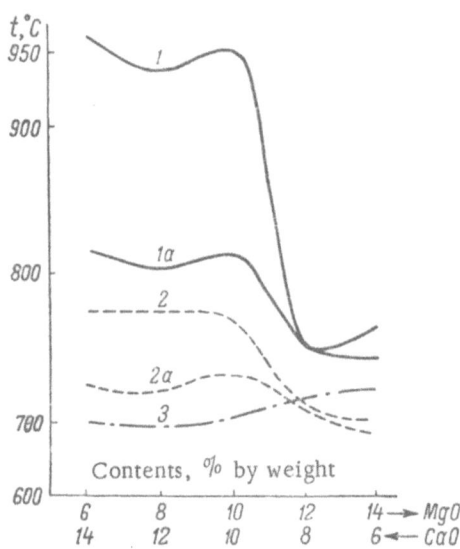

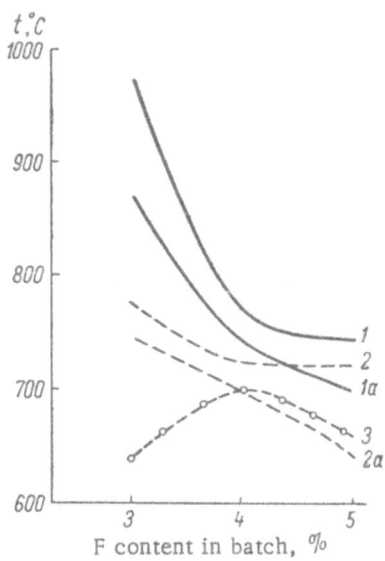

Fig. 1. Effect of replacement of CaO by MgO on the temperatures at which the glasses begin to crystallize and to deform under load. 1, 1a) Intensive crystallization for 2 and 4 h, respectively; 2, 2a) weak crystallization for 2 and 4 h, respectively; 3) deformation of the glass under load.

Fig. 2. Effect of added fluoride on the temperatures at which the glasses begin to crystallize and to deform under load. 1, 1a) Intensive crystallization for 2 and 4 h, respectively; 2, 2a) weak crystallization for 2 and 4 h, respectively; 3) deformation of the glass under load.

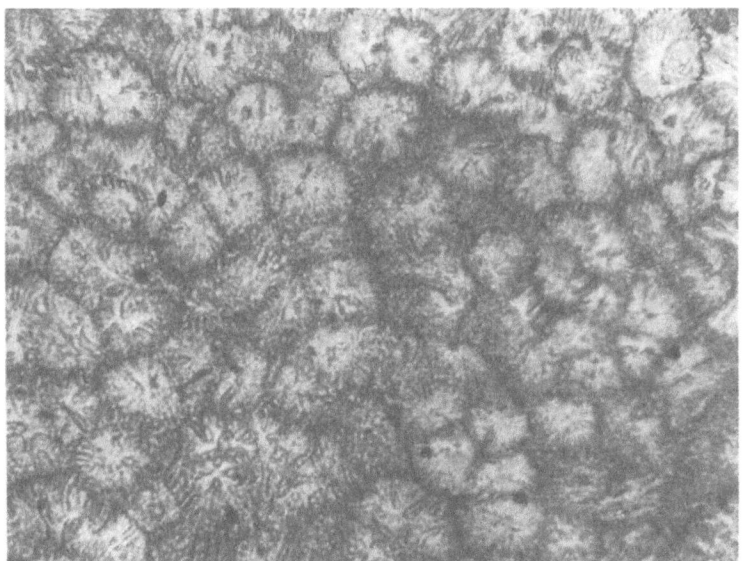

Fig. 3. Photomicrograph of a polished section of crystallized glass with addition of 6% fluoride (magnification 600).

If the amount of fluoride is lowered below the optimum, crystallization is weak and irregular. With a higher amount of fluoride, large spherulites are formed and the final product contains a gaseous phase in interstices, favoring the formation of interstitial cristobalite crystals (Fig. 3). This results in a loose structure and has an adverse effect on the mechanical strength and other properties of the glassceramic.

By regulation of glass crystallization by appropriate selection of the composition and heat-treatment conitions it is possible to obtain glassceramics with predetermined physicochemical properties and to obtain prod-.cts from them without deformation.

## LITERATURE CITED

. St. Lungu and D. Popescu-Has, Industria Usoara 5(2):63, 1954.
. S. I. Sil'vestrovich and É. M. Rabinovich, Zhur. Vsesoyuzn. Khim. Obshch. im. Mendeleeva 5(2):186, 1960.

# DEPENDENCE OF CERTAIN PHYSICAL PROPERTIES OF GLASSES OF THE $BaO-CaO-Al_2O_3-SiO_2$ SYSTEM ON HEAT TREATMENT

## I. S. Kachan and Z. I. Shalimo

In this paper we present the preliminary results of one part of an investigation of the relationship between the structure, heat treatment, and properties of crystallized glasses, conducted in the Glass Problems Laboratory of the Belorussian Polytechnic Institute. The $BaO-CaO-Al_2O_3-SiO_2$ system was chosen because materials based on glasses in this system should have good electrical properties after heat treatment.

At the first stages of the investigation we were interested in the kinetic conditions of crystallization. There are many possible methods for studying the mechanism of crystallization. We studied the influence of crystallization on the coefficient of thermal expansion and on Young's modulus, as these quantities are highly sensitive to structural changes in glasses. The optimum kinetic conditions of crystallization were investigated at different heat-treatment temperatures, the range of which differed from the softening range by 25, 50, 75, and 100°.

The glasses were thoroughly annealed and then tested in the form of rods 4.5-5 mm in diameter and 80 mm long. X-ray structural and electron microscope analyses showed that up to 400° the structure of the glass remains virtually unchanged; accordingly, in this temperature range the heating rate was over 40° per hour. Above 400° the temperature was raised at various rates: 40, 30, 20, and 10° per hour. The coefficient of thermal expansion was measured in the 20-400° range with the aid of a quartz dilatometer designed by the State Institute of Glass.

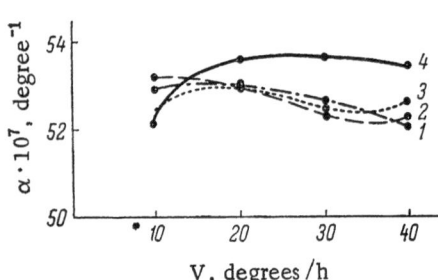

Fig. 1. Dependence of the coefficient of thermal expansion of the glasses on rate of heating and on heat-treatment temperature. 1) 700°; 2) 725°; 3) 750°; 4) 775°.

The variation of the coefficient of thermal expansion depends significantly on the heating rate and on the final temperature of heat treatment (Fig. 1).

The strength properties of the glass and of the materials obtained from it were investigated, as stated above, by determinations of Young's modulus and bending strength. Young's modulus characterizes the resistance of a material to elastic deformation under extension. Since the coefficient of linear expansion and Young's modulus are closely related to the elastic properties of glasses, it is to be expected that the heating rate should also influence the Young's modulus of glass.

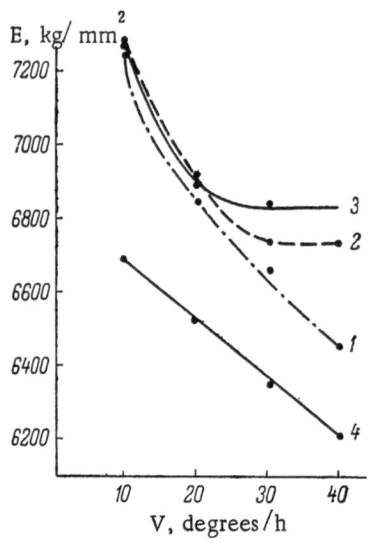

Fig. 2. Dependence of the Young's modulus of the glasses on the rate of heating and on the final heat-treatment temperature. 1) 725°; 2) 750°; 3) 775°; 4) 800°.

Young's modulus was studied by the bending method. This is the simplest of the known methods and is sufficiently accurate.

198

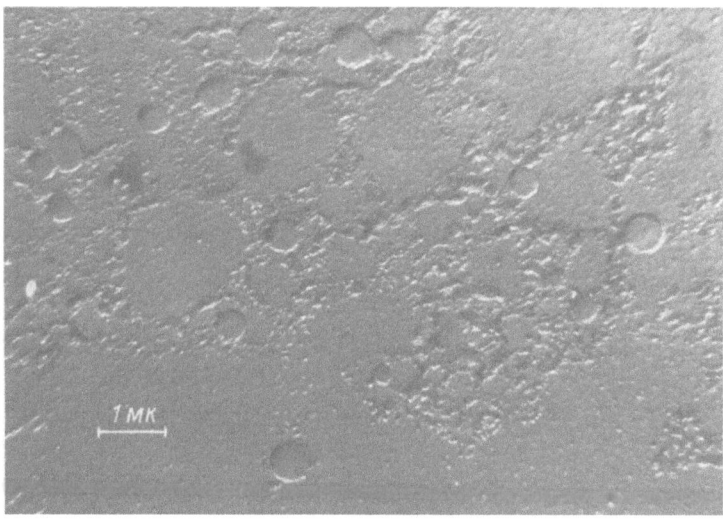

Fig. 3. Glass structure with phase microheterogeneity.

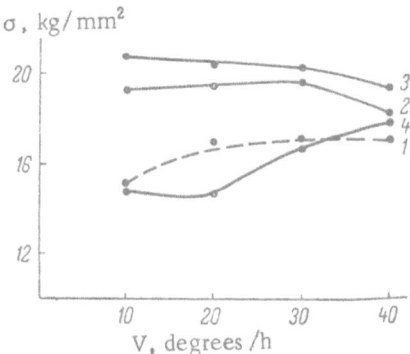

Fig. 4. Dependence of the static bending strength on the rate of heating and final temperature of heat-treatment of glasses in the $BaO-CaO-Al_2O_3-SiO_2$ system. 1) 700°; 2) 725°; 3) 750°; 4) 775°.

The bending of the rods was measured with the IZV-1 instrument; the deflection was measured to a precision of 0.2 mm from the difference between readings on the comparator scale before and after application of the load.

As was to be expected (Fig. 2), when the heat-treatment temperature is raised from 725 to 775°, Young's modulus also changes considerably as a result of differences in the heating rate and the final temperature of heat treatment.

Figure 3 shows that in the 700-750° range the chosen glass compositions in the $BaO-CaO-Al_2O_3-SiO_2$ system exhibit phase microheterogeneity.

Comparison of the temperature ranges in which phase microheterogeneity occurs and the nature of the dependence of the coefficient of thermal expansion and Young's modulus on the heating rate alters suggests that these effects are caused by demixing of the glass, changes in its phase composition.

In distinction from the coefficient of thermal expansion and Young's modulus, which are associated with volume properties of glasses, a quantity such as the bending strength depends to a large extent on their surface properties. However, despite this distinction, it is to be expected that the static bending strength must depend on the heating rate and on the final temperature of heat treatment, as demixing of the glass in the ranges indicated above leads to a change of the interfacial surface energy, which must naturally affect the mechanical properties of the glass.

We investigated static bending strength in relation to the heating rate and the final temperature of heat treatment. The strength was measured on a tensile machine of the RPM-50U type, which gives a uniformly increasing force up to destruction of the specimen. It follows from Fig. 4 that in the 725-775° range the surface properties also undergo changes.

The following generalization follows from the experimental results: the demixing of glasses that occurs in the 725-775° range leads to a change in the glass structure, and this influences the manner in which the coefficient of thermal expansion, Young's modulus, and the bending strength depend on the rate of heating and the final temperature of heat treatment.

# INVESTIGATION OF THE ELECTRICAL PROPERTIES OF CERTAIN ALKALI-FREE CRYSTALLINE GLASS MATERIALS IN RELATION TO THE COMPOSITION AND HEAT-TREATMENT CONDITIONS

G. A. Pavlova and V. G. Chistoserdov

Investigation of the influence of crystallization on the electrical properties of alkali-free glasses is of practical importance for solving the problem of production of crystalline glass materials with good insulating properties. It is also of theoretical interest for interpretation of the fine structure of crystalline glass materials. We investigated crystalline glass materials based on the $MgO-Al_2O_3-SiO_2-TiO_2$ and $BaO-B_2O_3-Al_2O_3-SiO_2$ systems; compositions of such materials are given in [1].

The glasses were synthesized from technical materials, so that they contained from 0.3 to 1.0% alkaline oxides.

Measurements of $\tan \delta$ and $\varepsilon$ at audio and radio frequencies were made with the MLE-1 and IP-3A instruments, the KV-1 Q-meter, and the 36-I instrument; the MOM-4 instrument was used for measurement of the volume resistivity $\rho_V$.

In order to elucidate how the electrical properties of glassceramics of the $MgO-Al_2O_3-SiO_2-TiO_2$ system depend on the conditions of heat treatment, specimens made from the same glass melt were crystallized at 800, 950, 1100, and 1200° and held for 4-7 h at these temperatures.

It follows from Fig. 1 that $\rho_V$ of crystallized materials is lower by 2-3 orders of magnitude than that of the original glass. The activation energy of conductivity falls from 2.7 to 1.6 eV. The tangent of the loss angle of materials crystallized at temperatures above 1000° rises sharply at relatively low frequencies (Fig. 2), falls with increase of frequency, and at $10^{10}$ cps becomes lower than for the original glass.

It follows from Figs. 2 and 3 than $\tan \delta$ in the $10^3-10^6$ cps range increases with rise of the crystallization temperature, and also begins to increase at lower temperatures. The dependence of $\tan \delta$ on temperature of the specimen crystallized at 1200° has a distinct maximum (Fig. 3, curve 4). This

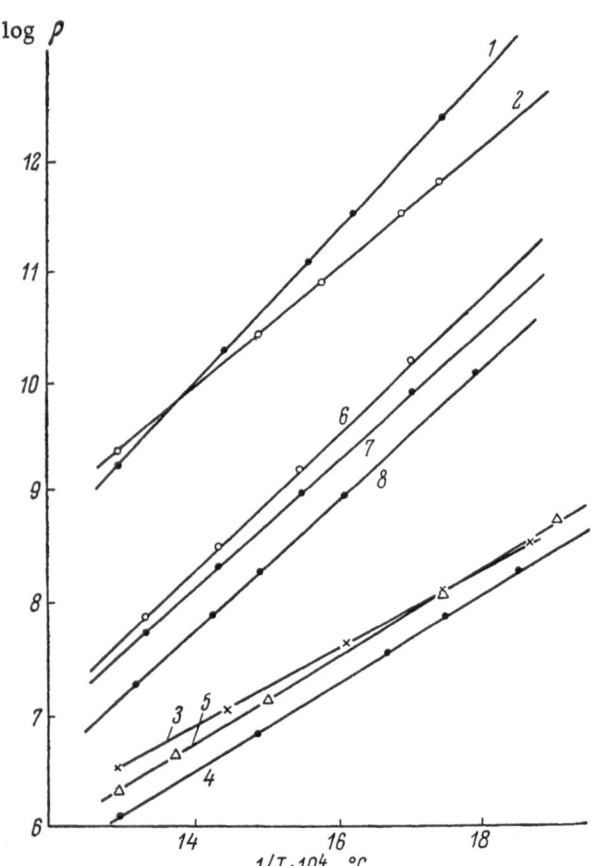

Fig. 1. Dependence of resistivity on temperature. 1) Original glass of the $MgO-Al_2O_3-SiO_2-TiO_2$ system; specimens crystallized at: 2) 800°; 3) 950°; 4,5) 1200°; 6) original glass of the $BaO-B_2O_3-Al_2O_3-SiO_2$ system; specimens crystallized at: 7) 1100°; 8) 1200°.

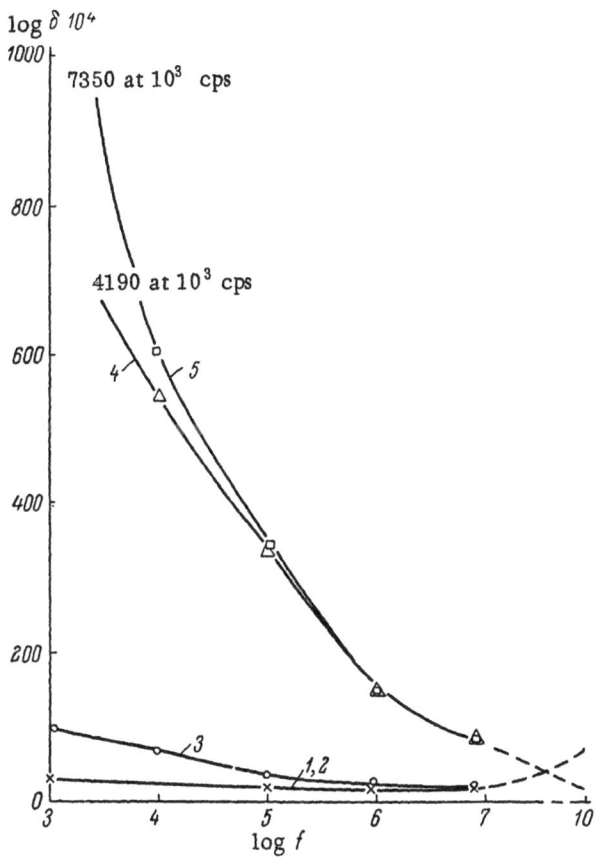

Fig. 2. Dependence of tan δ on frequency. 1) Original glass of the MgO-Al₂O₃-SiO₂-TiO₂ system; specimens crystallized at: 2) 800°; 3) 950°; 4,5) 1200°.

maximum shifts toward higher temperatures with increase of frequency (Fig. 4). This is indicative of the relaxational character of the losses. The energy of the relaxation process with maxima at 70 and 180° is about 0.6 eV.

Measurements of $\tan \delta$ and $\varepsilon$ (at $10^{10}$ cps frequency and room temperature) for specimens crystallized at various temperatures showed than $\tan \delta$ and $\varepsilon$ decrease with rise of the maximum crystallization temperature to 1100-1200°; for the best specimens, $\tan \delta = (7-13) \cdot 10^{-4}$ and $\varepsilon = 5.6$-$5.9$. The dependence of $\tan \delta$ on temperature at $10^{10}$ cps for these specimens is shown in Fig. 4. It must be pointed out that considerable scattering of the results was obtained with specimens crystallized at 1100-1200° and containing distinctly visible cords, and with specimens made from glasses of different meltings.

Specimens of crystalline glass materials containing the same components as our material of the MgO-Al₂O₃-SiO₂-TiO₂ system, but of considerably modified composition, were passed for testing to the State Electroceramics Research Institute (Moscow). The results, published in [2], showed that these materials gave similar maxima for the dependence of dielectric losses on frequency and temperature, but the maxima were at somewhat higher temperatures.

The nature of the variations of dielectric losses with frequency and temperature in crystalline glass materials of the MgO-Al₂O₃-SiO₂-TiO₂ system (crystallized at 1100-1200°) suggests that it is the crystalline phase and not the residual glassy phase which has the predominant influence on the electrical properties of these materials.

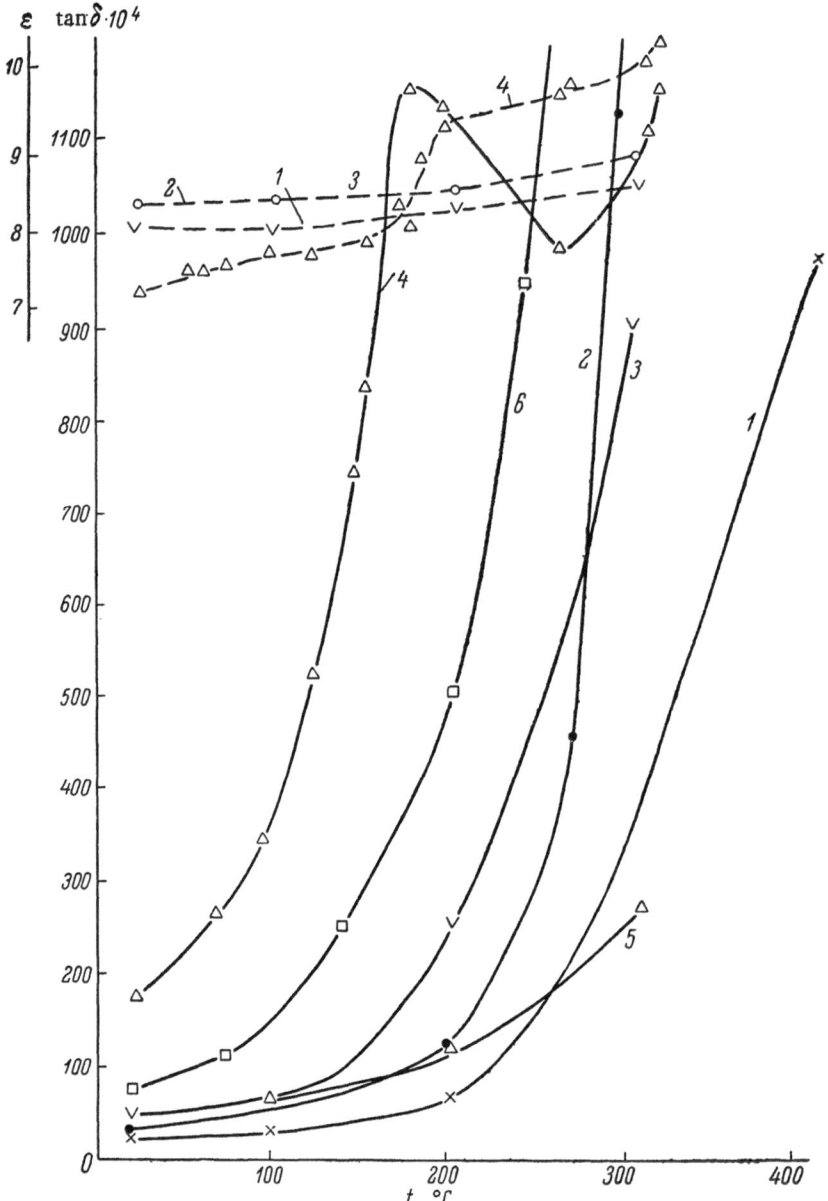

Fig. 3. Dependence of tan δ and ε on temperature for glass and crystalline glass materials based on the MgO-Al$_2$O$_3$-SiO$_2$-TiO$_2$ system, at 10$^6$ cps frequency. 1) Original glass; specimens crystallized at: 2) 800°; 3) 950°; 4) 1200°; 5) cordierite synthesized from technical oxides [4]; 6) glass of the MgO-Al$_2$O$_3$-SiO$_2$ system, containing α-cordierite as the main crystalline phase, crystallized at 1200°.

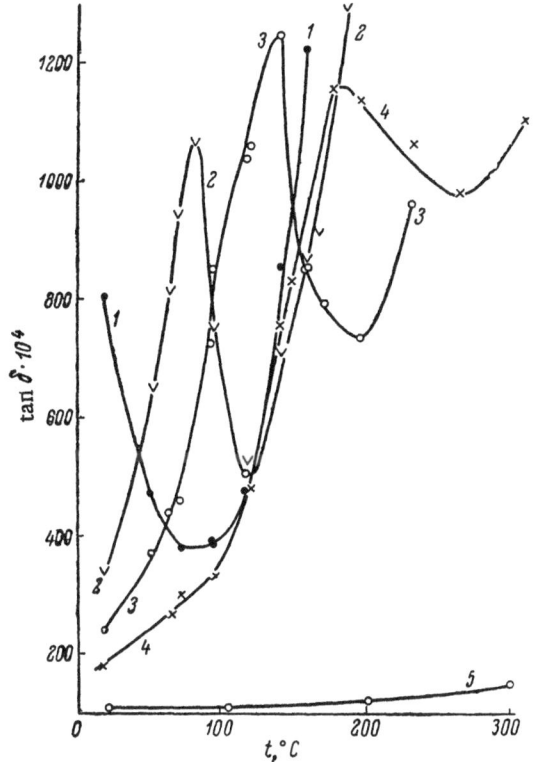

Fig. 4. Dependence of tan δ on temperature for crystalline glass material based on the $MgO-Al_2O_3-SiO_2-TiO_2$ system and crystallized at 1200°, at the following frequencies: 1) $10^3$ cps; 2) $10^4$ cps; 3) $10^5$ cps; 4) $10^6$ cps; 5) $10^{10}$ cps.

The results of x-ray diffraction and crystallographic optical studies [3] indicate the presence of α-cordierite and geikielite, $MgO \cdot TiO_2$, as the main crystalline phase in these materials when crystallized at 1200°. If the crystallization is performed at 1000° or lower, the predominant crystalline phase is μ-cordierite.

We also investigated the electrical properties of crystallized three-component glasses of the $MgO-Al_2O_3-SiO_2$ system, in which cordierite is the predominant crystalline phase. Figure 3 shows the dependence of tan δ on temperature for one of these glasses (curve 6) and for cordierite made by the ceramic process (curve 5).

Figure 3 shows that materials which do not contain $TiO_2$ and which have cordierite as the main crystalline phase do not give maxima of tan δ as a function of temperature. It follows that the maxima of tan δ with temperature must be attributed to a crystalline phase containing $TiO_2$.

It was shown in [4] that after irradiation with a neutron flux of $10^{18}$ cm$^{-2}$, pure crystalline $MgO \cdot TiO_2$ exhibits pronounced frequency and temperature maxima which are relaxational in character. Vodop'-yanov [5] attributes these changes of the dielectric properties to formation of defects of the vacancy and ion dislocation types. It seems that similar relaxational losses are possible when extraneous ion impurities are present in crystalline magnesium titanate.

The electrical properties of crystalline glass materials based on the $BaO-B_2O_3-Al_2O_3-SiO_2$ system differ little from those of the original glasses (Fig. 1, curves 6-8; Fig. 5). The nature of the variations of tan δ and ε with frequency and temperature for crystallized materials is similar to that for glasses. These materials have better insulating properties at elevated temperatures than crystallized materials of the $MgO-Al_2O_3-SiO_2-TiO_2$ system. Increase of the maximum crystallization temperature from 1100 to 1200° leads to a somewhat sharper increase of tan δ at temperatures above 300° and to a decrease of tan δ by about 0.3 of an order of magnitude. At $10^{10}$ cps frequency and 25°, tan δ = $(15-20) \cdot 10^{-4}$ and ε = 5.5.

We consider that the character of the variations of the electrical properties of crystallized materials of this type is determined primarily by the residual glassy phase.

The following conclusions may be drawn from this work.

1. Investigation of the dielectric losses of alkali-free crystalline glass materials and glasses over a wide range of frequencies showed that tan δ of crystalline glass materials can be either lower or higher than that of the original glasses.

2. The resistivity of the crystallized alkali-free glasses investigated either remains approximately equal to that of the original glasses or decreases considerably.

3. Either the residual glassy phase or the crystalline phase may have the predominant influence on the nature of the variations of the electrical properties of crystalline glass materials with frequency and temperature.

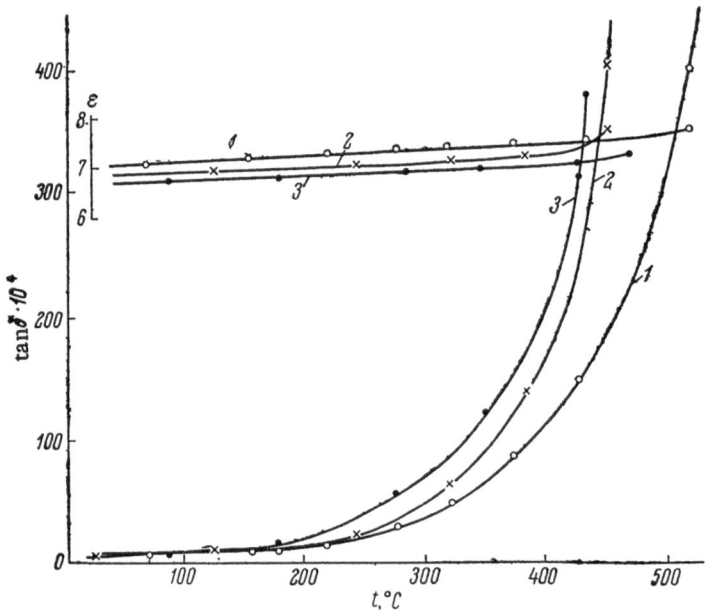

Fig. 5. Dependence of tan δ and ε on temperature for glass and crystalline glass materials based on the BaO-B$_2$O$_3$-Al$_2$O$_3$-SiO$_2$ system, at $10^6$ cps
frequency. 1) Original glass; specimens crystallized at: 2) 1100°; 3) 1200°.

4. The considerable variations of the electrical characteristics of crystalline glass materials (in the MgO-
Al$_2$O$_3$-SiO$_2$-TiO$_2$ system) in accordance with the conditions of heat treatment suggest that the structure of crystalline glass materials also influences their electrical properties.

## LITERATURE CITED

1. W. Hinz and L. Baiburt, Silikattechnik 11(10):455, 1960.
2. M. D. Mashkovich, Voprosy Radioelektroniki, Ser. IV, No. 7, 77, 1961.
3. V. G. Chistoserdov, N. A. Shmeleva, and A. N. Serdyuk, this collection, p.172.
4. N. P. Bogoroditskii and I. D. Fridberg, Élektrichestvo No. 5, 52, 1951.
5. L. K. Vodop'yanov, Fizika Tverd. Tela 3(8):2331, 1961.

# DISCUSSION

A. F. Borisov and V. I. Zadumin reported on a method developed in the Saratov Branch of the Institute of Glass for continuous control of the structure of oxide materials. If two glass plates of the same chemical composition, one of which is crystallized, are connected, placed between platinum electrodes, and heated, an emf arises between them; its magnitude reaches hundreds of millivolts. As the second glass crystallizes, the emf gradually decreases until the relative proportions of the crystalline and glassy phases become the same in both specimens. The internal resistance of the galvanic cell can be reduced and the current in the circuit increased by decrease of the thickness of the specimen. Experiments showed that specimens 0.5-0.7 mm thick have sufficient conductivity for such measurements at about 600°.

They presented curves representing thermal effects for two types of glasses. In a glass of the first type the emf decreases at 730° owing to separation of the first microcrystalline phase, which then dissolves to a considerable degree when the temperature is raised. In the 790-825° range there is another sharp decrease of emf, accompanied by formation of a crystalline phase in the specimen, as confirmed by differential thermal analysis. The emf increases with rise of temperature up to 1225°; crystal growth then occurs and the emf decreases sharply.

Differential thermal analysis of glass of the second type reveals a small effect at 750°, corresponding to separation of the first crystalline phase in the glass. On the emf − temperature curve, the effect at 750° is much more distinct. Crystal growth at 870-880° is also distinctly seen.

I. D. Tykachinskii and G. G. Zainulin reported the results of experiments on the production of a material with a tangled fibrous structure similar to the natural mineral nephrite.

Native minerals corresponding to nephrite in composition contain OH groups and have not been synthesized previously. It was shown by D. P. Grigor'ev that amphiboles may be obtained by replacement of the $OH^-$ group by $F^-$. Therefore, on the basis of the chemical composition of nephrite, the glasses were made with the $CaO$-$MgO$-$SiO_2$ system as a basis, with additions of small amounts of $Fe_2O_3$ and $F^-$ instead of $OH^-$. $TiO_2$, $ZrO_2$, and $Li_2O$ were used as mineralizers. Twelve different compositions were made (see table).

Batch Composition

| $SiO_2$ | CaO | MgO | $Fe_2O_3$ | $F^-$ | $TiO_2$ | $Li_2O$ | $ZrO_2$ |
|---|---|---|---|---|---|---|---|
| 57.69 | 13.46 | 28.85 | — | — | — | — | — |
| 57.69 | 13.46 | 28.85 | — | 4.0 | — | — | — |
| 57.69 | 13.46 | 28.85 | 0.5 | — | — | — | — |
| 57.69 | 13.46 | 28.85 | 1.0 | — | — | — | — |
| 57.69 | 13.46 | 28.85 | 4.0 | — | — | — | — |
| 57.69 | 13.46 | 24.85 | 4.0 | — | — | — | — |
| 57.69 | 13.46 | 27.85 | 0.5 | 4.0 | — | — | — |
| 57.69 | 13.46 | 24.85 | 4.0 | 4.0 | — | — | — |
| 57.69 | 13.46 | 24.85 | 4.0 | 4.0 | 4.0 | — | — |
| 57.69 | 13.46 | 22.85 | — | — | — | — | 6.0 |
| 57.69 | 13.46 | 22.85 | — | 4.0 | — | — | 6.0 |
| 57.69 | 13.46 | 22.85 | — | — | — | 3.0 | 3.0 |
| 57.69 | 13.46 | 22.85 | — | 4.0 | 4.0 | 4.0 | 3.0 |

These glasses crystallized at temperatures from 750° to 1300°. Microscope studies of the crystallized glasses showed that they begin to crystallize from the outside at 900°. Microscopic examination of the glasses revealed absence of a structure analogous to nephrite in the crystalline phase. It was shown by x-ray diffraction analysis that the main phases in the crystallized glasses are wollastonite, clinoenstatite, and enstatite. There is less diopside and magnetite.

V. G. Chistoserdov spoke on the separation of metallic crystallization centers. In glasses irradiated by ultraviolet light at room temperature silver nuclei are formed at 400-480°; in glasses irradiated at 100°, silver nuclei form at 350°. In glasses irradiated at 220°, large metallic silver centers form immediately. After irradiation at −180°, nucleation occurs only in melts at temperatures above 600°.

It was also pointed out that the precrystallization period occupies a wide range and depends on a number of factors. Additions of fluoride, chromium, boron, manganese, and titanium have different effects on the character of crystallization; cooling of the original glass melt has a considerable effect on crystallization. In an answer to a question concerning the possibility of obtaining glassceramics from photosensitive glasses without irradiation, it was stated that glass made under reducing conditions developed well, as under such conditions crystallization centers are formed.

L. Baiburt pointed out that the course of crystallization in the cordierite system depends on the composition. Further, he pointed out the desirability of elucidating the influence of the valence of titanium on the dielectric properties of glass. It can hardly be said that titanium dioxide is a crystallization catalyst; the amount is so large that it is a component. Baiburt considers that it is not possible to give a general answer to the question whether the formation of a glassceramic is preceded by phase separation or nucleation, as there are no universal schemes and terms for the numerous systems studied.

G. T. Petrovskii stressed that it is possible only as an exception to apply the thermal-emf method to studies of glass crystallization. The magnitude of the thermal emf is influenced by crystallization of the glass and changes in the coordination of aluminum and titanium; therefore we cannot explain the origins of the thermal emf.

R. S. Shevelevich noted that the following conclusions may be drawn from the papers presented at the seminar. 1) The sharp alteration of the physical properties of crystalline glass materials resulting from changes in the conditions of heat treatment probably indicates considerable changes in bonding; 2) it is likely that the chemical bonds in glassceramics are considerably more heterodynamic in character than in the original glass, since there is distinct microheterogeneity. These changes can be detected by the methods of radiospectroscopy (nuclear resonance). Assessment of the difference between the resonance frequencies of a given nucleus in the amorphous and crystalline phases gives an indication of changes in the degree of ionicity of the corresponding chemical bonds. The information obtained by radiospectroscopic methods can be used in conjunction with x-ray diffraction data for calculation of the lattice energy of glassceramics. Determination of the cohesive lattice energy of the glassceramic for a given chemical composition of the original glass and a given type of heat treatment involves the solution of a multidimensional variation problem.

V. I. Shelyubskii emphasized the special significance of electron microscopy and proposed the organization of a more restricted colloquium for discussion of the finer details of this method. Further, he pointed out the gulf between studies of the mechanism of formation of glassceramics and active control of the production process.

K. G. Bondarev discussed the crystallization mechanism of glasses of the $CaO-Al_2O_3-SiO_2-Na_2O$ system. Fluoride is used as the catalyst in glasses of this system. Transition of glasses of this system into the crystalline state can be represented as follows: when the glass is cooled to a temperature 150° above the annealing temperature, the catalyst gives rise to crystallization centers. This process is preceded by separation of the glass into two phases. When the temperature is raised, crystallization of the main phase begins on the crystallization centers. Formation of the crystallization centers, consisting of catalyst microcrystals, leads to formation of a rigid framework within the glass as a result of bond formation between the catalyst microcrystals. This framework strengthens the glass. Thus he disagreed with E. V. Podushko with regard to this aspect of the mechanism of glass crystallization, as he does not consider the formation of prenucleation groups reported by Avgustinik to be a physical reality. He considers that his view is confirmed by the fact that much more than 1-2% of catalyst must be added to the glass to ensure good crystallization.

Further, Bondarev discussed the significance of the production of glassceramics and especially the importance of producing expanded glassceramics. Glassceramics are being produced on the pilot scale, but not yet as standard products. The theory of the crystallization process which has been evolved should be used as a basis for development of a technological process for industry; for this it is necessary to increase the number of people engaged in technological development work. The production of large amounts of homogeneous glass, and in particular of cordierite glass, for glassceramics presents great technological problems. The relationship between the chemical composition, structure, and dielectric properties of glassceramics is not clear.

E. A. Porai-Koshits (chairman of the final session), in summarizing the work of the symposium on catalyzed crystallization of glasses, noted the active participation of representatives of three of the largest glass plants. The work of their factory laboratories is conducted at a very high scientific level and provides a vivid example of the benefits of close cooperation between science and industry.

"At the symposium," said Porai-Koshits, "we discussed work in two directions; the first is associated with the nature of the glassy state, the processes which occur in glass during heat treatment and as a result of changes in composition, and the second relates to the theory of formation of glassceramics or the processes of catalyzed crystallization leading to the formation of glassceramics with the optimal properties. The most important result is the establishment of a distinct connection between the two directions and of the fact that the state and structure of the glass in the precrystallization period (before the start of mass crystallization) play an important and sometimes a decisive part in the formation of the necessary properties of the material during crystallization and the formation of a glassceramic.

"In my view, the discussion revealed a lack of well-developed methods for studying the dependence of the course of crystallization on the state of the glass and on the processes which occur during the precrystallization period; therefore, the members of the symposium must direct special attention in their future work on development and perfection of methods for investigation of the precrystallization state and, in particular, methods for investigating the critical point of transition from the glassy to the crystalline state, the instant of formation of the first crystal nuclei. As regards investigation of the precrystallization state of glass, this should probably begin with investigations of the structure of melts.

"An extremely important result of the work of the symposium is the establishment of a distinct, if not yet detailed connection between the crystallization theory of V. N. Filipovich and many experimental facts. This work contains the foundations of the physical principles of the theory of heterogeneous equilibria, which has hitherto been essentially a phenomenological theory; i.e., it contains the fundamentals of verification, refinement, and calculation of phase diagrams by application of the physical theory to actual compositions, thermal conditions, and chemical interactions prevailing during cooling of melts and heat treatment of glass. It is possible with the aid of the theory to explain and predict the complex 'submicroheterogeneity' effect, the appearance and disappearance of the associated opalescence, and other experimental phenomena.

"Another important experimental fact, now generally accepted, is the existence of chemical differentiation into submicroscopic regions of different compositions in most complex glasses. Until recently, experimental detection of these regions was hindered by technical difficulties associated with the various structural methods. Most success has been achieved in this respect by the method of low-angle x-ray scattering which is being worked out at the Institute of Silicate Chemistry of the Academy of Sciences of the USSR. As was shown in the papers, this method can now easily reveal heterogeneity regions smaller than 50 A in sodium borosilicate glasses and about 100 A in lithium silicate glasses (where the difference between the electron densities of the 'submicrophases' is very small); however, further refinement of the technique is necessary for detection of regions 15-20 A in size. It is very important that this method can be used for determination of the degree of chemical heterogeneity from the absolute value of the low-angle x-ray scattering intensity. The electron microscope is extremely tempting for such investigations, but the replica method restricts the resolving power of the microscope.

"In his symposium paper on the mechanical properties and structure of glasses, F. K. Aleinikov presented the results obtained in a study of glasses with the aid of a Japanese electron microscope, without the use of replicas: very thin glass films made by blowing and etching were used. However, his results arouse suspicion, as he detected heterogeneity regions 50-80 A in size and larger in all glasses with more than one component. The existence of such regions is easily detected by the low-angle method, but the latter method gave a negative result for many glasses. The possible combination of selective dissolution of individual regions of the glass by acids and electron microscopy seems to me to be more promising.

"This accounts both for the attention which we must devote to the development of the most sensitive structural methods and for the caution with which the first results obtained by such methods must be interpreted. It is to be hoped that at the 4th All-Union Conference on the Glassy State, which is to be held in November, 1963, we will learn of new work in this direction.

"In conclusion, I would like to express, on behalf of the Organizing Committee, our thanks to the management of the Institute of Russian Literature of the Academy of Sciences of the USSR for the premises provided by them, to Deputy Director of the Institute of Silicate Chemistry of the Academy of Sciences of the USSR A. S. Gotlib for extensive and infallible assistance in organization of the symposium, and to the team of workers from the Leningrad Electrotechnical Institute, who ensured the successful demonstration of numerous diagrams for almost 50 papers by means of the epidiascope. I would also like to thank our guests from other cities, whose contributions enlivened the meetings, for their part in the development of the theory of the catalyzed crystallization of glass."